Meta-topolin: A Growth Regulator for Plant Biotechnology and Agriculture

Naseem Ahmad • Miroslav Strnad
Editors

Meta-topolin: A Growth Regulator for Plant Biotechnology and Agriculture

Springer

Editors
Naseem Ahmad
Department of Botany
Aligarh Muslim University
Aligarh, Uttar Pradesh, India

Miroslav Strnad
Laboratory of Growth Regulators
Institute of Experimental Botany of the Czech
Academy of Sciences and Palacký University
Olomouc, Czech Republic

ISBN 978-981-15-9045-0   ISBN 978-981-15-9046-7 (eBook)
https://doi.org/10.1007/978-981-15-9046-7

This Springer imprint is published by the registered company Springer Nature Singapore Pte Ltd.
The registered company address is: 152 Beach Road, #21-01/04 Gateway East, Singapore 189721, Singapore

Preface

In recent past, plant tissue culture technology and plant biotechnology have gained unassailable success in plant sciences for their application in various plant species, leading to its commercial application. Plant growth regulator (PGR) studies, plant tissue culture and plant biotechnology research are closely inter-related and reciprocally supportive. The manipulation of plant cell, tissue and organs in culture with important applications in propagation and modification of plants is highly dependent on the use of an appropriate PGR regime. Nowadays, a wide range of PGRs are available including different adenine and phenyl urea derivatives. Among them the most effective ones are the aromatic cytokinins, viz. kinetin, N^6-benzyladenine and topolins that show cytokinin activities and promote cell division and micropropagation. Topolins are a naturally occurring aromatic compound and are well accepted as an alternative for other frequently used cytokinins like isopentenyladenine and zeatin in in vitro culture of different plant species. Among the different topolins, *meta*-topolin [6-(3-hydroxybenzylamino)purine] is becoming the most popular and its use in plant tissue culture amplified swiftly. During the last few decades, a number of encouraging reports are available where *meta*-topolin not only proved to be a better cytokinin in micropropagation but also helps in alleviation of various physiological disorders, rooting and acclimatization of tissue culture raised plants. It is also used in a wide range of plant species from medicinal and food crops to plantation trees. As per our understanding, no report or comprehensive data is available on this interesting and valuable theme. Therefore, the present topic for an edited book has been taken into consideration.

The volume will consolidate the sprinkled information on the role of *meta*-topolin in plant physiology in general and plant tissue culture in particular.

Meta-topolin: A Growth Regulator for Plant Biotechnology and Agriculture is an assemblage of research and review papers in various aspects from history to uses of *meta*-topolin with advantageous inferences. The edited volume will provide a comprehensive literature for researchers and academicians across the globe who want to do further research in this particular subject matter.

We are thankful to each and every contributor who agreed and accepted our proposal and helped us in this venture.

The edited volume contains about 23 chapters by international dignitaries which give weightage to this book.

Aligarh, Uttar Pradesh, India
Olomouc, Czech Republic
June 02, 2020

Naseem Ahmad
Miroslav Strnad

Contents

Editors and Contributors

About the Editors

Naseem Ahmad is working as Assistant Professor in the D/o Botany, Aligarh Muslim University, Aligarh. He has also worked as Young Scientist and has more than 10 years of experience in Plant Tissue Culture and Molecular Biology. He has published about 44 research papers, 02 edited books entitled *Plant Tissue Culture: Propagation, Conservation and Crop Improvement* (Springer; 2016) and *Thidiazuron: From Urea Derivative to Plant Growth Regulator* (Springer, 2018). He holds life membership of many learned/scientific societies and is a fellow of Indian Botanical Society (FBS). He is recipient of "Yuva Vaigiyanik Samman (2009), CST, UP; Young Scientist (2010) by NESA, N. Delhi; Rashtriya Gaurav Award (2015) by IIFS, N. Delhi; Young Scientist (2015) by IFEE, Kolkata; and Prof. Hira Lal Chakravarty award (2019) by ISCA, Kolkata (India). He was also a Senior Research Fellow (SRF) of CSIR, N. Delhi.

Miroslav Strnad is the head of the joint Laboratory of Growth Regulators, Institute of Experimental Botany of the Czech Academy of Sciences and Palacký University in Olomouc, Czech Republic (rustreg.upol.cz). Strnad's current focus is on the research and development of a new generation of compounds with anti-viral, anti-proliferative, anti-angiogenic and anti-senescence properties, the molecular mechanisms of their action and the potential combinatory therapies based on these compounds, new phytohormone derived cosmetic as well as plant growth regulators for plant biotechnology and agriculture. He graduated in Phytotechnology from the Faculty of Agronomy, Mendel University, Brno, in 1982 (Ing.). In 2001 he was promoted to Professor of the Palacký University in Olomouc, Czech Republic. In 1998, he was awarded the Rhone-Poulenc Rorer Award by the Phytochemical Society of Europe (PSE), in recognition of his work on the identification, analysis and biochemistry of phytohormones topolins. Prof. Strnad has been widely publishing (11 chapters in books; more than 410 papers in recognized journals; 3 books; 8 Czech and 49 international patents; citations >12,500; Hirsch index: 56). He was PSE president (2014–2016), since 2018 vice-president.

Contributors

Naseem Ahmad Department of Botany, Aligarh Muslim University, Aligarh, Uttar Pradesh, India

Elena V. Ambros Central Siberian Botanical Garden, Siberian Branch of the Russian Academy of Sciences, Novosibirsk, Russian Federation

Mohammad Anis Plant Biotechnology Laboratory, Department of Botany, Aligarh Muslim University, Aligarh, Uttar Pradesh, India

N. V. Aswathi Department of Plant Science, Central University of Kerala, Periye (PO), Kasaragod, Kerala, India

Jana Bíbová Department of Chemical Biology and Genetics, Faculty of Science, Centre of the Region Haná for Biotechnological and Agricultural Research, Palacký University, Olomouc, Czech Republic
Laboratory of Growth Regulators, Institute of Experimental Botany of the Czech Academy of Sciences, Palacký University Olomouc, Olomouc, Czech Republic

Magdalena Bryksová Laboratory of Growth Regulators, Institute of Experimental Botany, AS CR, Palacký University, Olomouc, Czech Republic
Department of Chemical Biology, Faculty of Science, Palacký University, Olomouc, Czech Republic

Jean Carlos Cardoso Laboratory of Plant Physiology and Tissue Culture, Department of Biotechnology Plant and Animal Production, Centro de Ciências Agrárias, Universidade Federal de São Carlos, Araras, SP, Brazil

M. J. Cernadas Biotechnology and Forestry Improvement Group, Instituto de Investigaciones Agrobiológicas de Galicia, CSIC, Santiago de Compostela, Spain
Diputación de Lugo, Lugo, Spain

P. Chandran Department of Plant Science, Central University of Kerala, Periye (PO), Kasaragod, Kerala, India

M. Debenham The New Zealand Institute for Plant and Food Research Limited, Palmerston North, New Zealand

Karuna Dhiman Department of Biotechnology, Dr Y.S. Parmar University of Horticulture and Forestry, Solan, Himachal Pradesh, India

Karel Doležal Department of Chemical Biology and Genetics, Faculty of Science, Centre of Region Hanáfor Biotechnological and Agricultural Research, Palacký University, Olomouc, Czech Republic
Laboratory of Growth Regulators, Institute of Experimental Botany of the Czech Academy of Sciences, Palacký University, Olomouc, Czech Republic

Nigar Fatima Plant Biotechnology Laboratory, Department of Botany, Aligarh Muslim University, Aligarh, Uttar Pradesh, India

Geetika Gambhir Department of Biotechnology, Dr Y.S. Parmar University of Horticulture and Forestry, Solan, Himachal Pradesh, India
ICAR-Indian Agricultural Research Institute (IARI), New Delhi, Delhi, India

Saikat Gantait Crop Research Unit, Bidhan Chandra Krishi Viswavidyalaya, Kalyani, Nadia, West Bengal, India
Department of Genetics and Plant Breeding, Faculty of Agriculture, Bidhan Chandra Krishi Viswavidyalaya, Mohanpur, Nadia, West Bengal, India

Cemil Işlek Biotechnology Department, Art and Sciences Faculty, Nigde Omer Halisdemir University, Nigde, Turkey

Helena Janečková Department of Biophysics, Faculty of Science, Centre of the Region Haná for Biotechnological and Agricultural Research, Palacký University, Olomouc, Czech Republic

L. V. Janeiro Biotechnology and Forestry Improvement Group, Instituto de Investigaciones Agrobiológicas de Galicia, CSIC, Santiago de Compostela, Spain
Diputación de Lugo, Lugo, Spain

Dominika Kaczorová Department of Genetic Resources for Vegetables, Medicinal and Special Plants, Crop Research Institute, Centre of the Region Haná for Biotechnological and Agricultural Research, Olomouc, Czech Republic
Department of Phytochemistry, Faculty of Science, Centre of the Region Haná for Biotechnological and Agricultural Research, Palacký University Olomouc, Olomouc, Czech Republic

Esra Koç Department of Biology, Faculty of Science, Ankara University, Ankara, Turkey

David Kopečný Department of Protein Biochemistry and Proteomics, Faculty of Science, Centre of the Region Haná for Biotechnological and Agricultural Research, Palacký University Olomouc, Olomouc, Czech Republic

Radoslav Koprna Department of Chemical Biology and Genetics, Faculty of Science, Centre of the Region Haná for Biotechnological and Agricultural Research, Palacký University, Olomouc, Czech Republic

Pankaj Kumar CSIR-Institute of Himalayan Bioresource Technology (IHBT), Palampur, Himachal Pradesh, India
Department of Biotechnology, Dr Y.S. Parmar University of Horticulture and Forestry, Solan, Himachal Pradesh, India
Science and Engineering Research Board, Department of Science and Technology, Government of India, New Delhi, Delhi, India

Monisha Mitra Department of Agricultural Biotechnology, Faculty of Agriculture, Bidhan Chandra Krishi Viswavidyalaya, Mohanpur, Nadia, West Bengal, India

Jaroslav Nisler Laboratory of Growth Regulators, Institute of Experimental Botany of the Czech Academy odf Sciences, Palacký University, Olomouc, Czech Republic

Ondřej Novak Laboratory of Growth Regulators, Institute of Experimental Botany of the Czech Academy of Sciences, Palacký University, Olomouc, Czech Republic

Tatyana I. Novikova Central Siberian Botanical Garden, Siberian Branch of the Russian Academy of Sciences, Novosibirsk, Russian Federation

R. Pathirana The New Zealand Institute for Plant and Food Research Limited, Palmerston North, New Zealand
Plant and Food Research Australia Pvt Ltd., Urrbrae, SA, Australia

Ondřej Plíhal Laboratory of Growth Regulators, Institute of Experimental Botany of the Czech Academy of Sciences, Palacký University, Olomouc, Czech Republic

Lucie Plíhalová Department of Chemical Biology, Faculty of Science, Palacký University, Olomouc, Czech Republic
Laboratory of Growth Regulators, Institute of Experimental Botany ASCR, Olomouc, Czech Republic

M. C. San José Biotechnology and Forestry Improvement Group, Instituto de Investigaciones Agrobiológicas de Galicia, CSIC, Santiago de Compostela, Spain
Diputación de Lugo, Lugo, Spain

Carolina Sánchez-Romero Departamento de Botánica y Fisiología Vegetal, Universidad de Málaga, Málaga, Spain

Ishani Shaunak Department of Biotechnology, Dr Y.S. Parmar University of Horticulture and Forestry, Solan, Himachal Pradesh, India

Lukáš Spíchal Department of Chemical Biology and Genetics, Faculty of Science, Centre of the Region Haná for Biotechnological and Agricultural Research, Palacký University, Olomouc, Czech Republic

Dinesh Kumar Srivastava Department of Biotechnology, Dr Y.S. Parmar University of Horticulture and Forestry, Solan, Himachal Pradesh, India

Miroslav Strnad Laboratory of Growth Regulators, Institute of Experimental Botany of the Czech Academy of Sciences and Palacký University, Olomouc, Czech Republic

Petr Tarkowski Department of Genetic Resources for Vegetables, Medicinal and Special Plants, Crop Research Institute, Centre of the Region Haná for Biotechnological and Agricultural Research, Olomouc, Czech Republic
Department of Phytochemistry, Faculty of Science, Centre of the Region Haná for Biotechnological and Agricultural Research, Palacký University Olomouc, Olomouc, Czech Republic

T. Dennis Thomas Department of Plant Science, Central University of Kerala, Periye (PO), Kasaragod, Kerala, India

Bengu Turkyilmaz Unal Biotechnology Department, Art and Sciences Faculty, Nigde Omer Halisdemir University, Nigde, Turkey

Jiří Voller Laboratory of Growth Regulators, Institute of Experimental Botany of the Czech Academy of Sciences, Palacký University, Olomouc, Czech Republic

Krishna Vrundha Department of Plant Science, Central University of Kerala, Periye (PO), Kasaragod, Kerala, India

Stefaan P. O. Werbrouck Laboratory for Applied In Vitro Plant Biotechnology, Faculty of Bioscience Engineering, Ghent University, Ghent, Belgium

Marek Zatloukal Department of Chemical Biology and Genetics, Faculty of Science, Centre of the Region Haná for Biotechnological and Agricultural Research, Palacký University, Olomouc, Czech Republic

Yulianna G. Zaytseva Central Siberian Botanical Garden, Siberian Branch of the Russian Academy of Sciences, Novosibirsk, Russian Federation

History of *Meta*-Topolin and the Aromatic Cytokinins

1

Miroslav Strnad

Contents

Abstract

Following the discovery of kinetin (Miller et al., J Am Chem Soc 78 (7):1375–1380, 1956), there was a period of intensive synthesis of new cytokinins, which led to the preparation of the highly efficient and easily prepared aromatic cytokinin 6-benzylaminopurine (BAP). A huge number of biological experiments were then performed with this cytokinin derivative, which significantly contributed to our understanding of cytokinin functions in plants. Findings of naturally occurring BAP subsequently led to the search for other aromatic cytokinins. In order to rapidly detect these substances in plants, a unique procedure based on screening of HPLC fractions using specific ELISAs with antisera against individual aromatic cytokinins has been introduced. Isolation of these substances and their subsequent identification by mass spectrometry confirmed the natural occurrence of highly active endogenous ARCK *meta*-topolin (6-(3-hydroxybenzylamino)purine), but also a less active *ortho*-topolin analogue (6-(2-hydroxybenzylamino)purine) and their metabolites. The structure of topolins suggests biosynthetic and metabolic biosynthetic pathways that are likely to be similar to closely related isoprenoid cytokinins like zeatins. It also seems likely that the molecular mechanisms of action will be based on very closely related principles, although differences certainly exist, at least at the level of the signals and their recognitions carried by these phytohormones. This review

M. Strnad (✉)
Laboratory of Growth Regulators, Institute of Experimental Botany of the Czech Academy of Sciences and Palacký University, Olomouc, Czech Republic
e-mail: miroslav.strnad@upol.cz

N. Ahmad, M. Strnad (eds.), *Meta-topolin: A Growth Regulator for Plant Biotechnology and Agriculture*, https://doi.org/10.1007/978-981-15-9046-7_1

attempts to summarise the current state of knowledge about ARCK and to point out its practical use in agriculture and biotechnology.

Keywords

Aromatic cytokinins · History · Occurrence · *Meta*-topolin · Topolins

Abbreviations

ARCK	Aromatic cytokinin
BAP	6-Benzylaminopurine
CBP	Cytokinin-binding protein
ELISA	Enzyme-linked immunosorbent assay
HPLC	High-performance liquid chromatography
iP	N^6-Isopentenyladenine
ISCK	Isoprenoid cytokinin
K	Kinetin
*m*T	*Meta*-topolin
*m*TR	*Meta*-topolin riboside
*o*T	*Ortho*-topolin
*o*TR	*Ortho*-topolin riboside
*p*T	*Para*-topolin
*p*TR	*Para*-topolin riboside
*t*Z	*Trans*-zeatin

The history of cytokinins (ARCK) began in 1955, when Miller et al. isolated the first biologically active member of this group of natural phytohormones, 6-furfurylaminopurine (Fig. 1.1), from autoclaved herring sperm DNA. Due to its ability to stimulate cytokinesis in various plant tissues, this compound was named kinetin (K) (Miller et al. 1955a, b, 1956, Fig. 1.1). K thus became the first

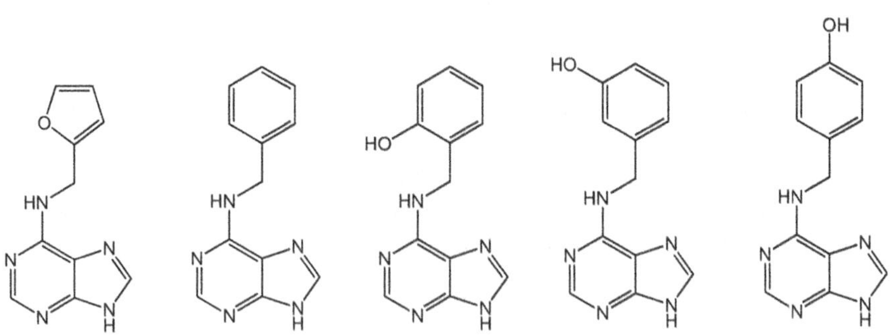

Fig. 1.1 Chemical structure of aromatic cytokinins. From left to right: kinetin (K, 6-furfurylaminopurine; BAP, 6-benzylaminopurine; *ortho*-topolin (*o*T, 6-(2-hydroxybenzylamino)purine); *meta*-topolin (*m*T, 6-(3-hydroxybenzylamino)purine); *para*-topolin, (*p*T, 6-(4-hydroxybenzylamino)purine)

representative of plant hormones, which were subsequently named cytokinins and which can be characterised as N^6-substituted adenine derivatives (Skoog et al. 1965). Therefore, the first attempts to prepare cytokinins (CKs) started with kinetin, followed by another cytokinin substituted by an aromatic ring at the N^6-position 6-benzylaminopurine (BAP, Miller et al. 1956). 6-(3-Hydroxybenzylamino)purine (*meta*-topolin) was synthesised a bit later. It was described for the first time by Okumura et al. (1959). The authors prepared BAP derivatives bearing hydroxyl, methyl, methoxy, amino, nitro and sulfonic acid groups on the benzyl ring. These functional groups were introduced into *ortho*-, *meta*- and *para*-positions of the N^6-benzyl substituent (Okumura et al. 1956, 1959, Fig. 1.1).

Cytokinin research is concentrated mainly on members of the isoprenoid class (ISCKs) represented by *cis*- and *trans-zeatin*, N^6-isopentenyladenine, dihydrozeatin and their metabolites. On the other hand, ARCKs are represented by BAP; *ortho*-, *meta*- and *para*-topolins (hydroxylated BAPs); and also kinetin. Aromatic cytokinins have long been considered purely synthetic cytokinins. For example, kinetin was considered an artificial product of DNA rearrangement (Hall and De Ropp 1955), although some scientists thought that it could be formed in vivo through DNA degradation (Skoog 1994). The first evidence of its natural occurrence came in 1996, when K was identified in extracts from nodes of the root of *Casuarina equisetifolia* infected with bacteria *Frankia* (Raman and Elumalai 1996). Barciszewski et al. (1996) discovered K in plant cell samples, in DNA isolated from in vitro cultured human fibroblasts and in commercial samples of calf thymus DNA (Barciszewski et al. 1996, 1997a, b). Subsequently, K was also found in the urine of cancer patients (10–100-fold less in healthy patients; Barciszewski et al. 2000), in coconut milk (Ge et al. 2005) and in the brain and liver tissue of transgenic mouse with familial dysautonomia (Shetty et al. 2011). In many cases, K has not however been detected in different tissues and organisms. Our efforts to prove the occurrence of K have so far been quite unsuccessful. For example, we have not demonstrated its occurrence in extracts from *Caenorhabditis elegans* and *Escherichia coli* (Kadlecová et al. 2018). Orr and colleagues were also unable to find K in native mouse and rat brain tissue (Orr et al. 2017). We have also failed to demonstrate the occurrence of K in many plant samples of different evolutionary origin (unpublished data), suggesting that the question of the natural occurrence of kinetin is probably more complicated than previously thought.

Cytokinins with a benzyl side chain have also been detected and identified in different plant tissues, but occur also much more rarely than their isoprenoid counterparts which are commonly present in all plant species. For example, Horgan et al. (1973, 1975) identified the first ARCK from mature poplar leaves and identified it as 6-(2-hydroxybenzylamino)-9-β-D-ribofuranosylpurine (*ortho*-topolin riboside, *o*TR). This ARCK metabolite was also discovered in *Zantedeschia aethiopica* fruits together with its 2-methylthio-9-glucofuranosyl derivative (Chaves das Neves and Pais 1980a, b). The glucosyl moiety of the last compound was assigned a furanosyl ring, but a 9-glucopyranoside structure is more probable by analogy with other cytokinin 9-glucosides (see MacLeod et al. 1976). The discoveries were followed by isolation and tentative identification of BAP and its metabolites in different plant tissues (Ernst et al. 1983; Nandi et al. 1989a, b). From the published data, we concluded that hydroxybenzylaminopurines could occur naturally as analogues of the commonly used cytokinin BAP (Strnad et al. 1992a, b). We were also disappointed with the results from studies on cytokinin-binding proteins (CBP) that ARCKs must be more widespread phytohormones. The first ARCK-binding protein CBF-1, studied in detail by Fox and his co-workers (Fox and Erion 1975, 1977; Erion and Fox 1981; Brinegar and Fox 1985a, b, 1987; Brinegar et al. 1985), is probably the best characterised of all plant hormone-binding proteins. It has amino acid similarity to vicilin-type seed storage proteins (Brinegar et al. 1988). It is unclear why this CBP has relatively high affinity ($K_d = 10^{-7}$ M) for BAP, but much lower affinity for natural ISCKs, nonactive cytokinins and adenines (Keim et al. 1981). In one case, the biological activity of various ARCKs in callus bioassay even agreed closely with their relative order of affinity for the binding protein (Sussman and Kende 1978). These data further supported our idea that cytokinins bearing a benzyl ring at the N^6-position must be more widespread in plant tissues than appeared.

For accurate and fast detection of the aromatic cytokinins (ARCKs) in plant tissues, we developed a very new approach based on combined high-performance liquid chromatography-enzyme-linked immunosorbent assay (HPLC-ELISA) which showed high selectivity for BAP; *ortho*-, *meta*- and *para*-hydroxybenzylaminopurine; and their 9-substituted derivatives (Strnad et al. 1992a; Strnad 1996). A typical protocol most commonly used for the isolation of new substances of natural origin is generally based on bioassay-guided fractionation, which consists of the gradual fractionation of bioactive substances based on differences in their physicochemical properties and evaluation of certain types of biological activity; the separation and testing steps are repeated several times (Weller 2012). The new cytokinin analytical strategy applied here for screening different plant tissues was much faster, precise and more robust being based on selective antibody recognition properties. The extracts were fractionated by HPLC similar to classical natural product discovery techniques, but the fractions were analysed by group-specific ELISA assays (ELISA selective for BAP; *ortho*-, *meta*- and *para*-topolin; and their appropriate 9-riboside and 9-glucosides). Immunoactivity ELISA analysis of HPLC fractions detected major immunoactive peaks co-chromatographing with the ARCK standards. Using this approach, we have been able to discover a brand new family of endogenous aromatic cytokinins centred around the highly active compound, 6-(3-hydroxybenzylamino)purine (*meta*-topolin, *m*T; Strnad et al. 1997). Because this compound was first detected in poplar leaves, we adopted the trivial name '*meta*-topolin' (*m*T), derived from 'topol', the Czech word for a poplar (Strnad et al. 1997). This new family of cytokinin-related growth regulators include, apart from *m*T, the highly active BAP and the much less active 6-(2-hydroxybenzylamino)purine (*ortho*-topolin, oT), 6-(4-hydroxybenzylamino)purine (*para*-topolin, *p*T) and their metabolites. Other hydroxyl and methoxy derivatives as well as their sugar conjugates have also been identified in other plant species than poplar leaves (Strnad et al. 1992b, 1994;

Goicoechea et al. 1995, Jones et al. 1996; Doležal et al. 2002; Tarkowská et al. 2003).

A recent study reports that topolins are also produced by *Mycobacterium tuberculosis*, a human-exclusive pathogen. They are formed by the activity of Rv1205, a homologue of the plant enzyme LONELY GUY, from respective riboside 5-′-monophosphates (Samanovic et al. 2015). Furthermore, the intracellular accumulation of *para*-hydroxybenzaldehyde (*p*HBA) in the LOG-deficient mycobacterial strain, which is a breakdown product of *para*-topolin, is likely to be responsible for the sensitisation of proteasomal degradation mutants to NO. The addition of *p*HBA to mycobacteria was lethal selectively in the NO presence but had no effect on similarly treated cultures of *E. coli*, suggesting that this is probably the mechanism by which LOG and cytokinins in the absence of the proteasome lead to NO sensitivity (Zhu and Javid 2015). I would like to mention here that Log-like homologues were identified in several other bacterial species, including the important human and animal pathogens *Bordetella* spp. and *Staphylococcus aureus*. It can be assumed that elucidation of the role of cytokinins in the bacterial pathophysiology of humans, animals and plants will undoubtedly provide more insights into their future important functions.

Targeted perturbation of this pathway by, for example, cytokinin analogues, may provide a novel therapeutic approach to tuberculosis treatment. As LONELY GUY homologues are present in the genomes of other bacterial pathogens, including *Staphylococcus aureus*, such an approach could have a significant impact on the treatment of bacterial infections. *Para*-topolin riboside (*p*TR) has been studied mainly in connection with neuroprotection. It was identified as one of the cytoprotective substances in plant *Gastrodia elata* that is used in Chinese traditional medicine for the treatment of headaches, dizziness and epilepsy (Huang et al. 2007). Overall, the most recent data show that *p*TR has promising activity in the treatment of neurological diseases (Huang et al. 2011; Visentin et al. 2013; Chou et al. 2015; Hong et al. 2019).

It is also interesting to note that a recent 12-week clinical study on 39 subjects (Garcia et al. 2018) demonstrated beneficial effects of topical *para*-topolin application (*p*T, designated 4HBAP in this study). At the concentration of 0.1%, *p*T improved the appearance of fine and coarse wrinkles and reduced skin roughness and hyperpigmentation. It also increased skin hydration and had a positive effect on both facial erythema and non-inflammatory acne lesions.

Collectively, *m*T, *o*T, pT and BAP and their metabolites constitute natural aromatic cytokinins, distinguishable from the isoprenoid cytokinins both chemically and because of their spectrum of biological activities. The aromatic cytokinins found in *P.* × *canadensis* were recently remeasured by sophisticated UHPLC-MS/MS analyses of leaf extracts from the same tree, from 12 gene bank accessions as well as during the season (Jaworek et al. 2019, 2020). While oT derivatives were found, those of mT were not. Currently, we cannot exclude the possibility that topolins are not products of poplar metabolism but endophyte products (Wang et al. 2019). Observed fluctuations could account for the changes in endophyte growth. This hypothesis is at least partially supported by identification of endogenous ARCKs in different algae strains (Stirk et al. 2003; Ordog et al. 2004) and also by identification of *o*TR in poplar tRNA (Jaworek et al. 2019). In algae, all three isomers of topolins are present (*ortho*-, *meta*- and *para*-), with *o*T occurring at higher concentrations than the other isomers. For O-glucosides, *meta*-isomers (*meta*-topolin-O-glucoside and *meta*-topolin riboside-O-glucoside) were present in higher concentrations than other isomers. It is important to note that no N-glucosides were detected in micro- and macroalgae (Stirk et al. 2003; Ordog et al. 2004). The main difference between the two groups of algae (micro- and macroalgae) was that the BAP-type cytokinins were found in higher percentages in microalgae (1%–28%) than in macroalgae (below 1%) (Stirk et al. 2003). The origin of topolins in plants remains unclear and needs to be further investigated in detail.

To conclude, the ARCKs as a group of plant growth regulators are already of immerse economic importance. The micropropagation industry is based on the ability of cytokinin to relax apical dominance so that axillary buds grow outwards, proliferating the numbers of shoots for cultivation. The future of plant biotechnology rests with genetically engineered plants, for which cytokinin-induced shoot formation is an obligatory step. For both of these processes, unwanted side effects of BAP treatment and the more or less permanent suppression of root formation and growth, for instance, are serious problems (Aremu et al. 2012). Evidence for the unique biological activity of *meta*-topolin in many plant in vitro systems (Werbrouck et al. 1996) has been followed up by a synthetic and biochemical study of new, more efficient ARCK derivatives as clearly seen from the many reviews in this book.

The important role of ARCKs in plant cell growth and differentiation has also attracted the attention of researchers working with animal and human cells. Here, cytokinins and their derivatives were demonstrated to either promote or inhibit the division of mammalian cells depending on the cytokinin and experimental system used (Strnad 1997; Voller et al. 2017). Some also induce differentiation of various cell lines, including keratinocytes and certain leukemias (see Voller et al. 2017). Many other pharmacological activities are reported, including neuroprotective, immunomodulatory and anti-angiogenic (e.g. see Voller et al. 2019).

The fact that plants and also animals have two related groups of cytokinins with overlapping spectra of biological activity gives us a fascinating insight into the potential complexity of developmental control. What is clear at this stage is that ARCKs and ISCKs are not merely alternative forms of the same signal. Clear differences are emerging. First, although the biosynthetic pathway of ARCKs is not firmly established, the evidence points to an entirely separate pathway for ARCKs, implying separate control of biosynthesis. It is also possible that the endogenous levels of compounds in the two groups are functionally linked in some way, either directly or inversely. Second, there would seem to be a difference in the nature of the receptors for ARCKs and ISCKs. It is an extraordinary fact that ARCK-specific binding proteins that do not bind ISCKs have been rejected for years as a puzzling artefact. Three cytokinin receptors, AHK2, AHK3 and CRE1/AHK4, have been described in *Arabidopsis* (Inoue et al. 2001; Suzuki et al. 2001; Yamada et al. 2001). Spíchal et al. (2004) showed that the most active ISCK *trans*-zeatin (*t*Z) is a strong activator of the cytokinin receptors, while *m*T activates preferably the receptor AHK3. The relative activity of *m*T at AHK3 and CRE1/AHK4 receptors was found to be 80% and 30% of *t*Z activity, respectively. *m*TR did not activate these receptors, but it was able to activate the expression of ARR5 (*Arabidopsis* cytokinin response regulator 5) in *Arabidopsis* plants (Spíchal et al. 2004). It has also been shown that cytokinin receptors from different plant species bind *m*T and with higher affinity than BAP and *o*T (Spíchal et al. 2004; Kuderová et al. 2014; Jaworek et al. 2020). To resolve the relationships between ARCKs and ISCKs at the receptor level will necessitate further controlled experimental manipulation of the levels of hormonal signals and receptor activity.

Acknowledgements The author of this review is grateful to Alexander Oulton, PhD., for his critical reading of the chapter. Moreover, this review is supported by grant no. 18-07563S from the Czech Science Foundation and ERDF project "Plants as a tool for sustainable global development" (No. CZ.02.1.01/0.0/0.0/16_019/0000827).

References

Aremu AO, Bairu MW, Doležal K, Finnie JF, Van Staden J (2012) Topolins: a panacea to plant tissue culture challenges? Plant Cell Tiss Org Cult 108:1–16

Barciszewski J, Siboska GE, Pedersen BO, Clark BF, Rattan SI (1996) Evidence for the presence of kinetin in DNA and cell extracts. FEBS Lett 393(2–3):197–200

Barciszewski J, Siboska GE, Pedersen BO, Clark BF, Rattan SI (1997a) Furfural, a precursor of the cytokinin hormone kinetin, and base propenals are formed by hydroxyl radical damage of DNA. Biochem Biophys Res Commun 238(2):317–319

Barciszewski J, Siboska GE, Pedersen BO, Clark BF, Rattan SI (1997b) A mechanism for the in vivo formation of N^6-furfuryladenine, kinetin, as a secondary oxidative damage product of DNA. FEBS Lett 414(2):457–460

Barciszewski J, Mielcarek M, Stobiecki M, Siboska G, Clark BF (2000) Identification of 6-furfuryladenine (kinetin) in human urine. Biochem Biophys Res Commun 279(1):69–73

Brinegar AC, Fox JE (1985a) Resolution of the subunit composition of a cytokinin binding protein from wheat embryos. Biol Plant 27:100–109

Brinegar AC, Fox JE (1985b) The developmental expression and molecular cloning of the gene for a wheat embryo cytokinin-binding protein. In: Freeling M (ed) Plant genetics. UCLA symposium on molecular and cellular biology, vol 35. Alan R. Liss, New York, pp 147–149

Brinegar AC, Fox JE (1987) Immunocytological localization of a wheat embryo cytokinin-binding protein and its homology with proteins in other cereals. In: Klambt D (ed) Plant hormone receptors. NATO ASI series H, cell biology, vol 10. Springer-Verlag, Berlin, pp 177–183

Brinegar AC, Stevens A, Fox JE (1985) Biosynthesis and degradation of a wheat embryo cytokinin-binding protein during embryogenesis and germination. Plant Physiol 79:706–710

Brinegar AC, Shabanowitz J, Hauer C, Cooper G, Stevens A, Fox JE (1988) Characterization of a benzyladenine binding with peptide isolated from a wheat cytokinin-binding protein: sequence analysis and identification of a single affinity—labeled histidine residue by mass spectrometry. Proc Natl Acad Sci U S A 85:5927–5931

Chaves das Neves HJ, Pais MSS (1980a) A new cytokinin from the fruits of *Zantedeschia aethiopica*. Tetrahedron Lett 21:4387–4390

Chaves das Neves HJ, Pais MSS (1980b) Identification of a spathe regreening factor in *Zantedeschia aethiopica*. Biochem Biophys Res Commun 95:1387–1392

Chou AH, Chen YL, Chiu CC et al (2015) T1-11 and JMF1907 ameliorate polyglutamine-expanded ataxin-3-induced neurodegeneration, transcriptional dysregulation and ataxic symptom in the SCA3 transgenic mouse. Neuropharmacology 99:308–317

Doležal K, Åstot C, Hanuš J et al (2002) Identification of aromatic cytokinins in suspension cultured photoautotrophic cells of *Chenopodium rubrum* by capillary liquid chromatography/frit–fast atom bombardment mass spectrometry. Plant Growth Regul 36:181–189

Erion JL, Fox JE (1981) Purification and properties of a protein which binds cytokinin active 6-substituted purines. Plant Physiol 64:156–162

Ernst D, Schafer W, Oesterhelt D (1983) Isolation and identification of a new naturally occurring cytokinin (6-benzylaminopurine riboside) from an old anise cell culture (*Pimpinella anisum* L.). Planta 159:222–225

Fox JE, Erion JL (1975) A cytokinin binding protein from higher plant ribosomes. Biochem Biophys Res Commun 64:694–700

Fox JE, Erion JL (1977) Cytokinin binding proteins in higher plants. In: Pilet PE (ed) Plant growth regulation, Proceedings of the 9th International Conference on plant growth substances. Springer Verlag, Berlin, pp 139–147

Garcia R, McCullough J, Reece B, Massino F (2018) Novel topical cytokinin improves coarse wrinkles and skin roughness without skin irritation. J Drugs Dermatol 17(7):772–779

Ge L, Yong JWH, Goh NK, Chia LS, Tan SN, Ong ES (2005) Identification of kinetin and kinetin riboside in coconut (*Cocos nucifera* l.) water using a combined approach of liquid chromatography–tandem mass spectrometry, high performance liquid chromatography and capillary electrophoresis. J Chromatogr B 829(1–2):26–34

Goicoechea N, Doležal K, Antolín MC, Strnad M, Sánchez-Díaz M (1995) Influence of mycorrhizae and *Rhizobium* on cytokinin content in drought-stressed alfalfa. J Exp Bot 46:1543–1549

Hall RH, De Ropp R (1955) Formation of 6-furfurylaminopurine from DNA breakdown products. J Am Chem Soc 77(23):6400–6400

Hong SI, Peyton L, Chern Y, Choi DS (2019) Novel adenosine analog, N^6-(4-hydroxybenzyl)-adenosine, dampens alcohol drinking and seeking behaviors. J Pharmacol Exp Ther 371:260–267

Horgan R, Hewett EW, Purse JG, Wareing PF (1973) A new cytokinin from *Populus robusta*. Tetrahedron Lett 30:2827–2828

Horgan R, Hewett EW, Horgan JM, Purse JG, Wareing PF (1975) A new cytokinin from *Populus x robusta*. Phytochemistry 14:1005–1008

Huang N-K, Chern Y, Fang J-M et al (2007) Neuroprotective principles from Gastrodia elata. J Nat Prod 70:571–574

Huang NK, Lin JH, Lin JT, Lin CI, Liu EM, Lin CJ, Chen WP, Shen YC, Chen HM, Chen JB et al (2011) A new drug design targeting the adenosinergic system for Huntington's disease. PLoS One 6(6):e20934

Inoue T, Higuchi M, Hashimoto Y et al (2001) Identification of CRE1 as a cytokinin receptor from *Arabidopsis*. Nature 409:1060–1063

Jaworek P, Kopečný D, Zalabák D et al (2019) Occurrence and biosynthesis of cytokinins in poplar. Planta 250:229–244

Jaworek P, Tarkowski P, Hluska T et al (2020) Characterization of five CHASE-containing histidine kinase receptors from *Populus × canadensis* cv. *Robusta* sensing isoprenoid and aromatic cytokinins. Planta 251:1

Jones LH, Martínková H, Strnad M, Hanke DE (1996) Occurrence of aromatic cytokinins in oil palm (*Elaeis guineensis* Jacq.). J Plant Growth Regul 15:39–49

Kadlecová A, Jirsa T, Novák O, Kammenga J, Strnad M, Voller J (2018) Natural plant hormones cytokinins increase stress resistance and longevity of *Caenorhabditis elegans*. Biogerontology 19(2):109–120

Keim P, Erion J, Fox JE (1981) The current status of cytokinin-binding moieties. In: Guern J, Péaud-Lenoël C (eds) Metabolism and molecular activities of cytokinins. Springer-Verlag, Berlin, pp 179–190

Kuderová A, Gallová L, Kuricová K et al (2014) Identification of AHK2- and AHK3-like cytokinin receptors in *Brassica napus* reveals two subfamilies of AHK2 orthologues. J Exp Bot 66:339–353

MacLeod JK, Summons RE, Letham DS (1976) Mass spectrometry of cytokinin metabolites. Per (trimethylsilyl) and permethyl derivatives of glucosides of zeatin and 6-benzylaminopurine. J Org Chem 41:3959–3967

Miller CO, Skoog F, Okumura FS, Von Saltza MH, Strong F (1955a) Structure and synthesis of kinetin. J Am Chem Soc 77(9):2662–2663

Miller CO, Skoog F, Von Saltza MH, Strong F (1955b) Kinetin, a cell division factor from deoxyribonucleic acid. J Am Chem Soc 77(5):1392–1392

Miller CO, Skoog F, Okumura F, Von Saltza M, Strong F (1956) Isolation, structure and synthesis of kinetin, a substance promoting cell division. J Am Chem Soc 78(7):1375–1380

Nandi SK, Palni LMS, Letham DS, Wong OC (1989a) Identification of cytokinins in primary crown gall tumors of tomato. Plant Cell Environ 12:273–283

Nandi SK, Letham DS, Palni LMS, Wong OC, Summons RE (1989b) 6-Benzylaminopurine and its glycosides as naturally occurring cytokinins. Plant Sci 61:189–196

Okumura FS, Masumura M, Motoki T, Takahashi T, Kuraishi S (1956) Syntheses of kinetin analogues I. Short Commun 30:194–195

Okumura FS, Kotani Y, Ariga T, Masumura M, Kuraishi S (1959) Syntheses of kinetin analogues II. Short Commun 32:883–884

Ordog V, Stirk WA, van Staden J, Novák O, Strnad M (2004) Endogenous cytokinins in three genera of microalgae from the chlorophyta. J Phycol 40(1):88–95

Orr AL, Rutaganira FU, de Roulet D, Huang EJ, Hertz NT, Shokat KM, Nakamura K (2017) Long-term oral kinetin does not protect against α-synuclein-induced neurodegeneration in rodent models of Parkinson's disease. Neurochem Int 109:106–116

Raman N, Elumalai S (1996) Presence of cytokinin in the root nodules of *Casuarina equisetifolia*. Indian J Exp Biol 34:577–579

Samanovic MI, Tu S, Novák O, Iyer LM, McAllister FE, Aravind L, Gygi SP, Hubbard SR, Strnad M, Darwin KH (2015) Proteasomal control of cytokinin synthesis protects *Mycobacterium tuberculosis* against nitric oxide. Mol Cell 57:984–994

Shetty RS, Gallagher CS, Chen YT, Hims MM, Mull J, Leyne M, Pickel J, Kwok D, Slaugenhaupt SA (2011) Specific correction of a splice defect in brain by nutritional supplementation. Hum Mol Genet 20:4093–4101

Skoog F (1994) A personal history of cytokinin and plant hormone research. In: Mok DWS, Mokeds MC (eds) Cytokinin: chemistry, activity and function. CRC, Boca Raton, FL, pp 1–14

Skoog F, Strong F, Miller CO (1965) Cytokinins. Science 148(3669):532–533

Spíchal L, Rakova NY, Riefler M et al (2004) Two cytokinin receptors of *Arabidopsis thaliana*, CRE1/AHK4 and AHK3, differ in their ligand specificity in a bacterial assay. Plant Cell Physiol 45:1299–1305

Stirk WA, Novák O, Strnad M, van Staden J (2003) Cytokinins in macroalgae. Plant Growth Regul 41:13–24

Strnad M (1996) Enzyme immunoassays of N^6-benzyladenine and N^6-(*meta*-hydroxybenzyl)adenine cytokinins. J. Plant Growth Regul 15:179–188

Strnad M (1997) The aromatic cytokinins. Physiol Plant 671:674–688

Strnad M, Vereš K, Hanuš J, Siglerová V (1992a) Immunological methods for quantification and identification of cytokinins. In: Kamínek M, Mok DWS, Zažímalová E (eds) Physiology and biochemistry of cytokinins in plants. SPB Academic, The Hague, pp 437–446

Strnad M, Peters W, Beck E, Kamínek M (1992b) Immunodetection and identification of N^6-(o-hydroxybenzylamino)purine as naturally occurring cytokinin in *Populus x canadensis* cv Robusta leaves. Plant Physiol 99:74–80

Strnad M, Peters W, Beck E (1994) Ortho-topolin-9-glucoside, an aromatic cytokinin from *Populus x canadensis* cv *Robusta* leaves. Phytochemistry 37:1059–1062

Strnad M, Hanuš J, Vaněk T et al (1997) Meta-topolin, a highly active aromatic cytokinin from poplar leaves (*Populus* × *canadensis* moench., cv. *Robusta*). Phytochemistry 45:213–218

Sussman MR, Kende H (1978) In vitro cytokinin binding to a particulate fraction of tobacco cells. Planta 140:251–259

Suzuki T, Miwa K, Ishikawa K et al (2001) The Arabidopsis sensor His-kinase, AHK4, can respond to cytokinins. Plant Cell Physiol 42:107–113

Tarkowská D, Doležal K, Tarkowski P et al (2003) Identification of new aromatic cytokinins in Arabidopsis thaliana and *Populus* × *can*adensis leaves by LC-(+)ESI-MS and capillary liquid chromatography/frit–fast atom bombardment mass spectrometry. Physiol Plant 117:579–590

Visentin S, De Nuccio C, Bernardo A, Pepponi R, Ferrante A, Minghetti L, Popoli P (2013) The stimulation of adenosine A2A receptors ameliorates the pathological phenotype of fibroblasts from Niemann-Pick type C patients. J Neurosci 33(39):15388–15393

Voller J, Maková B, Kadlecová A, Gonzalez G, Strnad M (2017) Plant hormone cytokinins for modulating human aging and age-related diseases. In: Hormones in ageing and longevity. Springer, Berlin, pp 311–335

Voller J, Béres T, Zatloukal M, Džubák P, Hajdúch M, Doležal K, Schmulling T, Strnad M (2019) Anti-cancer activities of cytokinin ribosides. Phytochem Rev 18(4):1101–1113

Wang Y, Zhang W, Ding C et al (2019) Endophytic communities of transgenic poplar were determined by the environment and niche rather than by transgenic events. Front Microbiol 10:588

Weller MG (2012) A unifying review of bioassay-guided fractionation, effect-directed analysis and related techniques. Sensors (Basel) 12(7):9181–9209

Werbrouck SPO, Strnad M, Van Onckelen HA, Debergh PC (1996) Meta-topolin, an alternative to benzyladenine in tissue culture? Physiol Plant 98:291–297

Yamada H, Suzuki T, Terada K et al (2001) The Arabidopsis AHK4 histidine kinase is a cytokinin-binding receptor that transduces cytokinin signals across the membrane. Plant Cell Physiol 42:1017–1023

Zhu JH, Javid B (2015) Tuberculosis: hey there, Lonely guy! Mol Cell 57:951–952

Synthesis and Chemistry of Meta-Topolin and Related Compounds

2

Lucie Plíhalová

Contents

Abstract

The chapter illustrates the gradual way in which the molecule of the active aromatic cytokinin, *meta*-topolin, was created. The historical development of the synthetic method from the first attempts to prepare aromatic cytokinins to professional organic synthesis is described here. The chapter also covers the preparation of second-generation aromatic derivatives of *meta*-topolin, e.g. C2- and/or N9-derivatives. The added value of such cytokinin derivatives over the original *meta*-topolin molecule is highlighted.

The stability and chemistry of N9-substituted *meta*-topolin derivatives were studied, and the mechanisms of action have been suggested. The benefits of these non-toxic compounds as well as the potential of specific functionalized derivatives of value to plants are discussed.

L. Plíhalová (✉)
Department of Chemical Biology, Faculty of Science, Palacký University, Olomouc, Czech Republic

Laboratory of Growth Regulators, Institute of Experimental Botany ASCR, Olomouc, Czech Republic
e-mail: lucie.plihalova@upol.cz

N. Ahmad, M. Strnad (eds.), *Meta-topolin: A Growth Regulator for Plant Biotechnology and Agriculture*, https://doi.org/10.1007/978-981-15-9046-7_2

Keywords

Aromatic cytokinin synthesis · Meta-topolin synthesis development · Modern synthetic approach

Abbreviations

ARCK	Aromatic cytokinins
BAP	6-Benzylaminopurine, N^6-benzyladenine
CK	Cytokinin
KIN	6-Furfurylaminopurine, N^6-furfuryladenine
mTR	*Meta*-topolin riboside, 6-(3-hydroxybenzylamino)purine riboside
oTR	*Ortho*-topolin riboside, 6-(2-hydroxybenzylamino)purine riboside
THF	Tetrahydrofuran-2-yl
THP	Tetrahydropyran-2-yl
WLSA	Detached wheat leaf senescence assay

2.1 From the First Artificial Aromatic Cytokinins to *Meta*-Topolin

The synthesis of the first cytokinins was initiated by the discovery of 6-furfurylaminopurine (kinetin, KIN) in autoclaved DNA samples in 1955. The compound was identified as a factor with the ability to influence cytokinesis (Miller et al. 1955, 1956). Despite the currently known fact that most purine-based cytokinins occurring naturally in plants are more likely substituted by an isoprenoid side chain at the N6 position of the adenine moiety, the first cytokinin discovered was accompanied by an aromatic furfuryl substituent. Therefore, the first attempts to prepare cytokinins (CKs) in the laboratory started with KIN, followed by another CK substituted by an aromatic benzyl moiety at the N6 position/6-benzylaminopurine (BAP, Miller et al. 1956).

Miller and his colleagues prepared KIN by the condensation of furfuryl chloride with adenine under alkaline conditions using sodium bicarbonate (Miller et al. 1956). BAP was prepared differently by a method based on the reaction of 6-methylthiopurine with benzylamine at a temperature of 140 °C for 16 h. The resulting yield was approximately 40–60% (Miller et al. 1955).

However, the process of aromatic cytokinin (ARCK) preparation was time consuming, and the resulting yields were not very satisfactory. Hence, over the years, a number of synthetic improvements were implemented. Currently, using nucleophilic substitution of SN_2 with 6-chloropurine by an appropriate amine under alkaline conditions is apparently provided by an excess of the appropriate amine or by the addition of trimethylamine. The reaction was for the first time published by Christensen and Daly for BAP synthesis (Daly and Christensen 1956). The reaction time was gradually shortened to approximately 3 h. The yields of the prepared CKs increased to 80% by further improvements of the reaction (Mik et al. 2011a). The excess of trimethylamine to provide the alkaline medium was used against excess of the appropriate amine. Finally, the reaction time of such synthesis has been recently shortened to several minutes due to the implementation of modern technologies such as microwave reactor synthesis. The yield was also increased to approximately 98% (Huang et al. 2007; Plíhalová et al. 2016). Of course, it must be noted that the reaction itself lasts for several minutes, but an hour of reaction condition tuning is necessary, and after the reaction, there are procedures linked with the purification of the product. The development of current synthetic chemistry tools led to phenomenal progress, especially in conversion rates.

Both ARCKs, KIN and BAP became very useful in biotechnologies, in particular in plant tissue culture (PTC), and the optimization of BAP preparation caused massive use of such CKs in PTC of many economically important plants such as banana (Arinaitwe et al. 2000; Aremu et al. 2012a), roses (Pati et al. 2006), strawberries (Borkowska 2001), apples (Dobránszki and Teixeira da Silva 2010), melons (Milazzo et al. 1999) and medicinal plants (Fajinmi et al. 2014; Bairu et al. 2007, 2009; Werbrouck et al. 1995). Several described BAP side effects such as inhibition of root growth and rooting of cultured explants provoked interest in research and development of new, improved BAP derivatives.

6-(3-Hydroxybenzylamino)purine (*meta*-topolin) was prepared for the first time by Okumura and his co-workers in 1959 (Okumura et al. 1959). They prepared BAP derivatives containing hydroxyl, methyl, methoxy, amino, nitro and sulfonic acid groups and introduced these functional groups into *ortho*-, *meta*- and *para*-positions of the benzyl ring (Okumura et al. 1956, 1959). The synthesis was performed by the condensation of 6-(methylmercapto)purine with the appropriate amines in a sealed and/or open tube at temperatures higher than 125 °C (Okumura et al. 1959). In order to gain more knowledge of the biological properties of CK derivatives containing the N^6 aromatic side ring, BAP derivatives substituted at the *ortho*-, *meta*- and *para*-positions on the benzyl ring, including *o*-, *m*- and *p*-topolin, were systematically prepared (Doležal et al. 2006). Their biological activity was tested using classical cytokinin bioassays, such as detached wheat leaf senescence assay (WLSA), Amaranthus bioassay and tobacco callus bioassay. The synthesis of *meta*-topolin and related analogues was based on the previously described procedures for BAP, in which the selected amine (3-hydroxybenzylamine in the case of *meta*-topolin) was mixed with trimethylamine and heated to reflux in a suitable alcohol, most often butanol, for 3 to 4 h (Fig. 2.1).

A total of 38 BAP derivatives were prepared and tested, including hydroxy (topolins) and methoxy groups (methoxy-topolins). Some of the methoxy-topolins had very strong antisenescence activity, especially *ortho*- and *meta*-methoxy-topolins (Doležal et al. 2004, 2006). It was found that the position of functional groups on the benzyl ring can significantly influence the biological properties of the CK derivatives. The biological activities were strongly dependent on the bioassay used (Doležal et al. 2004, 2006, 2007; Szüčová et al. 2009). The rapid development in the synthesis of the ARCKs enabled the construction of a large library of such substances. These derivatives were tested not only in a number of plant bioassays but

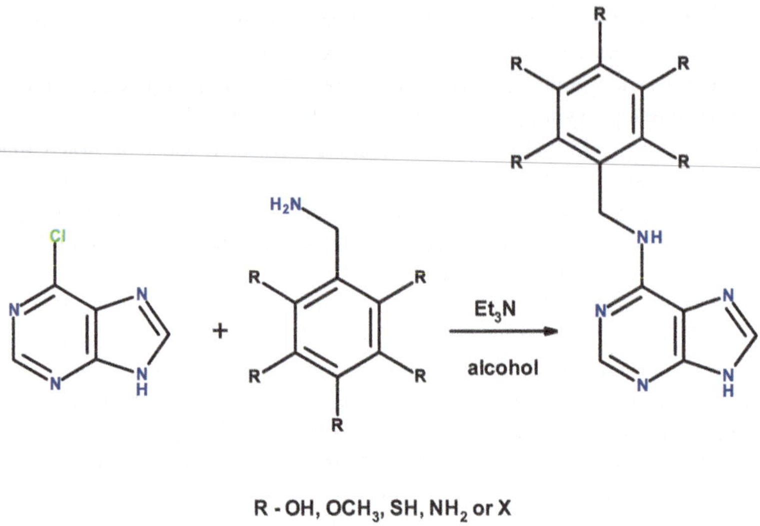

Fig. 2.1 Recent procedure, SN2 used for preparation of aromatic cytokinins

also in several biotechnological and agricultural applications (Plíhalová et al. 2016; Koprna et al. 2016—UV).

2.2 9-Substituted Topolin Derivatives

Substitutions at the N9 purine atom in CKs are common in plants. They are represented by a variety of natural conjugates, such as nucleosides, nucleotides and alanine derivatives (Davies 2007). CK-9-glucosides are highly abundant metabolites in plants. They are considered to be inactive forms of CK for their very low or zero CK activity. The binding of glucose to the N9 purine atom is considered irreversible in plants (Davies 2007; Kotek et al. 2010; Zahajská et al. 2019; Sakakibara 2006). In addition, naturally occurring 9-ribosides of ARCKs were discovered (Horgan et al. 1975; Strnad et al. 1997). Hydroxylated BAP derivatives substituted by ribose (BAPRs) were the first ARCK metabolites discovered in poplar leaves (*Populus x canadensis* Moench., cv. Robusta) (Horgan et al. 1975; Strnad 1997). Naturally occurring BAPRs were uniquely identified as *ortho-* and *meta-*topolin ribosides (oTR and mTR; 6-(2- and 3-hydroxybenzylamino)purine riboside) which were also recognized as active CK in standard bioassays such has tobacco callus, *Amaranthus* or WLSA. Especially in WLSA, ARCK ribosides exhibited significant biological activity. Although oTR was less active than BAPR, mTR was the most active of the three tested cytokinins (Holub et al. 1998; Zhang and Letham 1989). Bases of these nucleosides were named "topolins" according to the poplar or "topol" in the Czech language. A number of ARCK ribosides were later prepared by Doležal et al and their synthesis as well as biological activity, especially antisenescence and anticancer activity, was protected by several patents (Doležal et al. 2007, 2012). ARCK ribosides are active CK with significant biological activity despite the fact that most 9-substituted CK derivatives are unable to activate CK receptors (Doležal et al. 2007; Zhang and Letham 1989; Kende and Tavares 1968). On the other hand, it is interesting to note that these substances often trigger a CK response regardless of the level of CK receptor recognition and often have considerable biological activity in plant bioassays and in plants (Podlešáková et al. 2012; Spíchal et al. 2004; Vylíčilová et al. 2016).

Mok and Mok (2001) found that if the N9 substitution of CK is made by special protection groups, such as tetrahydropyran-2-yl (THP), tetrahydrofuran-2-yl (THF), 4-chlorbutyl or methyl, the arising 9-substituted CK derivatives show even higher biological activity than the original free bases. Especially in case of THP and THF, they mimic the structure of sugars, and the comparison with ribosides comes into consideration. On the other hand, it was shown that neither THP nor THF derivatives are toxic as comparable molecules accompanied by ribose (Doležal et al. 2007; Szüčová et al. 2007, 2009). These findings are supported by many other reports (Mok and Mok 2001; Fleysher et al. 1969; Szüčová et al. 2009; Young and Letham 1969; Weaver et al. 1965; Kende and Tavares 1968). It appears that 9-substitution is related to block of the purine moiety position for naturally occurring 9-glucosylation that inactivates CKs. Bearing in mind that the majority of 9-substituted CK still possess CK activity, blocking the N9 atom by suitable protecting group can serve as a barrier to the in situ formation of 9-glucosides and prolongs CK persistence in living plant systems (Bairu et al. 2011; Podlešáková et al. 2012).

As mentioned above, 9-substitution by THP or THF group can significantly increase the antisenescence activity of topolins (Young and Letham 1969; Szüčová et al. 2009). Combination of the appropriate 9-substitution with the presence of hydroxyl or methoxy group in the benzyl ring enables these second-generation ARCK molecules to avoid 9-glucosylation and promote reversible but less often O-glucosylation of CKs. 9-THP and 9-THF derivatives of *meta*-topolin and *meta*-methoxy-topolin in particular are revealed to be interesting compounds for use in plant biotechnologies and agriculture.

The first attempts for the preparation of 6,9-disubstituted derivatives resulted in the mixtures of 6,7- and 6,9-derivatives, which generally contained preferentially 9-isomers. For this reason, the synthetic procedure had to be subjected to further modification, and later, due to the implementation of modern BOP and DIPEA catalysts, it was possible to prepare regional-specific 6,9-isomers (Wan et al. 2005). N9 atom substitution can be performed in two steps using 6-chloropurine as the starting material. In the first step, the N9 atom of purine is substituted by the chosen protecting group, such as THP or ribose. The second step is realized via nucleophilic substitution of N6 atom by the appropriate amine, similar to the preparation of free bases (Szüčová et al. 2009, Fig. 2.2).

In addition to CK activities, some of the 6,9-disubstituted derivatives were cytotoxic to human tumour cells, e.g. ribosides, and hence, there was considerable interest in their development (Legraverend et al. 2000).

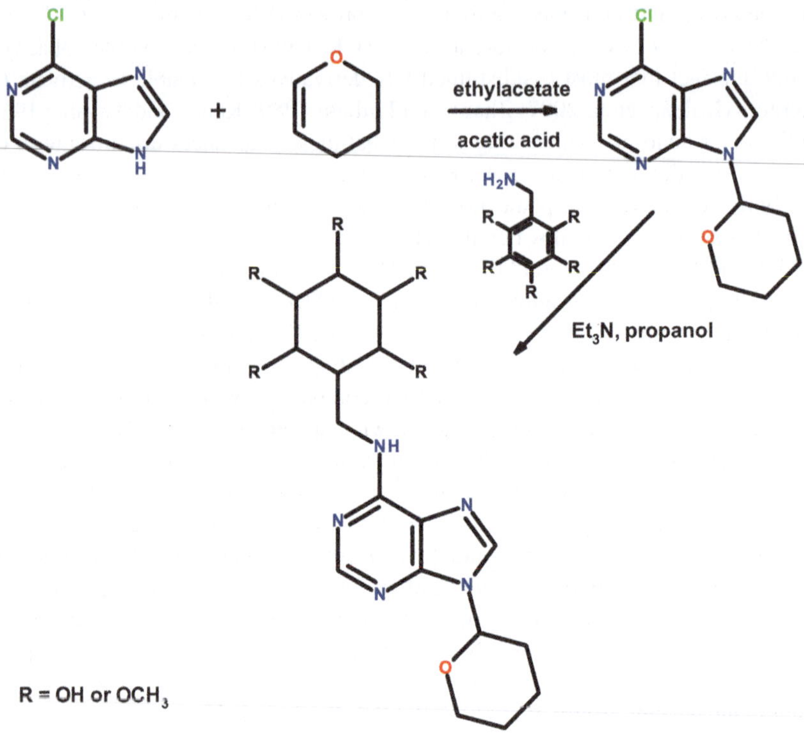

Fig. 2.2 Preparation of N9-(tetrahydropyran-2-yl)purine derivatives of substituted BAPs (Szüčová et al. 2009)

This aside, these compounds have a number of benefits for plant tissue culture, such as stimulation of root formation and its architecture in explants (Bairu et al. 2009), improvement of acropetal transport of CK and regulation of leaf senescence (Plíhal et al. 2013; Szüčová et al. 2011). 9-Substituted topolins also resist cytokinin oxidase/dehydrogenase (CKX) enzymatic degradation for a longer period of time (Podlešáková et al. 2012). In addition, 9-substituted KIN derivatives have been shown to actively protect lipid membranes against reactive oxygen species (ROS, Mik et al. 2011b).

2.3 2,6-Disubstituted and 2,6,9-Trisubstituted Aromatic Cytokinin Derivatives

2-Methyl-BAP was recognized as a compound with the ability to protect green plant material from harm, and its synthesis was firstly mentioned in the patent of Shell Oil Company nearly 60 years ago (Shell Patent 1966; Van Overbeek 1961).

Today, 2,6-dichloropurine or 2-fluoro-6-chloropurine (Zatloukal et al. 2008) is used as a starting material for preparing 2-substituted CK derivatives. 2-Substituted

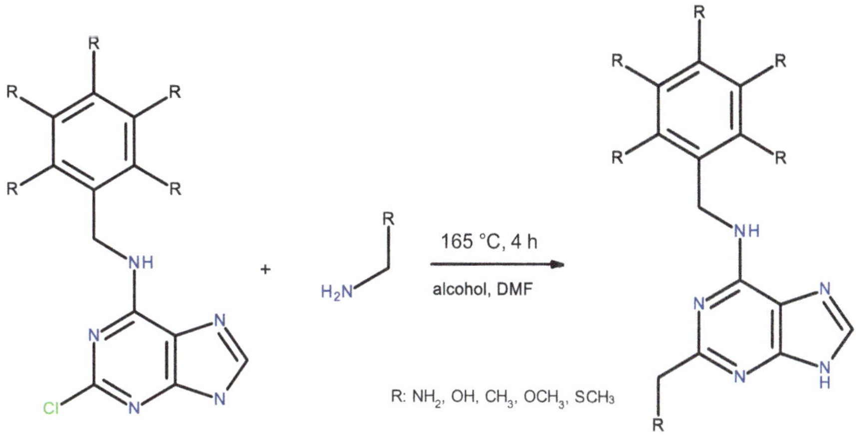

Fig. 2.3 Reaction scheme for the preparation of ARCKs substituted on purine C2

derivatives of *meta*-topolin were prepared from 2,6-dichloropurine or 2-fluoro-6-chloropurine (Zatloukal et al. 2008). The first step in the synthesis is the substitution of the C6-attached chlorine by suitable nucleophile, e.g. amine. In a subsequent step, C2-halogen can be suitably substituted with another functional group, e.g. amine, hydroxyl, methyl or methylthio group. However, not all functional groups can be added with equal efficiency, and the conversion rate and product purity may vary (Langli et al. 1996; Zatloukal et al. 2008) (Fig. 2.3).

2,6-Disubstituted derivatives have been shown to be biologically active in plant bioassays. For example, very interesting and highly biologically active *meta*-methoxy-topolin derivatives have been prepared by shortening the bond between the benzyl ring and the N^6-adenine by one carbon atom (Zatloukal et al. 2008) and by the addition of halogen atom to the C2 atom of the purine moiety. Specifically, 2-chloro-6-(3-methoxyphenylamino)purine and 2-fluoro-6-(3-methoxyphenylamino)purine were prepared and described as highly active inhibitors of the enzyme cytokinin oxidase/dehydrogenase (AtCKX2) from *A. thaliana* (Zatloukal et al. 2008; Spíchal et al. 2004, 2012). The inhibition of such enzyme can postpone the degradation of endogenously present CK in plants (Spíchal et al. 2004). From the example of these two derivatives, it is self-evident how the discovery of the molecular mechanisms of CK action can effectively be used to prepare fertilizers capable of significantly influencing plant growth and yield and enable plants to overcome abiotic stress. The compound, inhibitor of cytokinin degradation (INCYDE), described above, is capable of significantly affecting the response of plants to saline or mineral (cadmium) stress at a concentration starting from 10 nM and improving the yield of a range of crops and vegetables such as tomatoes, *Bulbine natalensis* Baker (Asphodelaceae) and *Rumex crispus* L. (Polygonaceae) (Gemrotová et al. 2013). In addition, it is also used in the micropropagation of some plants, for example, *Eucomis autumnalis* (Aremu et al. 2015) and banana trees (Aremu et al. 2012b).

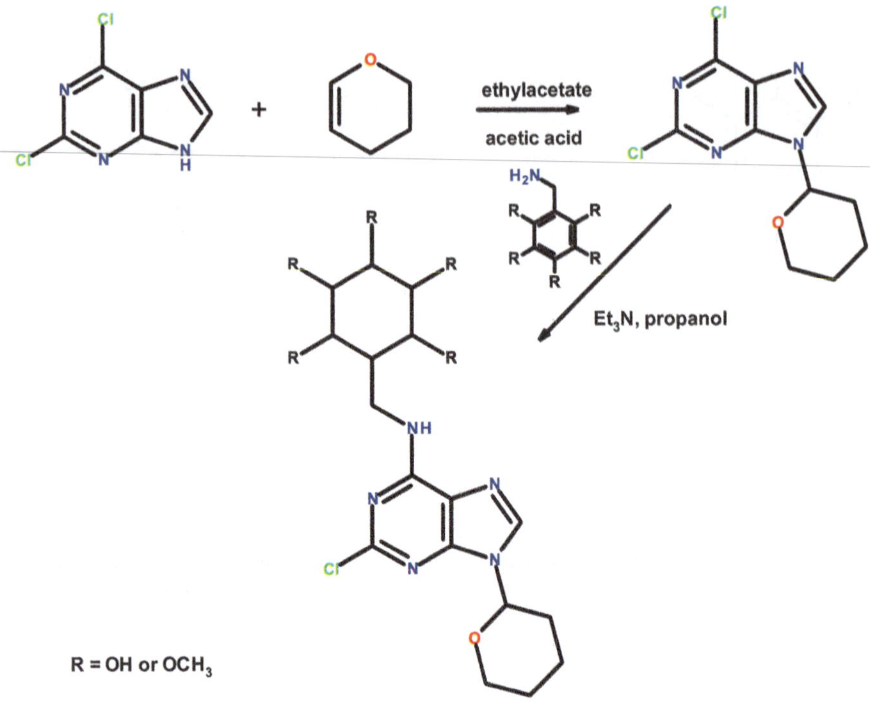

Fig. 2.4 The preparation of 2,6,9-trisubstituted ARCKs containing tetrahydropyran-2-yl (THF) at N9

It is also possible to prepare 2,6,9-trisubstituted purine derivatives using similar procedures as described above for the preparation of 6,9-disubstituted ARCKs (Szüčová et al. 2009). With this procedure, it is also possible to prepare very effective ARCKs, e.g. *meta*-topolin derivatives, which are viable as antisenescence agents (Fig. 2.4).

2.4 Stability of Meta-Topolin and N9-Substituted Meta-Topolin Derivatives

Although *meta*-topolin is a relatively stable compound, its 9-substituted derivatives, especially those covered by ribose, THP or THF groups, may undergo pH-dependent disintegration (Szüčová et al. 2009). It is for this reason that ribose or THP/THF are commonly used as protective groups in organic synthesis when further substitution of the molecule is needed and can be elegantly removed under controlled acidic conditions (Szüčová et al. 2009). The stability of THP/THF derivatives including *meta*-topolin-THP was described previously, and it was shown that the compounds are stable in pH around 5 and higher. Under this pH, gradual disintegration occurs and the free base of *meta*-topolin is released (Szüčová et al. 2009).

2.5 Summary

The preparation of ARCKs, including *meta*-topolin, has evolved over the past few decades when yields of prepared ARCKs were increased, syntheses were improved to produce higher HPLC purity products, while reaction time periods were gradually decreased. Improved syntheses over the past three decades have made ARCKs very readily available molecules that can be prepared in one- or two-step syntheses. More than 60 years after the preparation of the first ARCK, there has also been a significant development in the recognition of their effects and metabolism, and therefore more potent derivatives could be prepared that followed the natural molecular mechanisms of action. Key knowledge and understanding of the mechanism of action of O- and N9-glucosylation enabled the preparation of effective ARCKs, but also biologically active 9-substituted derivatives, which we now call second-generation ARCKs. All topolins and BAPs including their 2- or 9-substituted derivatives combined with the functionalization of the benzyl ring are protected by international patent applications (PCT) all over the world for use in agriculture and plant biotechnology, including plan tissue culture applications (Doležal et al. 2004). These derivatives effectively prevent the formation of N9-glucosides and thereby reduce ARCK inhibitory effects on root growth. Together with their high antisenescence activity and the prevention of unwanted side effects, these molecules form a new generation of ARCKs more and more utilized in plant tissue cultures of many crops and economically important plants.

Acknowledgement We would like to thank Mr. Alex Oulton for the careful reading of the manuscript. The work was supported by an ERDF project entitled "Development of Pre-Applied Research in Nanotechnology and Biotechnology" (No. CZ.02.1.01/0.0/0.0/17_048/0007323). The author of this review is grateful to Alexander Oulton, PhD., for the critical reading of the chapter.

References

Aremu AO, Bairu MW, Szüčová L, Doležal K, Finnie JF, Van Staden J (2012a) Assessment of the role of *meta*-topolins on *in vitro* produced phenolics and acclimatization competence of micropropagated "Williams" banana. Acta Physiol Plant 34:2265–2273

Aremu AO, Bairu MW, Szüčová L, Finnie JF, Van Staden J (2012b) The role of meta-topolins on the photosynthetic pigment profiles and foliar structures of micropropagated "Wiliams" bananas. J Plant Physiol 169:530–1541

Aremu AO, Stirk WA, Masondo NA, Plačková L, Novák O, Pěnčík A, Zatloukal M, Nisler J, Spíchal L, Doležal K, Finnie JF, van Staden J (2015) Dissecting the role of two cytokinin analogues (INCYDE and PI-55) on in vitro organogenesis, phytohormone accumulation, phytochemical content and antioxidant activity. Plant Sci 238:81–94

Arinaitwe G, Rubaihayo PR, Magambo MJ (2000) Proliferation rate effects of cytokinins on banana (*Musa* spp.) cultivars. Sci Hortic 86:13–21

Bairu MW, Stirk WA, Doležal K, Van Staden J (2007) Optimizing the micropropagation protocol for the endangered *Aloe polyphylla*: can *meta*-topolin and its derivatives serve as replacement for BA and zeatin? Plant Cell Organ Cult 90:15–23

Bairu MW, Jain N, Stirk WA, Doležal K, Van Staden J (2009) Solving the problem of shoot-tip necrosis in *Harpagophytum procumbens* by changing the cytokinin types, calcium and boron concentrations in the medium. S Afr J Bot 75:122–127
Bairu MW, Novák O, Doležal K, van Staden J (2011) Changes in endogenous cytokinin profiles in micropropagated Harpagophytum procumbens in relation to shoot-tip necrosis and cytokinin treatments. Plant Growth Regul 63:105–114
Borkowska B (2001) Morphological and physiological characteristics of micropropagated strawberry plants rooted *in vitro* or *ex vitro*. Sci Hortic 89:195–206
Daly JW, Christensen BE (1956) Purines. VI: the preparation of certain 6-substituted and 6,9-disubstituted purines. J Org Chem 21:177–179
Davies PJ (2007) The plant hormones: their nature, occurrence, and functions. In: Davies PJ (ed) Plant hormones, biosynthesis, signal transduction, action! Kluwer Academic, Dordrecht, pp 1–8
Dobránszki J, Teixeira da Silva JA (2010) Micropropagation of apple—a review. Biotechnol Adv 28:462–488
Doležal K, Popa I, Holub J, et al (2004) Heterocyclic compound based on N^6-substituted adenine, methods of their preparation, their use for preparation of drugs, cosmetic preparations and growth regulators, pharmaceutical preparations, cosmetic preparations and growth regulators containing these compounds. ÚEB AV ČR, CN1556808, AU2008203838, CA2455972, CZ294535, EP1419157, RS52389, US7279482
Doležal K, Popa I, Kryštof V, Spíchal L, Fojtíková M, Holub J, Lenobel R, Schmülling T, Strnad M (2006) Preparation and biological activity of 6-benzylaminopurine derivatives in plants and human cancer cells. Bioorg Med Chem 14:875–884
Doležal K, Popa I, Hauserová E, Spíchal L, Chakrabarty K, Novák O, Kryštof V, Voller J, Holub J, Strnad M (2007) Preparation, biological activity and endogenous occurrence of N6-benzyladenosines. Bioorg Med Chem 15:3737–3747
Doležal K, Popa I, Zatloukal M, Lenobel R, Hradecká D, Vojtěšek B, Uldrijan S, Mlejnek P, Werbrouck S, Strnad M (2012) Substitution derivatives of n6-benzyladenosine, methods of their preparation, their use for preparation of drugs, cosmetic preparations and growth regulators, pharmaceutical preparations, cosmetic preparations and growth regulators containing these compounds, US8119614
Fajinmi OO, Amoo SO, Finnie JF, Van Staden J (2014) Optimization of *in vitro* propagation of Coleonema album, a highly utilized medicinal and ornamental plant. S Afr J Bot 94:9–13
Fleysher MH, Bloch A, Hakala MT, Nichol CA (1969) Synthesis and biological activity of some new N6-substituted purine nucleosides. J Med Chem 12:1056–1061
Gemrotová M, Kularni MG, Stirk WA, Strnad M, van Staden J, Spíchal L (2013) Seedlings of medicinal plants treated with either a cytokinin antagonist (PI-55) or an inhibitor of cytokinin degradation (INCYDE) are protected against the negative effects of cadmium. Plant Growth Regul 71:137–145
Holub J, Hanuš J, Hanke DE, Strnad M (1998) Biological activity of cytokinins derived from ortho- and meta-hydroxybenzyladenine. Plant Growth Regul 26:109–115
Horgan R, Hewett EW, Horgan JM, Purse J, Wareing PF (1975) A new cytokinin from *Populus x robusta*. Phytochemistry 14:1005–1008
Huang LK, Cherng YC, Cheng YR, Jang JP, Chao YL, Cherng YJ (2007) An efficient synthesis of substituted cytosines and purines under focused microwave irradiation. Tetrahedron 63:5323–5327
Kende H, Tavares JE (1968) On the significance of cytokinin incorporation into RNA. Plant Physiol 43:1244–1248
Koprna R, De Diego N, Dundálková L, Spíchal L (2016) Use of cytokinins as agrochemicals. Bioorg Med Chem 24:484–492
Kotek V, Chudíková N, Tobrman T, Dvořák D (2010) Selective synthesis of 7-substituted purines via 7,8-dihydropurines. Org Lett 12:5724–5727
Langli G, Gundersen LL, Rise F (1996) Tetrahedron 52:5625–5638

Legraverend M, Tunnah P, Noble M, Ducrot P, Ludwig O, Grierson DS, Leost M, Meijer L, Endicott J (2000) Cyclin-dependent kinase inhibition by new C-2 Alkynylated purine derivatives and molecular structure of a CDK2 inhibitor complex, J. Med Chem 43:1082–1292

Mik V, Szüčová L, Spíchal L, Strnad M, Zahajská L (2011a) N^6-[(3-methylbut-2-en-1-yl)amino] purine derivatives and their biological activity in selected cytokinin bioassays. Bioorg Med Chem 19:7244–7251

Mik V, Szüčová L, Šmehilová M, Zatloukal M, Doležal K, Nisler J, Grúz J, Galuszka P, Strnad M, Spíchal L (2011b) N9-substituted derivatives of kinetin: effective anti-senescence agents. Phytochemistry 72:821–831

Milazzo MC, Zheng Z, Kellett G, Haynesworth K, Shetty K (1999) Stimulation of benzyladenine—induced *in vitro* shoot organogenesis and endogenous proline in melon (*Cucumis melo* L) by fish protein hydrolysates in combination with proline analogs. J Agric Food Chem 47:1771–1775

Miller CO, Skoog F, Okumura FS et al (1955) Kinetin, a cell division factor from deoxyribonucleic acid. J Am Chem Soc 77:1392

Miller CO, Skoog F, Okumura FS, Von Saltza MH, Strong FM (1956) Isolation, structure and synthesis of kinetin, a substance promoting cell division. J Am Chem Soc 78:1375–1380

Mok DWS, Mok MC (2001) Cytokinin metabolism and action. Annu Rev Plant Physiol Plant Mol Biol 52:89–118

Okumura FS, Masumura M, Motoki T, Takahashi T, Kuraishi S (1956) Syntheses of kinetin analogues I. Short Commun 30:194–195

Okumura FS, Kotani Y, Ariga T, Masumura M, Kuraishi S (1959) Syntheses of kinetin analogues II. Short Commun 32:883–884

Pati PK, Rath SP, Sharma M, Sood A, Ahuja PS (2006) *In vitro* propagation of rose—a review. Biotechnol Adv 24:94–114

Plíhal O, Szüčová L, Galuszka P (2013) N9-substituted aromatic cytokinins with negligible side effects on root development are an emerging tool for *in vitro* culturing. Plant Signal Behav 8:24392–24392

Plíhalová L, Vylíčilová H, Doležal K, Zahajská L, Zatloukal M, Strnad M (2016) Synthesis of aromatic cytokinins for plant biotechnology. N Biotechnol 33:614–624

Podlešáková K, Zalabák D, Čudejková M, Plíhal O, Szüčová L, Doležal K, Spíchal L, Strnad M, Galuszka P (2012) Novel cytokinin derivatives do not show negative effects on root growth and proliferation in submicromolar range. PLoS One 7:39293

Sakakibara H (2006) Cytokinins: activity, biosynthesis, and translocation. Annu Rev Plant Biol 57:431–449

Shell Internationale Research Maatschappij N.V. Company (1966) Compounds, compositions and method for promoting growth and/or preventing decay processes of plants or plant tissues. Patent GB 1,027,756

Spíchal L, Raková NY, Riefler M, Mizuno T, Romanov GA, Strnad M, Schmülling T (2004) Two cytokinin receptors of Arabidopsis thaliana, CRE1/AHK4 and AHK3, differ in their ligand specificity in a bacterial assay. Plant Cell Physiol 45:1299–1305

Spíchal L, Gemrotová M, Zatloukal M, Frébortová J, Galuszka P, Werner T, Schmülling T, Doležal K, Strnad M (2012) Substituted 6-anilinopurine derivatives as inhibitors of cytokinin oxidase/dehydrogenase and preparations containing these derivatives. US patent 8,222,260

Strnad M (1997) The aromatic cytokinins. Physiol Plant 101:674–688

Strnad M, Hanuš J, Vaněk T, Kamínek M, Ballantine JA, Fussell B, Hanke DE (1997) Meta-Topolin, a highly active aromatic cytokinin from poplar leaves (*Populus canadensis* Moench., cv. *Robusta*). Phytochemistry 45:213–218

Szüčová L et al (2007) 6,9-disubstituted purine derivatives and their use as cosmetics and cosmetic compositions, IEB AS CR, patents application WO2009086457

Szüčová L, Spíchal L, Doležal K, Zatloukal M, Greplová J, Galuszka P, Kryštof V, Voller J, Popa I, Massino FJ, Jorgensen JE, Strnad M (2009) Synthesis, characterization and biological activity of ring-substituted 6-benzylamino-9-tetrahydropyran-2-yl and 9-tetrahydrofuran-2-ylpurine derivatives. Bioorg Med Chem 17:1938–1947

Szüčová L, Zatloukal M, Spíchal L, Doležal K, Strnad M, Voller J, Massino FJ (2011) 6,9-Disubstituted purine derivatives and their use as cosmetics and cosmetic compositions. US patent 7,960,397

Van Overbeek J (1961) Method of preserving edible plant material and product. US patent 3,013,885

Vylíčilová H, Husičková A, Spíchal L, Srovnal J, Doležal K, Plíhal O, Plíhalová L (2016) C2-substituted aromatic cytokinin sugar conjugates delay the onset of senescence by maintaining the activity of the photosynthetic apparatus. Phytochemistry 122:22–33

Wan ZK, Binnu E, Wilson DP, Lee J (2005) A highly facile and efficient one-step synthesis of N6-adenosine and N6-2′- deoxyadenosine derivatives. Org Lett 22:5877–5880

Weaver RJ, Van Overbeek J, Pool RM (1965) Induction of fruit set in *Vitis vinifera* L. by a Kinin. Nature 206:952–953

Werbrouck SPO, Van der Jeugt J, Dewitte W, Prinsen E, Van Onckelen HA, Debergh PC (1995) The metabolism of benzyladenine in *Spathiphyllum floribundum* Śchott petité in relation to acclimatisation problems. Plant Cell Rep 14:662–665

Young H, Letham DS (1969) 6-(substituted amino)purines: synthesis by a new method and cytokinin activity. Phytochemistry 8:1199–1203

Zahajská L, Nisler J, Kadlocová A, Zatloukal M, Grúz J, Voller J, Doležal K, Strnad M (2019) 6,8-disubstituted-9-(heterocyclyl)purines, composition containing these derivatives and their use in cosmetic and medicinal applications, EP3233861I

Zatloukal M, Gemrotová M, Doležal K, Havlíček L, Spíchal L, Strnad M (2008) Novel potent inhibitors of *Arabidopsis thaliana* cytokinin oxidase/dehydrogenase. Bioorg Med Chem 16:9268–9275

Zhang R, Letham DS (1989) Cytokinin biochemistry in relation to leaf senescence. III. The senescence-retarding activity and metabolism of 9-substituted 6-benzylaminopurines in soybean leaves. J Plant Growth Regul 8:181–197

Cytokinin Properties of *Meta*-Topolin and Related Compounds

3

Jaroslav Nisler

Contents

Abstract

6-Benzylaminopurine, *meta*-topolin and their derivatives have been and are being used in and tested for the micropropagation of plants using tissue culture techniques. *Meta*-topolin and *meta*-methoxy-topolin were synthesized and tested for cytokinin activity in the late 1960s, but interest in these substances has accelerated since the 1990s. The family of topolin and methoxy-topolin cytokinins was first discovered in plants. Synthetic derivatives were then prepared, and this enabled the cytokinin activity of a wide variety of aromatic cytokinins to be evaluated in various bioassays. The latter revealed that aromatic cytokinins and most notably their *N9*-ribosides were very effective in retarding chlorophyll degradation, i.e. possessing strong senescence delaying effect. This led to the development of topolin derivatives with other *N9*-substituents. Such substances, in addition to their anti-senescence activity, show a milder negative effect on root growth and development than classical cytokinins. A combination of these two properties was crucial for choosing topolins for use and testing in plant tissue cultures. The uniqueness of the aromatic cytokinins is being gradually revealed using the molecular techniques of today. This chapter chronologically overviews what has been discovered so far about the cytokinin activity of topolins and compounds related to them.

J. Nisler (✉)
Laboratory of Growth Regulators, Institute of Experimental Botany, AS CR & Palacký University, Olomouc, Czech Republic

N. Ahmad, M. Strnad (eds.), *Meta-topolin: A Growth Regulator for Plant Biotechnology and Agriculture*, https://doi.org/10.1007/978-981-15-9046-7_3

Keywords

Meta-topolin · Topolin · Cytokinin · Anti-senescence · Plant tissue culture

3.1 The History of *Meta*-Topolin Activity Discovery

The history of the discovery of *meta*-topolin is closely related to that of kinetin and its cell division stimulatory activity (Miller et al. 1955). When this was first described, Japanese researchers synthesized a wide range of kinetin analogues which included among others the *ortho*-, *meta*- and *para*-substituted 6-benzylaminopurine (BAP) derivatives (Kuraishi 1959; Okumura et al. 1959). Their series contained *meta*-hydroxy-BAP which was later called *meta*-topolin (*m*T, Strnad et al. 1997). The compounds were tested as possible cytokinetic agents in the radish (*Raphanus sativus* L.) leaf disc expansion assay (Kuraishi and Okumura 1956). *Ortho*-methyl-BAP and *ortho*-chloro-BAP were slightly more active than unsubstituted BAP which was a little more active than *ortho*- and *meta*-hydroxy-BAP (this refers to *ortho*- topolin (*o*T) and *m*T, respectively, Okumura et al. 1959). The effect of different substituents introduced into the benzyl ring of BAP on the promotion of the growth of tobacco callus was analysed by Iwamura et al. (1980), but no significant effect of the substitution was found. These results appeared to be consistent with the results reported by Keim et al. (1981) who showed that BAP, *o*T and *m*T all bind to CBF-1 protein with similar high affinity, although the affinity of *o*T was a little lower. CBF-1 is a soluble protein isolated from wheat germ which binds cytokinins with relatively high affinity and specificity (Fox and Erion 1975).

It was not until 28 years later from the first *m*T synthesis that Kaminek et al. (1987) reported significant biological activity of *meta*-hydroxy-benzyladenosine (*m*TR) over the standard *N*6-benzyladenosine (BAPR) in the tobacco callus and wheat leaf senescence assay (WLSA). In the discussion section of their paper, the authors noted that *m*T was slightly more active than *m*TR in the *Amaranthus* assay, while the two compounds exhibited about the same activity in WLSA. *Ortho*- and *para*-hydroxy-BAPRs showed significantly lower activity than BAPR in all bioassays (Kaminek et al. 1987). The latter was the first report showing that *meta*-hydroxylation of BAPR leads to a significant increase in cytokinin activity, while *ortho*- and *para*-hydroxylation of BAPR have the opposite effect. These results contrasted sharply with the results of the radish leaf disc expansion assay (Kuraishi and Okumura 1956). Interestingly, the cytokinin activity of *m*TR was as high as that of *trans*-zeatin in all tested bioassays, including the *Amaranthus* assay and pea bud formation assay (Kaminek et al. 1987) apart from the two aforementioned. In this report, *m*TR was first introduced and tested for cytokinin activity, which was the beginning of *m*TR exploration. In the same year, Kaminek and Vaněk (1987) patented *m*T and *m*TR for use in agriculture and horticulture as shooting- and branching-stimulating plant growth regulators. As is so far apparent from the overview, since the discovery of *m*T and its cytokinin properties, very little attention had been paid to this molecule for almost three decades. This was about to change in the 1990s.

Meta-topolin riboside (mTR) has been tested by Czech scientists as a crop yield-increasing substance in barley (Hradecká and Petr 1992a) and wheat (Trčková et al. 1992; Hradecká and Petr 1992b; Borkovec and Prochazka 1995). Relative plant growth rate, 1000-grain weight, number of grains per plant and other yield-forming parameters were improved after application of *m*TR. In 1991, Holub presented his teams' findings at the 14th IPGSA conference in Amsterdam that *m*T is slightly more active than *m*TR in the tobacco callus and WLSA, making it the most active compound of the hydroxylated aromatic cytokinins studied so far in these assays (Holub et al. 1991). Such observation led to the idea whether *m*T could be an alternative to BAP in tissue culture (Werbrouck et al. 1996). Strnad and his co-workers further developed an analytical HPLC-ELISA screening system specific for BAP, *o*T and *m*T and their N9-substituted derivatives (Strnad et al. 1992a; Strnad 1996). This enabled the screening of plant tissue extracts for these types of cytokinins. Shortly afterwards, a brand new family of endogenous aromatic cytokinins, including *m*T, was discovered in plants, characterized and labelled as topolins (Strnad et al. 1992b, 1994, 1997; Jones et al. 1996).

The structure-activity relationships of natural cytokinins, namely, *trans*-zeatin, BAP, *o*T, *m*T and their N9-ribosides and N9-glucosides, were then examined and compared in three cytokinin bioassays (Holub et al. 1998). The bioassays were based on the stimulation of tobacco callus growth (tobacco callus assay), retention of chlorophyll in excised wheat leaves (wheat leaf senescence assay, WLSA) and dark induction of betacyanin synthesis in *Amaranthus* cotyledons (*Amaranthus* assay). The results showed what could have been read between the lines, but was never shown in one publication. Hydroxylation of the benzyl ring in the *meta*-position significantly increased the activity of BAP and BAPR in WLSA. In this assay, the activity of *m*T as well as *m*TR was doubled compared to BAP and BAPR, respectively. In the tobacco callus assay, the activity was increased only in the case of the *m*T free base but not in the case of *m*TR. Interestingly, in the *Amaranthus* assay, *meta*-hydroxylation led to a decrease in BAP and BAPR activity. *Ortho*-hydroxylation reduced the activity of all compounds in all bioassays, and the N9-glucosylation inactivated all the compounds completely (Holub et al. 1998). The authors further reported that the N9-ribosylation had no significant effect on the cytokinin activity of the compounds; however, N9-ribosylation slightly diminished their activity in tobacco callus and *Amaranthus* assays but markedly improved their activity in WLSA. Even from the graphs in Holub's work, it is clear that N9-ribosides retained more of the chlorophyll in the detached wheat leaves than cytokinin free bases. This effect was later observed by other authors and further investigated. It can be definitively affirmed that the WLSA played an irreplaceable role in the future development of meta-topolin derivatives, their analogues and all aromatic cytokinins in general.

3.2 *Meta*-Topolin and Related Compounds in a New Millennium

For the family of topolin compounds, a new millennium began with the discovery of their very close relatives—methoxy-topolins, 6-(2-methoxybenzylamino)purine (*ortho*-methoxy-topolin, Me*o*T) and 6-(3-methoxybenzylamino)purine (*meta*-methoxy-topolin, Me*m*T). Their N9-ribosides (Me*o*TR and Me*m*TR, respectively) were found again in *Populus* x *canadensis* leaves and in *Arabidopsis thaliana* plants (Tarkowská et al. 2003). The latter evaluated the cytokinin activity of the four new hormones in the three abovementioned cytokinin bioassays. It is not very surprising that Me*m*T and Me*m*TR were highly active in the WLSA. Their *ortho*-analogues were highly active as well. All derivatives exhibited plus/minus double activity relative to BAP (Table 3.1). In the other two bioassays, the compounds had an equal or a slightly lower activity than BAP (Tarkowská et al. 2003). These results demonstrated that the cytokinin activity of the benzyl ring-substituted BAPs and BAPRs is not limited to their *meta*-derivatives. The need for a comprehensive review of the cytokinin activity of individual BAP and BAPR derivatives was accomplished by Doležal et al. (2006, 2007). Their systematic approach made it possible to identify key substituents for achieving high cytokinin and particularly high anti-senescence activity of BAP and BAPR derivatives. Of the substituted BAPs, two fluorinated compounds, 3F-BAP and 2F-BAP, showed the highest cytokinin activities (Doležal et al. 2006 and Table 3.1). Apropos BAPR derivatives, many of them showed high activity in WLSA (Doležal et al. 2007). From the available results (Holub et al. 1998; Doležal et al. 2007; Hönig et al. 2018), it is clear that the presence of ribose at the N9-position alone significantly increased the compound's anti-senescence activity. This is very well demonstrated in Table 3.1. Based on the disproportion in activity of aromatic cytokinins in WLSA and in the other cytokinin bioassays, Doležal et al. suggested that the effect of aromatic cytokinins including topolins is predominantly associated with protection of the photosynthetic apparatus. This is consistent with the work of other authors. It was shown that in *Rosa hybrida* cuttings, Me*m*TR exhibited the best anti-senescing effects of all tested cytokinins (Bogaert et al. 2006). On the other hand, the compounds 3F-BAP and 3F-BAPR showed the best shoot multiplication rate. Me*m*TR was also superior to BAP in the multiplication of *Petunia hybrida* (Bogaert et al. 2006). Further, *m*T was evaluated as the most effective cytokinin in stimulating chlorophyll and protein biosynthesis at low concentrations in cucumber cotyledons (*Cucumis sativus* L., Çağ et al. 2007). *Meta*-topolin was also very efficient in retarding the leaf senescence of *Pelargonium* cuttings with only marginal negative effect on the root development. This was in contrast with the use of thidiazuron which severely affected root growth (Mutui et al. 2012).

These observations correspond very well with the ability of each compound to activate cytokinin receptors in plants. Three cytokinin receptors, AHK2, AHK3 and CRE1/AHK4, have been described in *Arabidopsis* (Inoue et al. 2001; Suzuki et al. 2001; Yamada et al. 2001). It has been reported that the receptor CRE1/AHK4 is responsible for the root inhibitory activity of cytokinins in *Arabidopsis* (Riefler et al.

Table 3.1 Comparison of the cytokinin activities of BAP and BAPR derivatives in wheat leaf senescence assay and tobacco callus assay. The values for BAPR, *o*T, *m*T, *o*TR and *m*TR were calculated from estimated values extracted from the graphs published by Holub et al. (1998). Further, it is important to note that the activities of all compounds were evaluated by the authors only according to the activity of the substance in the highest concentration tested, and it was compared to BAP activity. The highest activity of BAP could differ slightly among these assays, but this value in % would strongly affect the activity of other substances (in % of BAP activity)

R = β-D-ribose

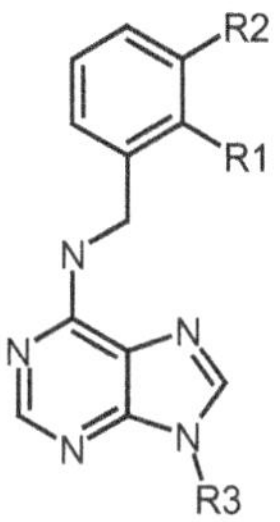

Compound abbreviation	R1	R2	R3	Senescence assay (% of BAP activity)	Tobacco callus a. (% of BAP activity)	Reference
BAP				100	100	
*o*T	OH	H	H	20	55	Holub et al. (1998)
*m*T	H	OH	H	210	130	Holub et al. (1998)
Me*o*T	OMe	H	H	203 ± 46	79 ± 5	Tarkowská et al. (2003)
Me*m*T	H	OMe	H	178 ± 16	76 ± 20	Tarkowská et al. (2003)
2F-BAP	F	H	H	169	111 ± 21	Doležal et al. (2006)
3F-BAP	H	F	H	200	135 ± 8	Doležal et al. (2006)
BAPR			**R**	205	108	Holub et al. (1998)
*o*TR	OH	H	**R**	120	40	Holub et al. (1998)
*m*TR	H	OH	**R**	320	110	Holub et al. (1998)
Me*o*TR	OMe	H	**R**	198 ± 12	108 ± 1	Tarkowská et al. (2003)
Me*m*TR	H	OMe	**R**	209 ± 8	92 ± 1	Tarkowská et al. (2003)
2F-BAPR	F	H	**R**	118 ± 39	100 ± 9	Doležal et al. (2007)
3F-BAPR	H	F	**R**	220 ± 16	91 ± 6	Doležal et al. (2007)

2006), while the AHK3 receptor is the key element in cytokinin-mediated leaf longevity (Riefler et al. 2006; Kim et al. 2006). Spíchal et al. (2004) showed that TDZ is a strong activator of both mentioned cytokinin receptors, while *m*T activates preferably the receptor AHK3. The relative activity of *m*T at AHK3 and CRE1/AHK4 receptors was 80% and 30% of *t*Z activity, respectively. *Meta*-topolin riboside did not activate these receptors, but it was able to activate the expression of ARR5 (Arabidopsis cytokinin response regulator 5) in *Arabidopsis* plants (Spíchal et al. 2004). Recently, it has also been shown that cytokinin receptors from *Brassica napus* bind *m*T and with higher affinity than BAP (Kuderová et al. 2014).

The continuous search for compounds with cytokinin activity but with limited root growth inhibitory activity led to the development of topolin derivatives with N9-substituents, such as tetrahydropyran-2-yl, 9-tetrahydrofuran-2-yl and 4-chlorobutyl (Szüčová et al. 2009; Plíhal et al. 2013; Podlešáková et al. 2012). Such compounds are reported to maintain high anti-senescence activity (Szüčová et al. 2009) while exhibiting negligible negative effects on root growth and development. It is assumed that the reasons for this are (1) the compounds do not activate the CRE1/AHK4 cytokinin receptor; and (2) they are preferably transported from roots to shoots, (3) where the free base is slowly released and takes action (Plíhal et al. 2013). It has been also speculated that such compounds may act as CRE1/AHK4 receptor antagonists, which would further favour their indifference to the root system (Plíhal et al. 2013). This is feasible given that two other derivatives of BAP were shown to be cytokinin receptor antagonists. A compound known as PI-55, which is (6-(2-hydroxy-3-methylbenzylamino)purine), is an antagonist of the CRE1/AHK4 receptor (Spíchal et al. 2009). The compound LGR-991 (2,5-dihydroxy-BAP) showed antagonistic effects on AHK3 and CRE1/AHK4 receptors (Nisler et al. 2010).

For the reasons described above, it is clear that the N9-substituted derivatives of aromatic cytokinins have great potential for improving plant in vitro regeneration techniques.

References

Bogaert I, Cauter S, Werbrouck SPO et al (2006) New aromatic cytokinins can make the difference. In: Fári MG, Holb I, Bisztray GD (eds) Acta horticulturae, vol 725. International Society for Horticultural Science (ISHS), Leuven, pp 265–270

Borkovec V, Prochazka S (1995) Proceedings of the Plant Growth Regulator Society of America, vol 22, pp 360–364

Çağ S, Gören-Sağlam N, Çıngıl-Barış Ç et al (2007) The effect of different concentration of epibrassinolide on chlorophyll, protein and anthocyanin content and peroxidase activity in excised red cabbage (*Brassica oleracea* L.) cotyledons. Biotechnol Biotechnol Equip 21:422–425

Doležal K, Popa I, Kryštof V et al (2006) Preparation and biological activity of 6-benzylaminopurine derivatives in plants and human cancer cells. Bioorg Med Chem 14:875–884

Doležal K, Popa I, Hauserová E et al (2007) Preparation, biological activity and endogenous occurrence of N6-benzyladenosines. Bioorg Med Chem 15:3737–3747
Fox JE, Erion JL (1975) A cytokinin binding protein from higher plant ribosomes. Biochem Biophys Res Commun 64:695–700
Holub J, Hanuš J, Vaněk T et al (1991) Cytokinin activities of some 9-substituted *ortho-* and *meta*-hydroxybenzyladenines. In: Abstract 14th IPGSA Conference. Agricultural University, Wageningen, Amsterdam, p 70
Holub J, Hanuš J, Hanke DE et al (1998) Biological activity of cytokinins derived from *Ortho-* and *Meta*-Hydroxybenzyladenine. Plant Growth Regul 26:109–115
Hönig M, Plíhalová L, Husičková A et al (2018) Role of cytokinins in senescence, antioxidant defence and photosynthesis. Int J Mol Sci 19:4045
Hradecká D, Petr J (1992a) Spring barley kernel weight after treatment with a cytokinin. In: SbornikVysokeSkolyZemedelske v Praze, FakultaAgronomicka, Rada A: RostlinnaVyroba, vol 54, pp 153–162
Hradecká D, Petr J (1992b) The effect of cytokinins on the yield of some cereal. In: Kaminek M, Mok WS, Zažímalová E (eds) Physiology and biochemistry of cytokinins in plants. SPB Academic, The Hague, pp 245–247
Inoue T, Higuchi M, Hashimoto Y et al (2001) Identification of CRE1 as a cytokinin receptor from *Arabidopsis*. Nature 409:1060–1063
Iwamura H, Fujita T, Koyama S et al (1980) Quantitative structure-activity relationship of cytokinin-active adenine and urea derivatives. Phytochemistry 19:1309–1319
Jones LH, Martinková H, Strnad M et al (1996) Occurrence of aromatic cytokinins in oil palm (*Elaeis guineensis Jacq.*). J Plant Growth Regul 15:39
Kaminek M, Vaněk T (1987) Agricultural and horticultural plant growth regulator compositions containing N^6-(*m*-hydroxybenzyl)adenosine and -adenine. From Czech.CS 238548.B1 19851113
Kaminek M, Vaněk T, Motyka V (1987) Cytokinin activities of N^6-Benzyladenosine derivatives hydroxylated on the side-chain phenyl ring. J Plant Growth Regul 6:113–120
Keim P, Erion J, Fox JE (1981) The current status of cytokinin-binding moieties. In: Guern J, Péaud-Lenoël C (eds) Metabolism and molecular activities of cytokinins. Proceedings in life sciences. Springer, Berlin
Kim HJ, Ryu H, Hong SH et al (2006) Cytokinin-mediated control of leaf longevity by AHK3 through phosphorylation of ARR2 in *Arabidopsis*. Proc Natl Acad Sci U S A 103:814–819
Kuderová A, Gallová L, Kuricová K et al (2014) Identification of AHK2- and AHK3-like cytokinin receptors in *Brassica napus* reveals two subfamilies of AHK2 orthologues. J Exp Bot 66:339–353
Kuraishi S (1959) Effect of kinetin analogs on leaf growth. Sci Pap Coll Gen Educ Univ Tokyo 9:67–104
Kuraishi S, Okumura FS (1956) Bot Mag (Tokyo) 69:300
Miller CO, Skoog F, Von Saltza MH et al (1955) Kinetin, a cell division factor from deoxyribonucleic acid. J Am Chem Soc 77:1392
Mutui TM, Mibus H, Serek M (2012) Effect of *meta*-topolin on leaf senescence and rooting in *Pelargonium* × *hortorum* cuttings. Postharvest Biol Technol 63:107–110
Nisler J, Zatloukal M, Popa I et al (2010) Cytokinin receptor antagonists derived from 6-benzylaminopurine. Phytochemistry 71:823–830
Okumura FS, Kotani Y, Ariga T, Kuraishi S (1959) Syntheses of kinetin-analogues. II. Bull Chem Soc Jpn 32:883–886
Plíhal O, Szüčová L, Galuszka P (2013) N9-substituted aromatic cytokinins with negligible side effects on root development are an emerging tool for *in vitro* culturing. Plant Signal Behav 8:6–9
Podlešáková K, Zalabák D, Čudejková M et al (2012) Novel cytokinin derivatives do not show negative effects on root growth and proliferation in submicromolar range. PLoS One 7:e39293

Riefler M, Novák O, Strnad M et al (2006) *Arabidopsis* cytokinin receptor mutants reveal functions in shoot growth, leaf senescence, seed size, germination, root development, and cytokinin metabolism. Plant Cell 18:40–54

Spíchal L, Rakova NY, Riefler M et al (2004) Two cytokinin receptors of *Arabidopsis thaliana*, CRE1/AHK4 and AHK3, differ in their ligand specificity in a bacterial assay. Plant Cell Physiol 45:1299–1305

Spíchal L, Werner T, Popa I et al (2009) The purine derivative PI-55 blocks cytokinin action *via* receptor inhibition. FEBS J 276:244–253

Strnad M (1996) Enzyme immunoassays of N^6-benzyladenine and N^6-(*meta*-hydroxybenzyl)adenine cytokinins. J Plant Growth Regul 15:179–188

Strnad M, Veres K, Hanus J et al (1992a) Immunological methods for quantification and identification of cytokinins. In: Kaminek M, Mok DWS, Zazimalova E (eds) Physiology and biochemistry of cytokinins in plants. SPB Academic, The Hague, pp 437–446. ISBN 90-5103-066-5

Strnad M, Peters W, Beck E et al (1992b) Immunodetection and identification of N^6-(*o*-hydroxybenzylamino)purine as a naturally occurring cytokinin in *Populus x canadensis* Moench cv. *Robusta* leaves. Plant Physiol 99:74–80

Strnad M, Peters W, Hanuš J et al (1994) Orthotopolin-9-glucoside, an aromatic cytokinin from *Populus x canadensis* cv. *Robusta* leaves. Phytochemistry 37:1059–1062

Strnad M, Hanuš J, Vaněk T et al (1997) *Meta*-Topolin, a highly active aromatic cytokinin from poplar leaves (*Populus x canadensis* Moencho cv. *Robusta*). Phytochemistry 42:213–218

Suzuki T, Miwa K, Ishikawa K et al (2001) The *Arabidopsis* sensor his-kinase, AHK4, can respond to cytokinins. Plant Cell Physiol 42:107–113

Szüčová L, Spíchal L, Doležal K et al (2009) Synthesis, characterization and biological activity of ring-substituted 6-benzylamino-9-tetrahydropyran-2-yl and 9-tetrahydrofuran-2-ylpurine derivatives. Bioorg Med Chem 17:1938–1947

Tarkowská D, Doležal K, Tarkowski P et al (2003) Identification of new aromatic cytokinins in *Arabidopsis thaliana* and *Populus* X *canadensis* leaves by LC-ESI-MS and capillary liquid chromatography/frit–fast atom bombardment mass spectrometry. Physiol Plant 117:579–590

Trčková M, Kamínek M, Zmrhal Z (1992) Grain formation and distribution of nutrients in wheat plants after application of synthetic cytokinin N6-(*meta*-hydroxybenzyl)adenosine. In: Kaminek M, Mok WS, Zažímalová E (eds) Physiology and biochemistry of cytokinins in plants. Kluwer Academic, The Hague, pp 241–244

Werbrouck SPO, Strnad M, Van Onckelen HA et al (1996) *Meta*-topolin, an alternative to benzyladenine in tissue culture? Physiol Plant 98:291–297

Yamada H, Suzuki T, Terada K et al (2001) The *Arabidopsis* AHK4 histidine kinase is a cytokinin-binding receptor that transduces cytokinin signals across the membrane. Plant Cell Physiol 42:1017–1023

Occurrence, Interconversion, and Perception of Topolins in Poplar

4

David Kopečný, Dominika Kaczorová, and Petr Tarkowski

Contents

Abstract

The poplar species *Populus* × *canadensis* cv. Robusta was the first organism found to contain aromatic cytokinins. Screening of the content of aromatic cytokinins in leaves of 12 *Populus* species revealed that the capacity to produce aromatic cytokinins is widespread among *Populus* accessions. The major aromatic metabolites are *ortho*-topolin and *ortho*-topolin riboside. Their levels transiently increase after daybreak and are much higher in older plants. Poplar species contain five genes coding for functional CHASE-containing histidine kinases acting as cytokinin receptors. Poplar genome further contains nine isopentenyl transferase genes coding for enzymes responsible for the biosynthesis of isoprenoid cytokinins, two genes coding for adenosine kinase, two genes of nucleoside *N*-ribohydrolase, and one gene encoding purine nucleoside phosphorylase. These enzymes contribute to interconversion of cytokinin ribosides.

D. Kopečný (✉)
Department of Protein Biochemistry and Proteomics, Faculty of Science, Centre of the Region Haná for Biotechnological and Agricultural Research, Palacký University Olomouc, Olomouc, Czech Republic
e-mail: david.kopecny@upol.cz

D. Kaczorová · P. Tarkowski
Department of Genetic Resources for Vegetables, Medicinal and Special Plants, Crop Research Institute, Centre of the Region Haná for Biotechnological and Agricultural Research, Olomouc, Czech Republic

Department of Phytochemistry, Faculty of Science, Centre of the Region Haná for Biotechnological and Agricultural Research, Palacký University Olomouc, Olomouc, Czech Republic

N. Ahmad, M. Strnad (eds.), *Meta-topolin: A Growth Regulator for Plant Biotechnology and Agriculture*, https://doi.org/10.1007/978-981-15-9046-7_4

Trans-Zeatin is the most abundant cytokinin in poplar and displays the highest variation in abundance. It shows the strongest affinity to all five cytokinin receptors and activates the cytokinin signaling via A-type response regulators. Among aromatic cytokinins, *meta*-topolin is efficiently bound to all receptors, while *ortho*-topolin binds only at micromolar concentrations. The origin of topolins in poplar remains unclear, and it is possible that they are not products of poplar metabolism but indeed endophyte-derived products.

Keywords

Activity · Aromatic cytokinin · *Ortho*-topolin · Poplar · Cytokinin receptor

Abbreviations

ADK	Adenosine kinase
BA	N^6-benzyladenine
*c*Z	*cis*-zeatin
HK	Histidine kinase
iP	N^6-isopentenyladenine
iPR	N^6-isopentenyladenosine
*m*T	*meta*-topolin
NRH	Nucleoside N-ribohydrolase
*o*T	*ortho*-topolin
PNP	Purine nucleoside phosphorylase
*t*Z	*trans*-zeatin
*t*ZR	Zeatin riboside

Plant signaling systems that integrate and control cellular responses are, at least partially, based on active chemical substances called phytohormones. Cytokinins represent a group of phytohormones defined as the compounds that in the presence of auxin (another phytohormone) induce cell division in a suitable assay material grown on a defined medium (Shaw 1994). Among other functions, they also regulate morphogenesis and cambial development; modulate the activity of root, shoot, and reproductive meristems; and inhibit leaf senescence. All native cytokinins are derivatives of adenine (free bases, sugar conjugates, or nucleotides) with at least one substituent at the exocyclic N^6 position. Depending on the side chain structure, two main subclasses of cytokinins are recognized – isoprenoid and aromatic. Isoprenoid cytokinins, comprising N^6-isopentenyladenine (iP), *trans*-zeatin (*t*Z), *cis*-zeatin (*c*Z) and dihydrozeatin, are ubiquitous in plants, whereas their aromatic counterparts, which consist of N^6-benzyladenine (BA) and *ortho*- and *meta*-topolin (*o*T and *m*T), are very rare. For this reason, much is known about the function, metabolism, and perception of isoprenoid cytokinins, while the origin and fate of aromatic cytokinins remain largely unexplained (Frébort et al. 2011).

Several aromatic cytokinins have been identified in crown gall tumors of tomato (Nandi et al. 1989a, b), calla (Chaves dasNeves and Pais 1980), palm oil (Jones et al. 1996), and red goosefoot (Doležal et al. 2002). However, poplar (*Populus* ×

canadensis cv. Robusta) appears to be the best model plant for studying aromatic cytokinins. Mainly hydroxylated derivatives of BAP, *o*T, and *m*T as well as their sugar conjugates have been identified in extracts of poplar leaves (Horgan et al. 1975; Strnad et al. 1992, 1994, 1997; Tarkowská et al. 2003). Recently, we performed screening for aromatic cytokinins in a collection of poplar genetic resources (Forestry and Game Management Research Institute, Kunovice, Czech Republic). UHPLC-MS/MS analyses of leaf extracts from 12 gene bank accessions revealed no BAP-, *methoxy-o*T-, *m*T-, or *para*-topolin-type cytokinins (Jaworek et al. 2019). On the other hand, *ortho*-topolin riboside (*o*TR) was present in all accessions and especially abundant in *P.* × *canadensis*, *P.* × *berolinensis*, *P. laurifolia*, *P. maximovici*, and *P.* × *oxford*. Surprisingly, apart from *P.* × *canadensis*, *o*T was present in low quantities or not at all in the accessions tested. A concentration of aromatic cytokinins is reported to vary according to the length of exposure to light and the quality of light (Hewett and Wareing 1973). Time-dependent accumulation of aromatic cytokinins was observed in *P.* × *canadensis*, with the highest levels of *o*T (253 pmol g^{-1} FW) and *o*TR (156 pmol g^{-1} FW) occurring 4 h after daybreak. Interestingly, only negligible fluctuations of *t*Z and its riboside (*t*ZR) were observed, with average concentrations of 2.2 pmol g^{-1} FW and 1.2 pmol g^{-1} FW, respectively (Jaworek et al. 2019). Annual and seasonal fluctuations were also observed in *P. tremula* (Edlund et al. 2017).

Three cytokinin moieties were identified in tRNA hydrolysates of *P.* × *canadensis* and *P.* × *deltoides*—N^6-isopentenyladenosine (iPR) (50 pmol g^{-1} FW), *cis*-zeatin riboside (*c*ZR) (19 pmol g^{-1} FW) and *o*TR (4 pmol g^{-1} FW) (Jaworek et al. 2019). While the dynamics of free *o*TR was observed in both *P.* × *canadensis* and *P.* × *deltoides*, there were no significant changes in levels of tRNA-bound *o*TR, either in a the short time (day) or throughout the season. In addition, tissues of mature *P.* × *canadensis* and *P.* × *deltoides* trees contained much higher free aromatic cytokinin levels than those detected in leaves from young suckers. It seems to be likely that hybrid *P.* × *canadensis* inherited the trait to synthesize aromatic cytokinins from *P.* × *deltoides* (Jaworek et al. 2019, 2020).

The presence of *o*TR and *o*T in various poplar cultivars raises the question on their mutual interconversion. Nucleosides can be synthesized or de novo or transported from the vacuole or apoplast into the cytosol after RNA degradation by members of the equilibrative nucleoside transporter (ENT) family exhibiting broad substrate specificity toward purine, pyrimidine, and cytokinin ribosides with affinities in the micromolar range (Wormit et al. 2004; Hirose et al. 2008; Girke et al. 2014). In the cytosol, purine ribosides can be hydrolyzed by calcium-dependent nucleoside N-ribohydrolases (NRHs, E.C. 3.2.2.-) to corresponding nitrogenous bases and ribose (Jung et al. 2009, 2011; Kopečná et al. 2013) or phosphorylated to monophosphates by adenosine kinases (ADK, E.C. 2.7.1.20, Moffatt et al. 2000; Schoor et al. 2011). Purine nucleoside phosphorylase in plants (PNP, E.C. 2.4.2.1, Chen and Petschow 1978; Bromley et al. 2014) preferentially catalyzes a ribosylation reaction of adenine/isoprenoid cytokinin with ribose-1-phosphate to release adenosine/cytokinin riboside and phosphate moiety. Although phosphorolytic reaction can also appear in the presence of phosphate, lower K_m

Table 4.1 A list of genes coding for a nucleosidase (NRH), adenosine kinase (ADK), and purine nucleoside phosphorylase (PNP) in *Populus trichocarpa*

Gene	Phytozome ID number	Citation
NRH		
PtNRH1	Potri.007G144600	–
PtNRH2	Potri.006G083400	–
AtNRH1	At2g36310	Jung et al. (2009)
AtNRH2	At1g05620	Jung et al. (2011)
ZmNRH1a	GRMZM2G029845	Kopečná et al. (2013)
ZmNRH1b	GRMZM2G134149	Kopečná et al. (2013)
ZmNRH2a	GRMZM2G085960	Kopečná et al. (2013)
ZmNRH2b	GRMZM2G015344	Kopečná et al. (2013)
ZmNRH3	GRMZM2G104999	Kopečná et al. (2013)
ADK		
PtADK1	Potri.010G224300	–
PtADK2	Potri.008G038100	–
AtADK1	At3G09820	Moffatt et al. (2000)
AtADK2	At5G03300	Moffatt et al. (2000)
PNP		
PtPNP	Potri.T096300	–
AtPNP	At4g28940	–
StPNP	Sotub08g016060	Bromley et al. (2014)

The table shows phytozome gene ID numbers and genes studied in detail are provided for comparison (*Pt* v *Populus trichocarpa*, *Zm Zea mays*, *At Arabidopsis thaliana*, *St Solanum tuberosum*)

values for bases favor the ribosylation reaction as shown for potato enzyme (Bromley et al. 2014). ADKs found in *Arabidopsis* and tobacco are known to catalyze phosphorylation of adenosine and isoprenoid cytokinins with K_m values in low micromolar range (Moffatt et al. 2000; Kwade et al. 2005). NRHs hydrolyze various ribosides including isoprenoid cytokinin ribosides. NRHs contain calcium ions in the active site which is coordinated by aspartate residues from the conserved N-terminal DXDXXXDD motif. Calcium ions are essential for the ribose moiety binding. Plant NRHs belong to nonspecific inosine-uridine containing NRHs as they are able to act on a wide range of ribosides. K_m values of *Arabidopsis* and maize NRHs for iPR and *t*ZR are in high micromolar range (Jung et al. 2009; Kopečná et al. 2013).

The genome of poplar (https://phytozome.jgi.doe.gov/pz/portal.html) contains at least two *NRH* genes, two *ADK* genes, and one *PNP* gene (Table 4.1). Two most active NRHs from maize, namely, ZmNRH2b and ZmNRH3, hydrolyze not only isoprenoid cytokinin ribosides (Kopečná et al. 2013) but also aromatic cytokinins including *o*TR to *o*T (Fig. 4.1a, b). Both isoforms display specific activities of 0.3 and 0.5 nkat mg^{-1} with iPR. Their activities with *o*TR are in similar range and attain values of 0.5 and 0.4 nkat mg^{-1}, respectively. For this reason, it is very likely that ADK and PNP reactions will also catalyze the conversion of *o*T and *o*TR in addition to isoprenoid cytokinins.

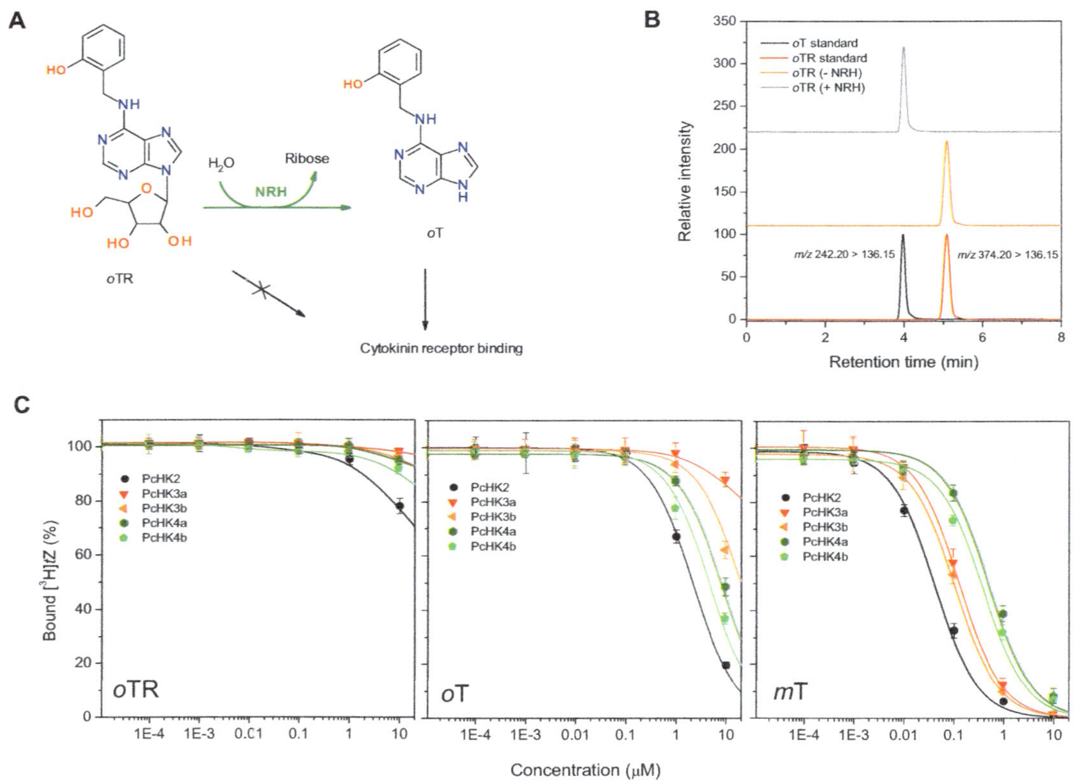

Fig. 4.1 (**a**) A scheme of the catalytic conversion of oTR to oT catalyzed by plant NRHs. (**b**) Confirmation of *o*TR conversion by ZmNRH2b using UHPLC-MS/MS with *m/z* transitions indicated for each peak. (**c**) Competitive binding curves for *o*TR, *o*T, and *m*T versus [^{3}H]-*t*Z using five poplar cytokinin receptors in live cell assay. Binding was studied at pH 7.0 (adapted from Jaworek et al. 2020)

The poplar genome contains five genes coding for CHASE-containing histidine kinases (HKs), which are known to function as cytokinin receptors. The abbreviation "CHASE" denotes a cyclase/histidine kinase-associated sensory extracellular domain, which comprises a region of ~270 amino acids and represents the cytokinin binding site located at the N-terminus of HK (Heyl et al. 2007). HKs from *Populus* × *canadensis* cv. Robusta, namely, PcHK2, PcHK3a, PcHK3b, PcHK4a and PcHK4b, display kinase activity and are able to bind both isoprenoid and aromatic cytokinins (Jaworek et al. 2020). All five PcHKs display strong *t*Z binding affinities ranging from 1.8 nM to 5.5 nM concentrations.

While both PcHK3a and PcHK3b display the strongest binding at pH 7.5, binding to both PcHK4 increases steadily toward pH 5.5 in line with their putative membrane localization in the endoplasmic reticulum and plasma membrane. Eventual intra- and intercellular transport of *o*T to the receptor can be mediated by several transporter families including purine permeases, ureide permeases, and nucleobase: cation symporter families 1 and 2 (reviewed in Girke et al. 2014). Of the tested aromatic cytokinins, *m*T binds more strongly than BAP. However, it is unlikely that these appear in vivo in poplar. K_i values for *m*T range between 23 nM and 300 nM (Jaworek et al. 2020). Conversely, *o*T, the metabolite found in poplar, is a much weaker ligand for PcHKs. The lowest K_i value of 1.1 μM and thus the highest sensitivity for *o*T are displayed by PcHK2 (Fig. 4.1c) (Jaworek et al. 2020). Its riboside is inactive and not bound below 10 μM concentrations. As it has been shown that leaves of mature *P.* × *canadensis* trees exhibit diurnally fluctuating levels of *o*T under physiological conditions and the concentration can peak at around 250 nM, it is possible that the local concentration may be even higher and trigger cytokinin signaling via PcHK2 (Jaworek et al. 2019).

A chlorophyll retention bioassay was used to assess the biological activity of *o*T and *m*T in poplar. The leaf discs floated on the micromolar cytokinin solution in the dark for 3 weeks at 25 °C. Data are presented in Fig. 4.2 as chlorophyll content relative to the initial value. Unlike the results outlined by Holub et al. (1998), the assay does not permit satisfactory determination of biological activity, mainly due to extremely high biological variability. While winter wheat plants were cultivated under controlled conditions in the original assay, poplar leaves were harvested from adult trees in an urban environment. The activity of *m*TR is higher and delays chlorophyll degradation. The effect of the *o*T is comparable with other three cytokinins *m*T, *t*Z, and BAP, and the values range between 60% and 80%. However, the differences are not significant enough from the control. Interestingly, chlorophyll degradation in poplar leaf discs is very slow compared with winter wheat. Bioassay with winter wheat requires 96 h, while bioassay with poplar needs 3 weeks or more.

To conclude, levels of aromatic cytokinins found in *P.* × *canadensis* and measured in leaves from the same tree are very variable between years and during the season (Jaworek et al. 2019, 2020). While *o*T derivatives are found, those of *m*T are not. Moreover, *m*T is more active in bioassays and activates cytokinin receptors at nanomolar concentrations. Currently, we cannot exclude the possibility that topolins are not products of poplar metabolism but indeed endophyte products

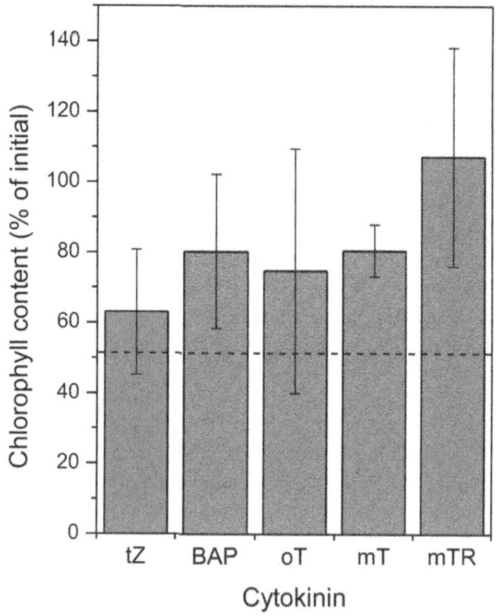

Fig. 4.2 A biological activity of topolins in poplar analyzed by a chlorophyll retention bioassay. The chlorophyll content was measured in leaf discs left in the dark for 3 weeks at 25 °C. It is shown relative to the initial value. The content in the absence of cytokinin is shown by a dashed line and corresponds to 51.5%

(Wang et al. 2019). Observed fluctuations could account for the changes in the endophyte growth. The origin of topolins in poplar remains to be identified.

Acknowledgments This study is supported by grant no. 18-07563S from the Czech Science Foundation and ERDF project "Plants as a tool for sustainable global development" (No. CZ.02.1.01/0.0/0.0/16_019/0000827).

References

Bromley JR, Warnes BJ, Newell CA et al (2014) A purine nucleoside phosphorylase in *Solanum tuberosum* L. (potato) with specificity for cytokinins contributes to the duration of tuber endodormancy. Biochem J 458:225–237

Chaves dasNeves HJ, Pais MSS (1980) Identification of a spathe greening factor in *Zantedeschia-Aethiopica*. Biochem Biophys Res Commun 95:1387–1392

Chen CM, Petschow B (1978) Metabolism of cytokinin: ribosylation of cytokinin bases by adenosine phosphorylase from wheat germ. Plant Physiol 62:871–874

Doležal K, Åstot C, Hanuš J et al (2002) Identification of aromatic cytokinins in suspension cultured photoautotrophic cells of Chenopodium rubrum by capillary liquid chromatography/frit–fast atom bombardment mass spectrometry. Plant Growth Regul 36:181–189

Edlund E, Novak O, Karady M et al (2017) Contrasting patterns of cytokinins between years in senescing aspen leaves. Plant Cell Environ 40:622–634

Frébort I, Kowalska M, Hluska T et al (2011) Evolution of cytokinin biosynthesis and degradation. J Exp Bot 62:2431–2452

Girke C, Daumann M, Niopek-Witz S et al (2014) Nucleobase and nucleoside transport and integration into plant metabolism. Front Plant Sci 5:443

Hewett EW, Wareing PF (1973) Cytokinins in *Populus* × *robusta* (Schneid): light effects on endogenous levels. Planta 114:119–129

Heyl A, Wulfetange K, Pils B et al (2007) Evolutionary proteomics identifies amino acids essential for ligand-binding of the cytokinin receptor CHASE domain. BMC Evol Biol 7:62

Hirose N, Takei K, Kuroha T et al (2008) Regulation of cytokinin biosynthesis, compartmentalization and translocation. J Exp Bot 59:75–83

Holub J, Hanuš J, Hanke DE et al (1998) Biological activity of cytokinins derived from *Ortho*- and *Meta*-Hydroxybenzyladenine. Plant Growth Regul 26:109–115

Horgan R, Hewett EW, Horgan JM et al (1975) A new cytokinin from *Populus × robusta*. Phytochemistry 14:1005–1008

Jaworek P, Kopečný D, Zalabák D et al (2019) Occurrence and biosynthesis of cytokinins in poplar. Planta 250:229–244

Jaworek P, Tarkowski P, Hluska T et al (2020) Characterization of five CHASE-containing histidine kinase receptors from *Populus × canadensis* cv. *Robusta* sensing isoprenoid and aromatic cytokinins. Planta 251:1

Jones LH, Martinková H, Strnad M et al (1996) Occurrence of aromatic cytokinins in palm oil (*Elaeis guineensis* Jacq). J Plant Growth Regul 15:39–49

Jung B, Flörchinger M, Kunz HH et al (2009) Uridine-ribohydrolase is a key regulator in the uridine degradation pathway of Arabidopsis. Plant Cell 21:876–891

Jung B, Hoffmann C, Möhlmann T (2011) Arabidopsis nucleoside hydrolases involved in intracellular and extracellular degradation of purines. Plant J 65:703–711

Kopečná M, Blaschke H, Kopečný D et al (2013) Structure and function of nucleoside hydrolases from Physcomitrella patens and maize catalyzing the hydrolysis of purine, pyrimidine, and cytokinin ribosides. Plant Physiol 163:1568–1583

Kwade Z, Swiatek A, Azmi A et al (2005) Identification of four adenosine kinase isoforms in tobacco By-2 cells and their putative role in the cell cycle-regulated cytokinin metabolism. J Biol Chem 280:17512–17519

Moffatt BA, Wang L, Allen MS et al (2000) Adenosine kinase of Arabidopsis. Kinetic properties and gene expression. Plant Physiol 124:1775–1785

Nandi SK, Letham DS, Palni LMS et al (1989a) 6-Benzylaminopurine and its glycosides as naturally occurring cytokinins. Plant Sci 61:189–196

Nandi SK, Palni LMS, Letham DS et al (1989b) Identification of cytokinins in primary crown gall tumours of tomato. Plant Cell Environ 12:273–283

Schoor S, Farrow S, Blaschke H et al (2011) Adenosine kinase contributes to cytokinin interconversion in Arabidopsis. Plant Physiol 157:659–672

Shaw G (1994) Chemistry of adenine cytokinins. In: Mok DWS, Mok MC (eds) Cytokinins: chemistry, activity and function. CRC, Boca Raton, FL, pp 15–34

Strnad M, Peters W, Beck E et al (1992) Immunodetection and identification of N6-(o-hydroxybenzylamino)purine as a naturally occurring cytokinin in *Populus × canadensis* Moench cv. *Robusta* leaves. Plant Physiol 99:74–80

Strnad M, Peters W, Hanuš J et al (1994) *Ortho*-topolin-9-glucoside, an aromatic cytokinin from *Populus × canadensis* cv. *Robusta* leaves. Phytochemistry 37:1059–1062

Strnad M, Hanus J, Vanek T et al (1997) *Meta*-topolin, a highly active aromatic cytokinin from poplar leaves (*Populus × canadensis* Moench., cv. *Robusta*). Phytochemistry 45:213–218

Tarkowská D, Doležal K, Tarkowski P et al (2003) Identification of new aromatic cytokinins in *Arabidopsis thaliana* and *Populus × canadensis* leaves by LC-(+)ESI-MS and capillary liquid chromatography/frit–fast atom bombardment mass spectrometry. Physiol Plant 117:579–590

Wang Y, Zhang W, Ding C et al (2019) Endophytic communities of transgenic poplar were determined by the environment and niche rather than by transgenic events. Front Microbiol 10:588

Wormit A, Traub M, Flörchinger M et al (2004) Characterization of three novel members of the *Arabidopsis thaliana* equilibrative nucleoside transporter (ENT) family. Biochem J 383:19–26

Cytokinin Signalling and Mechanism of Action of *Meta*-Topolin and Its Derivatives

5

Ondřej Plíhal

Contents

Abstract

The discovery of *ortho*-topolin, *meta*-topolin and its conjugates in the early 1990s prompted further research on the biological activity of these aromatic cytokinins. Structure–activity relationship studies (SAR) revealed that hydroxylation at the *meta*-position on the benzyl side chain of BAP and its riboside significantly improved the compound performance in standardized cytokinin bioassays compared to BAP. Elucidation of the molecular mechanism of the cytokinin signal transduction and identification of membrane-bound receptor histidine kinase (HK) led to deeper understanding of the function of topolins in planta. While aromatic cytokinins generally show weak interaction with *Arabidopsis* and other cytokinin-sensitive HKs, *meta*-topolin is able to provoke a strong response comparable to that observed with the most active isoprenoid free bases. In contrast, *ortho*-topolin and *para*-topolin are much less active, and it is possible that hydroxylation at these positions serves as a deactivation step. In contrast, derivatives of *meta*-topolin, although not directly interacting with cytokinin-sensitive HKs, have proven to be very useful in plant propagation techniques, and slow gradual release of the free active base allows for the potentiation of

O. Plíhal (✉)
Laboratory of Growth Regulators, Faculty of Science of Palacký University and Institute of Experimental Botany of the Czech Academy of Sciences, Olomouc, Czech Republic
e-mail: ondrej.plihal@upol.cz

N. Ahmad, M. Strnad (eds.), *Meta-topolin: A Growth Regulator for Plant Biotechnology and Agriculture*, https://doi.org/10.1007/978-981-15-9046-7_5

cytokinin effects. This chapter summarizes basic facts about cytokinin signalling and adds to our knowledge of signalling aspects and metabolism of aromatic cytokinins in various plant models.

Keywords

Meta-topolin · Cytokinin signalling · Aromatic cytokinin · AHK · PcHK · Response regulator

Abbreviations

[9G]	9-β-D-glucopyranosyl
[9R]	9-β-D-ribofuranosyl
3MeOBA9THPP	3-methoxy(6-benzylamino-9-tetrahydropyran-2-yl)purine
ARCK	Aromatic cytokinin
BAP	6-benzylaminopurine
CK	Cytokinin
*c*Z	*cis*-zeatin
HK	Receptor histidine kinase
iP	N^6-isopentenyladenine
*m*T	*meta*-topolin, 6-(3-hydroxybenzylamino)purine
*m*TR	*meta*-topolin riboside, 6-(3-hydroxybenzylamino)purine-9-riboside
*o*T	*ortho*-topolin, 6-(2-hydroxybenzylamino)purine
*o*TR	*ortho*-topolin riboside, 6-(2-hydroxybenzylamino)purine-9-riboside
SAR	Structure–activity relationship
THF	Tetrahydrofuran-2-yl
THP	Tetrahydropyran-2-yl
*t*Z	*trans*-zeatin
*t*ZR	*trans*-zeatin-9-riboside

5.1 Cytokinin Signal Transduction Via Multistep Phosphorelay

Cytokinins are a major class of plant hormones responsible for regulating various developmental processes, including coordination of cell division and differentiation, activity of shoot and root apical meristems, proportional organ formation and others (Kieber and Schaller 2018). CKs are perceived by a subfamily of sensor histidine kinases (HKs), which transmit the signal via a two-component phosphorelay cascade, a signalling platform sharing its origin with bacterial and yeast two-component signal transduction system (TCS) (To and Kieber 2008; Hwang et al. 2012).

Characterization of cytokinin-insensitive *cre1* (*cytokinin response 1*) mutants and overexpression of *CRE1* in a yeast mutant background led to identification of the CRE1 cytokinin receptor in *Arabidopsis* (Inoue et al. 2001). Other groups independently confirmed these observations and established the role of CRE1/AHK4 receptor in CK sensing (Suzuki et al. 2001; Ueguchi et al. 2001a; Yamada et al. 2001). The CK perception is initiated by ligand interaction with a sensory CHASE domain, resulting in subsequent autophosphorylation on a conserved histidine residue (His) in the transmitter domain and further intramolecular transfer to a conserved aspartate residue (Asp) in the receiver domain. The signal is then transmitted via AHP proteins (AHP1–AHP5) through the His-Asp phosphorelay signal transduction pathway to the type-B response regulators (RRs), which are localized in the nucleus. Their activation then leads to promoter binding and transcription of target genes (Hwang and Sheen 2001; Lohrmann et al. 2001; Hosoda et al. 2002; Yamada et al. 2004).

AHP proteins which act downstream of cytokinin HKs are partially redundant positive regulators of cytokinin signalling (Hutchison et al. 2006). Analyses of protein–protein interactions revealed that AHPs can interact with several upstream HKs as well as downstream RRs (Dortay et al. 2006). On the other hand, some HKs show preference for specific AHPs. The receiver domain of HK CKI1 has been shown to specifically interact with AHP2, AHP3 and AHP5 with different affinities (Pekárová et al. 2011). This particular HK preferentially interacts with AHP2 and AHP3, but only weakly with AHP5. Interestingly, the receptor domain of another hybrid HK, ETR1 receptor of the ethylene pathway, interacts with AHP1, AHP2, AHP3 and AHP5 providing evidence of a possible crosstalk mechanism between cytokinin- and ethylene-signalling pathways, presumably via ETR1 pathway-related phosphorylation of ARR10, a positive regulator of cytokinin signalling (Zdarska et al. 2019).

There are two types of RRs involved in the cytokinin pathway: type-A RRs and type-B RRs. Type-B RRs possess an N-terminal receiver domain and C-terminal output domain, which contains a DNA-binding GARP motif (named after GOLDEN2, *Arabidopsis* ARR type-B proteins and *Chlamydomonas* Psrl) defining these transcription factors as members of the MYB superfamily (Riechmann et al. 2000). In addition, the C-terminal portion of these proteins also contains nuclear localization sequences (NLSs), a prerequisite for playing a role of a transcription factor (Lohrmann et al. 2001).

In contrast, type-A RRs contain a receiver domain but lack the output domain necessary for transcription function. They respond rapidly to exogenous cytokinin and their expression is controlled by type-B RRs (Brandstatter and Kieber 1998; D'Agostino et al. 2000; Taniguchi et al. 2007). Transcript abundance of type-A RRs in response to CK is thus, at least in part, regulated on the transcriptional level. Analysis of *ARR5::GUS* transgenic plants revealed high expression in the root and shoot apical meristems, consistent with the idea that cytokinins are active in dividing cells (D'Agostino et al. 2000). The response to exogenous cytokinin is fast, and *ARR5* transcript levels together with those of other *RR* (*ARR4*, *ARR6*, *ARR7*, *ARR15* and *ARR16*) transcripts reach maximal induction 10 min following CK treatment. Generally, many type-A RRs are expressed particularly strongly in the root in response to changing CK levels providing a negative mechanism for attenuation of the cytokinin signalling output (To et al. 2004). The mechanism by which they exert their negative regulatory function is not well characterized, but likely involves phosphorylation-dependent interactions and a phosphocompetition relationship between type-A and type-B ARRs (To et al. 2007). Hence, it seems that the orchestrated action of type-B and type-A RRs mediates cell type-specific response in dividing meristematic cells and, in synergy with other signalling pathways, coordinates various aspects of root and shoot development. For example, for early phases of lateral root organogenesis (such as priming and initiation), the ectopic production of CK in the root basal meristem is required to secure development of lateral root primordia in root zones with a repressed cytokinin response (Bielach et al. 2012).

5.2 Aromatic Cytokinins Show Different Responses in Cytokinin Bioassays

Although aromatic cytokinins (ARCKs) were known for a relatively long time and initially assumed to occur as rare metabolic products in a limited number of plant species (Letham and Palni 1983), it was not until the 1990s that *m*T and its riboside (*m*TR) were detected in poplar leaves and their biological activity scrutinized (Strnad et al. 1992, 1997; Werbrouck et al. 1996; Holub et al. 1998). The surprising discovery that position isomers of topolin, *o*T and *m*T as well as their ribosides showed different biological properties stirred up a new interest in ARCKs and more detailed SAR studies. These results were centred around *m*T, the biologically active endogenous cytokinin form (Holub et al. 1998). While *meta*-position appeared to have a generally positive impact on the cytokinin signalling mediated by this ARCK, the corresponding *ortho*-derivative produced mostly negative effects.

Identification of cytokinin receptors and characterization of the cytokinin signalling pathway paved the way for the preparation of a new generation of synthetic CK derivatives and allowed us to generate new tools for precise quantification of the cytokinin signalling output. The *ARR5::GUS* was among the first reporter-based assays to allow for transcriptional output measurements elevated by CK (D'Agostino et al. 2000). Another assay based on the CRE1/AHK4 His-to-Asp phosphorelay system and assying its His-kinase activity in a mutant *E. coli* strain (*ΔrcsC*, *cps::lacZ*) was reported shortly afterwards (Suzuki et al. 2001; Yamada et al. 2001). This receptor together with *Arabidopsis* AHK2 and AHK3 are among the best characterized cytokinin receptors in the plant kingdom (Ueguchi et al. 2001b). In 2004, the bacterial assay previously used for the characterization of CRE1/AHK4 (Suzuki et al. 2001; Yamada et al. 2001) was utilized in the screening of ligand specificities of both CRE1/AHK4 and AHK3 receptors with a broad range of CKs and CK ribosides/ribotides (Spíchal et al. 2004). Consistent with the previous results, both receptors showed preference for the free bases of isoprenoid-type cytokinins (*t*Z and iP) but had different sensitivity to some ARCKs and CK ribosides and ribotides. Interestingly, both CRE1/AHK4 and AHK3 were considerably activated by *m*T, which produced responses comparable to those achieved by isoprenoid cytokinins. However, both receptors showed low sensitivity to *o*T. In addition, *m*T, but not *o*T, elicited high response in *ARR5::GUS* assay in *Arabidopsis* (Spíchal et al. 2004). These results have provided the first mechanistic evidence that *m*T interaction with *Arabidopsis* HKs, notably with AHK3, is most likely responsible for the high activity of this ARCK in the cytokinin bioassays.

Similarly, cytokinin-responsive HKs from maize (ZmHK1, ZmHK2 and ZmHK3a) were expressed in the above *E. coli* strain that conferred cytokinin-inducibility of *lacZ* expression on the transformed bacteria (Yonekura-Sakakibara et al. 2004). iP triggered strong response specifically in ZmHK1, while ZmHK2 showed higher sensitivity to *t*Z and *t*ZR. ZmHK3a had similar preference for *t*Z, *c*Z, iP and BAP in the bacterial assay. Direct comparison of CRE1/AHK4 and ZmHK1 in the bacterial receptor assay further confirmed that only the *m*T isomer interacted with *Arabidopsis* CRE1/AHK4 receptor, while both *m*T and *o*T were found to be equally good ligands for maize ZmHK1 (Mok et al. 2005). However, the most striking difference between HKs of monocots and dicots seems to be their different affinity for *c*Z—only maize receptors recognized this ligand with significant affinity (Yonekura-Sakakibara et al. 2004; Mok et al. 2005). These findings clearly show that maize cytokinin HKs differ in ligand preference and that *c*Z is an active form of CK at least in monocot species.

An improved version of a live cell-based assay using transgenic bacteria expressing CK receptors was developed (Romanov et al. 2005, 2006). In this assay, radiolabelled *t*Z was used to quantitatively assess various kinetic parameters of cytokinin binding to CRE1/AHK4 and AHK3 receptors (Romanov et al. 2006). Using intact bacteria expressing a functional receptor, it was possible to characterize important parameters of ligand–receptor interaction, including affinity and ligand specificity of the individual CK interaction partners. Both CRE1/AHK4 and AHK3 shared high affinity for *t*Z, but AHK3 had an approximately tenfold lower affinity to iP (Romanov et al. 2006). On the other hand, the AHK3 receptor showed genuine affinity for CK ribosides, *t*ZR and iPR, which are considered to represent cytokinin transport forms in the long-distance acropetal and basipetal cytokinin translocations between plant organs (Kudo et al. 2010). The receptor also recognized *c*Z and the non-adenine compound thidiazuron as its natural ligands. Affinities of both receptors to BAP were rather low, inconsistent with the view that this ARCK is one of the most potent cytokinins for *in planta* use; however, this was later put into a broader perspective by the same group using bacterially expressed CHASE domain of CRE1/AHK4 and AHK2 receptors (Stolz et al. 2011). These results together with known localization patterns of both receptors (*CRE1/AHK4* expression being predominantly detected in roots and *AHK3* in shoots) strongly suggested that AHK3 plays a specific function in root-to-shoot communications and that both time- and space-controlled gene expression and ligand preference are the key factors to specify the receptor activity.

5.3 Metabolism and Activity of Topolins and their Derivatives in Various Plant Models

A broad range of ARCK derivatives has been utilized in the past decades, and various substitutions on the purine backbone as well as on the benzyl side chain were designed and synthesized to test their specific modus operandi in both plants and human cells. In the first of these systematic structure–activity relationship studies, the authors prepared a chemical library of BAP derivatives with various substituents attached to the phenyl ring and tested their activity in standardized assays (Doležal et al. 2006, 2007). The high biological activity of many prepared compounds, notably their ability to delay the onset of senescence measured in detached wheat leaf senescence bioassay, contrasted with their relatively low responses in CRE1/AHK4 and AHK3 bacterial receptor assays. Several of these compounds also showed strong cytotoxic activity against various cancer cell lines while having negligible cytotoxicity to a normal murine fibroblast cell line (Doležal et al. 2007). Another SAR study firmly established the role of *meta*-substitution at the N^6-benzyl side chain of BAP. 6-Benzylamino-9-tetrahydropyran-2-ylpurine (THP) and 6-benzylamino-9-tetrahydrofuran-2-ylpurine (THF) derivatives, with variously positioned hydroxy and methoxy functional groups on the aromatic side chain, were prepared and tested (Szüčová et al. 2009). The biological activity of THP derivatives was shown to be correlated to these substitutions, as follows: *meta* > *ortho*> > *para*.

6-(3-Methoxybenzylamino)purine (*meta*-methoxy topolin, Me*m*T) derivatives with *meta*-methoxy substitution on the benzyl ring and tetrahydropyranyl or 4-chlorobutyl groups at the N9-position of the cytokinin purine backbone were further used in a detailed study on two model plants, *Arabidopsis* and maize (Podlešáková et al. 2012; Plíhal et al. 2013). These studies brought together data from model plants after exogenous treatments with detailed metabolic profiling and distribution of various CK forms throughout the plant body. It was found that THP substitution significantly enhanced acropetal transport of the parental CK and prevented unwanted N-glucosylation (Podlešáková et al. 2012). It is well-known that cytokinins including BAP actively regulate root meristem activity via modulation of cytokinin–auxin interactions by affecting auxin transport (Růžička et al. 2009) and that the cytokinin–auxin balance is also important for lateral root formation (Chang et al. 2013). More specifically, auxin influx carrier AUX1 is required for cytokinin-dependent changes in auxin levels in the lateral root cap associated with the control of cell elongation in concert with another plant hormone, ethylene (Street et al. 2016). It is possible that substitutions of BAP at N^6- and N9-positions may prevent some negative effects of glucosylation and different metabolic pathways then lead to the accumulation of metabolites with different CK activities. In the case of *m*T exogenous treatment in the in vitro propagation of *Spathiphyllum floribundum*, (OG)[9R]*m*T was shown to be the main metabolic product that was degraded rapidly during acclimatization (Werbrouck et al. 1996). In contrast, when BAP was used in place of *m*T in the plant acclimatization, the predominant metabolic product was non-active [9G]BAP, which may be responsible for affecting cytokinin–auxin balance and eventually some negative effects on the primary root development, inhibition of lateral root branching, etc. Similarly, in the case of Me*m*T and Me*m*TTHP treatments in maize, the authors observed fast metabolic turnover of these ARCKs leading to a noticeable accumulation of (OG)*m*T and deprotected *m*T as early as 1 h post treatment (Podlešáková et al. 2012). The elevated endogenous concentrations of the active base *m*T corresponded well with slow gradual increase of the transcript levels of type-A *RRs* that was significant already 30 min following the application of Me*m*T and a little delayed after Me*m*TTHP treatment.

A deeper understanding of the specific role played by ARCKs can be obtained in comparative studies with various plant models, bearing in mind that CK levels vary significantly in a time- and place-dependent manner (diurnal or seasonal cycles, etc.). Only recently, the poplar species *Populus × canadensis*, cv. Robusta, the first organism found to contain ARCKs, was confirmed to have five CHASE-containing histidine kinase isoforms: PcHK2, PcHK3a, PcHK3b, PcHK4a and PcHK4b (Jaworek et al. 2019). While all these receptor histidine kinases showed strong affinity for *t*Z—the most abundant cytokinin form in poplar—and some also to iP, the kinase activity in the presence of the CK ligand differed significantly. PcHK4a displayed over 400-fold higher kinase activity compared to other HKs, which suggests a major role in responses to changing CK levels (Jaworek et al. 2019). Interestingly, another recent study on cytokinin-specific multistep phosphorelays in poplar also showed involvement of the above cytokinin-related HKs together with 10 HPts in the suggested cytokinin signalling pathway (Héricourt et al. 2019). A yeast complementation study with the above HKs again confirmed the functionality of all these, but only HK4a, was shown to have kinase activity dependent on the presence of CK.

Both studies also showed that all poplar HKs had strong expression in all tested plant organs, but HK4a and HK4b, however, seem to show different distribution suggesting possible functional diversification of these two paralogous genes. Live-cell bacterial assay was employed for cytokinin-binding studies with radiolabelled *t*Z (Jaworek et al. 2019). The cytokinin-binding strengths of all cytokinin-responsive HKs declined in the following order: *t*Z > iP > *c*Z. PcHK3a and PcHK3b exhibited weaker binding to iP-type cytokinins, *c*Z, BAP and *o*T than the other HKs. PcHK2 showed relatively good affinity to *m*T, iPR and *t*ZR, which was significantly better than the affinity towards BAP. It seems that *m*T shows relatively good affinities to PcHK2, PcHK3a and PcHK3b isoforms, having K_i values in the range of approximately 20–60 nM, whereas *o*T represents a relatively poor ligand for all PcHK isoforms ($K_i > 1000$ nM). Additionally, PcHK4 receptors showed strong binding at pH 5.5, while PcHK3 receptors proved to be more active at pH 7.5, in line with their putative membrane localization at the plasma membrane and in the membrane of the endoplasmic reticulum, respectively.

Acknowledgement The work was supported by an ERDF project entitled “Development of Pre-Applied Research in Nanotechnology and Biotechnology” (No. CZ.02.1.01/0.0/0.0/17_048/0007323). The author of this review is grateful to Alexander Oulton, Ph.D., for his critical reading of the chapter.

References

Bielach A, Podlešáková K, Marhavý P, Duclercq J, Cuesta C, Müller B, Grunewald W, Tarkowski P, Benková E (2012) Spatiotemporal regulation of lateral root organogenesis in *Arabidopsis* by cytokinin. Plant Cell 24:3967–3981

Brandstatter I, Kieber JJ (1998) Two genes with similarity to bacterial response regulators are rapidly and specifically induced by cytokinin in *Arabidopsis*. Plant Cell 10:1009–1019

Chang L, Ramireddy E, Schmülling T (2013) Lateral root formation and growth of *Arabidopsis* is redundantly regulated by cytokinin metabolism and signalling genes. J Exp Bot 64:5021–5032

D'Agostino IB, Deruère J, Kieber JJ (2000) Characterization of the response of the Arabidopsis response regulator gene family to cytokinin. Plant Physiol 124:1706–1717

Doležal K, Popa I, Kryštof V, Spíchal L, Fojtíková M, Holub J, Lenobel R, Schmülling T, Strnad M (2006) Preparation and biological activity of 6-benzylaminopurine derivatives in plants and human cancer cells. Bioorg Med Chem 14:875–884

Doležal K, Popa I, Hauserová E, Spíchal L, Chakrabarty K, Novák O, Kryštof V, Voller J, Holub J, Strnad M (2007) Preparation, biological activity and endogenous occurrence of N^6-benzyladenosines. Bioorg Med Chem 15:3737–3747

Dortay H, Mehnert N, Bürkle L, Schmülling T, Heyl A (2006) Analysis of protein interactions within the cytokinin-signaling pathway of *Arabidopsis thaliana*. FEBS J 273:4631–4644

Héricourt F, Larcher M, Chefdor F, Koudounas K, Carqueijeiro I, Lemos Cruz P, Courdavault V, Tanigawa M, Maeda T, Depierreux C, Lamblin F, Glévarec G, Carpin S (2019) New insight into HPts as hubs in poplar cytokinin and osmosensing multistep phosphorelays: cytokinin pathway uses specific HPts. Plants (Basel) 8:591

Holub J, Hanuš J, Hanke DE, Strnad M (1998) Biological activity of cytokinins derived from *ortho*- and *meta*-hydroxybenzyladenine. Plant Growth Regul 26:109–115

Hosoda K, Imamura A, Katoh E, Hatta T, Tachiki M, Yamada H, Mizuno T, Yamazaki T (2002) Molecular structure of the GARP family of plant Myb-related DNA binding motifs of the Arabidopsis response regulators. Plant Cell 14:2015–2029

Hutchison CE, Li J, Argueso C, Gonzalez M, Lee E, Lewis MW, Maxwell BB, Perdue TD, Schaller GE, Alonso JM, Ecker JR, Kieber JJ (2006) The *Arabidopsis* histidine phosphotransfer proteins are redundant positive regulators of cytokinin signaling. Plant Cell 18:3073–3087

Hwang I, Sheen J (2001) Two-component circuitry in *Arabidopsis* cytokinin signal transduction. Nature 413:383–389

Hwang I, Sheen J, Müller B (2012) Cytokinin signaling networks. Annu Rev Plant Biol 63:353–380

Inoue T, Higuchi M, Hashimoto Y, Seki M, Kobayashi M, Kato T, Tabata S, Shinozaki K, Kakimoto T (2001) Identification of CRE1 as a cytokinin receptor from *Arabidopsis*. Nature 409:1060–1063

Jaworek P, Tarkowski P, Hluska T, Kouřil S, Vrobel O, Nisler J, Kopečný D (2019) Characterization of five CHASE-containing histidine kinase receptors from *Populus* x *canadensis* cv. *Robusta* sensing isoprenoid and aromatic cytokinins. Planta 251:1

Kieber JJ, Schaller GE (2018) Cytokinin signaling in plant development. Development 145:1–7

Kudo T, Kiba T, Sakakibara H (2010) Metabolism and long-distance translocation of cytokinins. J Integr Plant Biol 52:53–60

Letham DS, Palni LMS (1983) The biosynthesis and metabolism of cytokinins. Annu Rev Plant Physiol 34:163–197

Lohrmann J, Sweere U, Zabaleta E, Bäurle I, Keitel C, Kozma-Bognar L, Brennicke A, Schäfer E, Kudla J, Harter K (2001) The response regulator ARR2: a pollen-specific transcription factor involved in the expression of nuclear genes for components of mitochondrial complex I in *Arabidopsis*. Mol Genet Genomics 265:2–13

Mok MC, Martin RC, Dobrev PI, Vaňková R, Ho PS, Yonekura-Sakakibara K, Sakakibara H, Mok DW (2005) Topolins and hydroxylated thidiazuron derivatives are substrates of cytokinin *O*-glucosyltransferase with position specificity related to receptor recognition. Plant Physiol 137:1057–1066

Pekárová B, Klumpler T, Třísková O, Horák J, Jansen S, Dopitová R, Borkovcová P, Papoušková V, Nejedlá E, Sklenář V, Marek J, Zídek L, Hejátko J, Janda L (2011) Structure and binding specificity of the receiver domain of sensor histidine kinase CKI1 from *Arabidopsis thaliana*. Plant J 67:827–839

Plíhal O, Szüčová L, Galuszka P (2013) N9-substituted aromatic cytokinins with negligible side effects on root development are an emerging tool for in vitro culturing. Plant Signal Behav 8: e24392

Podlešáková K, Zalabák D, Čudejková M, Plíhal O, Szüčová L, Doležal K, Spíchal L, Strnad M, Galuszka P (2012) Novel cytokinin derivatives do not show negative effects on root growth and proliferation in submicromolar range. PLoS One 7:e39293

Riechmann JL, Heard J, Martin G, Reuber L, Jiang C, Keddie J, Adam L, Pineda O, Ratcliffe OJ, Samaha RR, Creelman R, Pilgrim M, Broun P, Zhang JZ, Ghandehari D, Sherman BK, Yu G (2000) *Arabidopsis* transcription factors: genome-wide comparative analysis among eukaryotes. Science 290:2105–2110

Romanov GA, Spíchal L, Lomin SN, Strnad M, Schmülling T (2005) A live cell hormone-binding assay on transgenic bacteria expressing a eukaryotic receptor protein. Anal Biochem 347:129–134

Romanov GA, Lomin SN, Schmülling T (2006) Biochemical characteristics and ligand-binding properties of *Arabidopsis* cytokinin receptor AHK3 compared to CRE1/AHK4 as revealed by a direct binding assay. J Exp Bot 57:4051–4058

Růžička K, Šimášková M, Duclercq J, Petrášek J, Zažímalová E, Simon S, Friml J, Van Montagu MC, Benková E (2009) Cytokinin regulates root meristem activity via modulation of the polar auxin transport. Proc Natl Acad Sci U S A 106:4284–4289

Spíchal L, Rakova NY, Riefler M, Mizuno T, Romanov GA, Strnad M, Schmülling T (2004) Two cytokinin receptors of *Arabidopsis thaliana*, CRE1/AHK4 and AHK3, differ in their ligand specificity in a bacterial assay. Plant Cell Physiol 45:1299–1305

Stolz A, Riefler M, Lomin SN, Achazi K, Romanov GA, Schmülling T (2011) The specificity of cytokinin signalling in *Arabidopsis thaliana* is mediated by differing ligand affinities and expression profiles of the receptors. Plant J 67:157–168

Street IH, Mathews DE, Yamburkenko MV, Sorooshzadeh A, John RT, Swarup R, Bennett MJ, Kieber JJ, Schaller GE (2016) Cytokinin acts through the auxin influx carrier AUX1 to regulate cell elongation in the root. Development 143:3982–3993

Strnad M, Peters W, Beck E, Kamínek M (1992) Immunodetection and identification of N^6-(*o*-hydroxybenzylamino) purine as a naturally occurring cytokinin in *Populus x canadensis* Moench cv. *Robusta* leaves. Plant Physiol 99:74–80

Strnad M, Hanuš J, Vaněk T, Kamínek M, Ballantine JA, Fussell B, Hanke DE (1997) *Meta*-topolin, a highly active aromatic cytokinin from poplar leaves (*Populus x canadensis* Moench, cv. *Robusta*). Phytochemistry 45:213–218

Suzuki T, Miwa K, Ishikawa K, Yamada H, Aiba H, Mizuno T (2001) The Arabidopsis sensor His-kinase, AHK4, can respond to cytokinins. Plant Cell Physiol 42:107–113

Szüčová L, Spíchal L, Doležal K, Zatloukal M, Greplová J, Galuszka P, Kryštof V, Voller J, Popa I, Massino FJ, Jørgensen JE, Strnad M (2009) Synthesis, characterization and biological activity of ring-substituted 6-benzylamino-9-tetrahydropyran-2-yl and 9-tetrahydrofuran-2-ylpurine derivatives. Bioorg Med Chem 17:1938–1947

Taniguchi M, Sasaki N, Tsuge T, Aoyama T, Oka A (2007) ARR1 directly activates cytokinin response genes that encode proteins with diverse regulatory functions. Plant Cell Physiol 48:263–277

To JP, Kieber JJ (2008) Cytokinin signaling: two-components and more. Trends Plant Sci 13:85–92

To JP, Haberer G, Ferreira FJ, Deruère J, Mason MG, Schaller GE, Alonso JM, Ecker JR, Kieber JJ (2004) Type-A Arabidopsis response regulators are partially redundant negative regulators of cytokinin signaling. Plant Cell 16:658–671

To JP, Deruère J, Maxwell BB, Morris VF, Hutchison CE, Ferreira FJ, Schaller GE, Kieber JJ (2007) Cytokinin regulates type-A Arabidopsis response regulator activity and protein stability via two-component phosphorelay. Plant Cell 19:3901–3914

Ueguchi C, Sato S, Kato T, Tabata S (2001a) The *AHK4* gene involved in the cytokinin-signaling pathway as a direct receptor molecule in *Arabidopsis thaliana*. Plant Cell Physiol 42:751–755

Ueguchi C, Koizumi H, Suzuki T, Mizuno T (2001b) Novel family of sensor histidine kinase genes in *Arabidopsis thaliana*. Plant Cell Physiol 42:231–235

Werbrouck SPO, Strnad M, Van Onckelen H, Debergh PC (1996) *Meta*-topolin, an alternative to benzyladenine in tissue culture? Physiol Plant 98:291–297

Yamada H, Suzuki T, Terada K, Takei K, Ishikawa K, Miwa K, Yamashino T, Mizuno T (2001) The Arabidopsis AHK4 histidine kinase is a cytokinin-binding receptor that transduces cytokinin signals across the membrane. Plant Cell Physiol 42:1017–1023

Yamada H, Koizumi N, Nakamichi N, Kiba T, Yamashino T, Mizuno T (2004) Rapid response of *Arabidopsis* T87 cultured cells to cytokinin through His-to-Asp phosphorelay signal transduction. Biosci Biotechnol Biochem 68:1966–1976

Yonekura-Sakakibara K, Kojima M, Yamaya T, Sakakibara H (2004) Molecular characterization of cytokinin-responsive histidine kinases in maize. Differential ligand preferences and response to cis-zeatin. Plant Physiol 134:1654–1661

Zdarska M, Cuyacot AR, Tarr PT, Yamoune A, Szmitkowska A, Hrdinová V, Gelová Z, Meyerowitz EM, Hejátko J (2019) ETR1 integrates response to ethylene and cytokinins into a single multistep phosphorelay pathway to control root growth. Mol Plant 12:1338–1352

Topolin Metabolism and Its Implications for In Vitro Plant Micropropagation

6

Karel Doležal and Magdalena Bryksová

Contents

Abstract

Topolins are a relatively recent discovery, following search for viable alternatives to 6-benzylaminopurine (BAP) which while effective and affordable has important disadvantages for certain crops. This chapter reviews some biochemical and technical aspects of topolin metabolism in relation to in vitro plant micropropagation.

Keywords

Meta-topolin · Topolin · Cytokinin · Metabolism · Plant tissue culture

6-Benzylaminopurine (BAP, also known as N^6-benzyladenine: BA) was long the most widely used cytokinin (CK) in micropropagation systems, due to its efficacy and affordability (Holub et al. 1998). However, its adverse effects on the growth, rooting, and acclimatization of some recalcitrant species and induction of other physiological disorders (Aremu et al. 2012a) prompted the search for and subsequent

K. Doležal (✉) · M. Bryksová
Department of Chemical Biology, Faculty of Science, Palacký University, Olomouc, Czech Republic

Laboratory of Growth Regulators, Institute of Experimental Botany, AS CR, Palacký University, Olomouc, Czech Republic
e-mail: karel.dolezal@upol.cz

N. Ahmad, M. Strnad (eds.), *Meta-topolin: A Growth Regulator for Plant Biotechnology and Agriculture*, https://doi.org/10.1007/978-981-15-9046-7_6

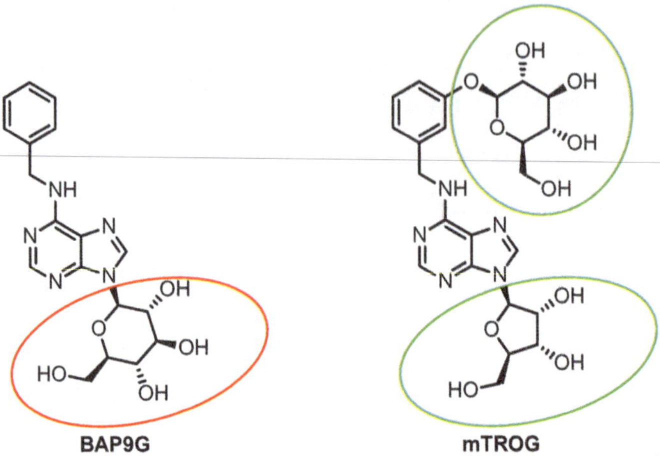

Fig. 6.1 Chemical structure of 6-benzylamino-9-β-D-glucopyranosylpurine (BAP9G) and 6-(3-*O*-β-D-glucopyranosyl)benzylaminopurine-9-β-D-riboside (*m*TROG)

discovery of viable alternatives: the choice of CK remains critical to the success or failure of any newly developed micropropagation protocol (Bairu et al. 2007; Werbrouck 2010; Aremu et al. 2012a; Valero-Aracama et al. 2010). The increasing importance of hydroxylated derivatives commonly referred to as topolins (Strnad et al. 1997) in micropropagation has been critically reviewed more recently (Aremu et al. 2012a, 2017).

Following description of the basic aspects of CK activity of topolin ribosides (Kaminek et al. 1987) and unambiguous identification of 6-(3-hydroxybenzylamino) purine (*meta*-topolin, *m*T) and its derivatives as naturally occurring CKs *in planta* (Strnad 1996, 1997; Strnad et al. 1997), pioneering work was published on the metabolism and in vitro effects of *m*T in micropropagated *Spathiphyllum floribundum* (Werbrouck et al. 1996). *m*T was compared to BAP, 6-benzylamino-9-(2-tetrahydropyranyl)purine (BAP9THP), and 6-benzylaminopurine-9-β-D-riboside (BAPR) using an HPLC separation system, coupled to a tandem quadrupole mass spectrometer (MS/MS) equipped with an electrospray interface (ESI). In vitro, BAP and its 9-substituted derivatives BAP9THP and BAPR were mainly converted into the stable and inactive metabolite, 6-benzylamino-9-β-D-glucopyranosylpurine (BAP9G) (Fig. 6.1), located mostly in the basal part of the micropropagated plant (Werbrouck et al. 1995). In contrast, 6-(3-*O*-β-D-glucopyranosyl) benzylaminopurine-9-β-D-riboside (*m*TROG) (Fig. 6.1) was identified as the main metabolite of *m*T. This new cytokinin-*O*-glucoside, which was present in all plant parts, was formed much faster than BAP9G during acclimatization process. The effect of BAP and *m*T on in vitro shoot and root production and ex vitro rooting was then compared (Werbrouck et al. 1996). Only *m*T combined sufficient shoot production with acceptable in vitro root formation. The plants developed on medium with 10 μM or more *m*T rooted better also during acclimatization in comparison with those grown on medium with comparable BAP concentrations.

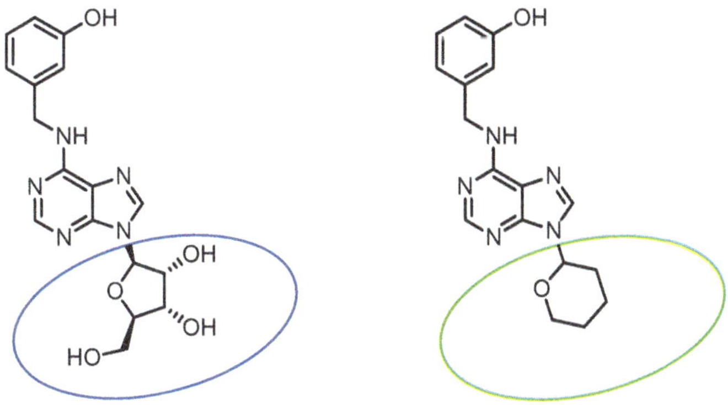

mTR **mT9THP**

Fig. 6.2 Chemical structure of 6-(3-hydroxybenzylamino)purine-9-β-D-ribofuranoside (*m*TR) and 6-(3-hydroxybenzylamino)-9-tetrahydropyran-2-ylpurine (*m*T9THP)

Very similar results were later obtained by Bairu et al. (2011) in micropropagated *Harpagophytum procumbens* tissues, where changes in endogenous CK profiles and the physiological implications of this in relation to shoot-tip necrosis (STN) and CK treatments were also studied. Generally, necrotic shoots contained more total CKs compared to normal shoots, and CK accumulation was higher in the basal section. More importantly, further analysis of structural and functional CK forms revealed the inability of BAP to form *O*-glucosides as well as excessive accumulation of 9-glucosides (irreversible deactivation product) in necrotic and basal callus like tissues of BAP-treated shoots (Bairu et al. 2011). The addition of IAA enhanced the formation of 9-glucosides in BAP-treated cultures. The symptom of STN could therefore be attributed to conversion of active CK to other forms such as 9-glucoside (Bairu et al. 2011). On the other hand, the presence of a hydroxyl group in their molecule gives topolins a structural advantage over BAP. This was reflected in the presence of a generous amount of *O*-glucosides in topolin-treated samples and hence little or no CK shortage (Bairu et al. 2011). Moreover, the level of irreversible inactivation (9-glucoside formation) of *m*T was found to be even lower when exogenously applied in the form of 9-riboside, compared with free base application.

As the most efficient plant growth regulator (PGR), 10 μM 6-(3-hydroxybenzylamino)purine-9-β-D-ribofuranoside (*m*TR, Fig. 6.2) treatment also produced the highest number of shoots (approximately five shoots per explant) during clonal propagation of *Lachenalia montana*, a species endemic to Southern Africa and extensively traded as ornamental plants in the international floriculture industry (Aremu et al. 2017b). Based on the concentrations of endogenous CKs subsequently determined, 10 μM *m*TR regenerants also had the highest CK levels which were mainly of the aromatic type (98%). In terms of the functional role of the CKs, *O*-glucosides were again the dominant CK metabolites in the regenerants of the 10 μM *m*TR treatment. On the other hand, the insufficient rooting, predominantly in regenerants of the BAP treatments, was closely related to the high accumulation of

N9-glucosides compared to regenerants from other treatments. These findings provided further evidence of the interrelationship between exogenous topolin application, positive phenotypic responses, and endogenous CK levels in the in vitro regenerants (Aremu et al. 2017b).

Based on these studies, a series of attempts was made to prepare other topolin derivatives substituted at the 9-position using various protective groups to improve the specific biological functions of the CKs already routinely used in the plant micropropagation industry. For example, inspired by discovery of the protective role of ribose in the 9-position (Bairu et al. 2009). Szüčová et al. (2009) prepared several substituted 6-benzylamino-9-tetrahydropyran-2-ylpurine (THPP) and 9-tetrahydrofuran-2-ylpurine (THFP) derivatives, with hydroxy and methoxy functional groups at various positions on the benzyl ring. The new compounds were synthesized by condensation of 6-chloropurine with 3,4-dihydro-2H-pyran or 2,3-dihydrofuran and then by the reaction of these intermediates with the corresponding benzylamines (in *n*-propanol or *n*-butanol, in the presence of triethylamine). Identity and purity of the prepared compounds were confirmed by CHN analysis, TLC, HPLC, melting point determinations, CI+ MS, and 1H NMR spectroscopy. The CK activity of the prepared derivatives was determined in three classical cytokinin bioassays (tobacco callus, wheat leaf senescence, and *Amaranthus* bioassay).

In another study (Podlešáková et al. 2012), in contrast to canonical CKs, the 9-tetrahydropyranyl derivative of *m*T (Fig. 6.2) and its methoxy counterpart showed negative effects on root development at only three orders of magnitude higher concentration. The methoxy derivative also demonstrated a positive effect on lateral root branching and leaf emergence in nanomolar concentration range in comparison with untreated plants. Tetrahydropyranyl substitution at the *N9*-position of CK purine ring was also found to significantly enhance acropetal transport of a given CK. Together with the methoxy substitution, inhibition of the formation of non-active CK glucosides in roots allows gradual release of the active base and has a significant effect on the distribution and amount of endogenous isoprenoid CKs in different plant parts (Podlešáková et al. 2012). These results provided a basis for anticipating that the use of novel aromatic CK derivatives could improve the expected hormonal effects in plant propagation methodology in the future.

This led to 9-THP topolin derivative (Fig. 6.2) being successfully used in various micropropagation systems, for example, two widely used medicinal plants, *Aloe arborescens* and *Harpagophytum procumbens* (Amoo et al. 2014). In terms of *A. arborescens* shoot multiplication, *m*TTHP and *m*T (at equimolar level) showed similar effects, and both were comparably better than the control and BAPR. In *H. procumbens*, *m*T-treated cultures were the most responsive to treatment at 2.5 μM compared to the control. At 5.0 μM concentration, *m*T9THP and *m*TR demonstrated a similar activity on shoot proliferation. Particularly at low concentrations, *m*T9THP had a better rooting stimulatory activity than the other CKs in both plant species. It is conceivable that *m*T9THP is another viable alternative topolin with the added advantage of inducing rooting at low concentrations (Amoo et al. 2014). Later (Amoo et al. 2015), this compound was also successfully used to improve micropropagated *Merwilla plumbea* shoot production without rooting inhibition as well as its positive carry-over effect on ex vitro growth. Unlike *m*TTHP treatments, an increase in concentration of *m*TR or TDZ, other tested CKs, beyond 0.5 μM resulted in a significant decrease in the concentrations of all the photosynthetic pigments quantified. Even after 6 months of ex vitro growth, regenerated plants of the 0.5 μM *m*TTHP treatment had the significantly higher total leaf area, total leaf fresh weight, and bulb size compared to all *m*TR- and TDZ-treated plants (Amoo et al. 2015).

In a similar study on *Merwilla plumbea* (Lindl.) Speta, a popular and highly sought-after South African medicinal plant with diverse therapeutic uses (Aremu et al. 2014), the effect of another *meta*-topolin derivative, 6-(3-methoxybenzylamino)-9-tetrahydropyran-2-ylpurine (Me*m*TTHP), was evaluated on the growth and level of endogenous CKs during micropropagation and acclimatization stages. A total of 37 (22 isoprenoid and 15 aromatic) CKs were determined in both in vitro and ex vitro acclimatized plants. Based on their metabolic function, these were separated into five different groups, including free bases, ribosides, ribotides, and *O*- and 9-glucosides. In addition to enhancing our understanding of the hormone physiology in *M. plumbea*, the current findings were discussed in line with the effect of exogenously applied CK on the observed differences in growth before and after the important stage of acclimatization. The observed dynamics in endogenous CK can provide an avenue to optimize the in vitro growth and development of investigated species.

Another successful attempt to enhance the anti-senescence properties of topolins was described by Doležal et al. (2017) and Matušková et al. (2020), by preparing their 9-β-D-arabinofuranosyl or 9-β-D-2′-deoxyribofuranosyl derivatives (Fig. 6.3) via a one-step reaction. The starting material, optically pure unprotected 9-(2′-deoxy-β-D-ribofuranosyl)hypoxanthine (2′-deoxy-9-β-D-inosine) or 9-(β-D-arabinofuranosyl)hypoxantine, and BOP [(benzotriazol-1yloxy)tris

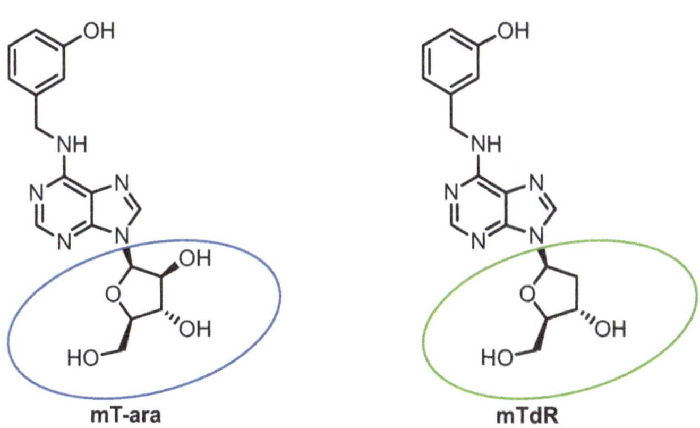

Fig. 6.3 Chemical structure of 6-(3-hydroxybenzylamino)purine-9-β-D-arabinofuranoside (*m*T-ara) and 6-(3-hydroxybenzylamino)purine-9-(2′-deoxy-β-D-ribofuranoside) (*m*TdR)

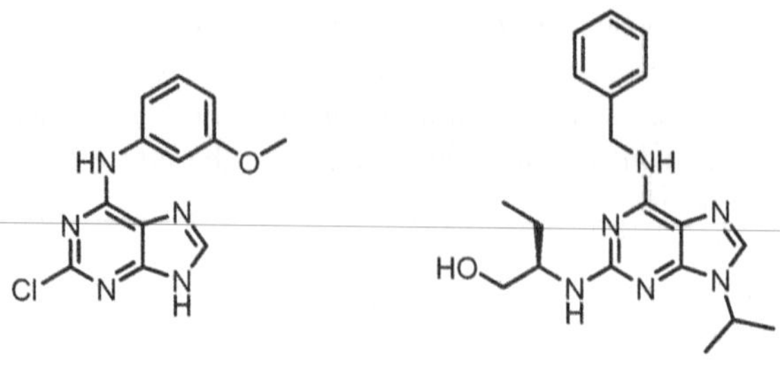

INCYDE roscovitine

Fig. 6.4 Chemical structure of 2-chloro-6-(3-methoxyphenyl) aminopurine (INCYDE) and 2-(1-ethyl-2-hydroxyethylamino)-6-benzylamino-9-isopropylpurine (roscovitine)

(dimethylamino)phosphonium hexafluorophosphate] were dissolved in dry DMF (3 mL) under nitrogen or argon atmosphere at 55–60 °C, and DIPEA was added, followed by appropriate substituted benzylamine (1.2 eq.) as the last component. The resulting white solid was isolated by filtration and re-crystallized from EtOH. The synthesized compounds were characterized by CHN and melting point analysis, analytical thin layer chromatography, high-performance liquid chromatography, ES^+ MS spectrometry, and ^1H NMR. The positive effect of selected derivatives on shoot proliferation in *Harpagophytum procumbens* and *Amelanchier alnifolia* as well as control of shoot-tip necrosis in in vitro cultures of the medicinal plant *Gymnosporia buxifolia* was then demonstrated (Doležal et al. 2017).

Modulating the CK status with inhibitors of CK perception and/or degradation may affect the general physiology of the plant (Dwivedi et al. 2010; Motte et al. 2013; Zatloukal et al. 2008). For this reason, regulation of phytohormone metabolism may be another potential way to improve plant growth and development during micropropagation.

The effect of supplementing either *m*T or BAP requiring cultures with INCYDE (2-chloro-6-(3-methoxyphenyl)aminopurine) (Fig. 6.4), an inhibitor of CK degradation (Spichal et al. 2012), on the endogenous CK profiles and physiology of banana in vitro was hence investigated (Aremu et al. 2012b).

Another interesting alternative approach for decreasing levels of irreversible topolin deactivation in banana tissue cultures was developed by Aremu et al. (2012b). An inhibitor of 9-glucosylation, roscovitine 2-(1-ethyl-2-hydroxyethylamino)-6-benzylamino-9-isopropylpurine (Fig. 6.4), which was previously discovered among a number of 2,6,9-trisubstituted purines, tested as potential *N*-glucosylation inhibitors (Blagoeva et al. 2004; Letham et al. 1977; Dwivedi et al. 2010). Aremu et al. (2012b) demonstrated that its application simultaneously with exogenous CK (s) in the cultivation media has the potential to change endogenous CK pools, thereby influencing the rooting and ex vitro acclimatization of in vitro-derived *Musa* spp. It was observed that plantlets regenerated from *m*T + roscovitine media produced the most shoots. They also had the highest total CK content (661 pmol/g FW) with the roots having approximately 68-fold more than the shoots (Aremu et al. 2012b).

A general trend observed was that the addition of roscovitine and/or INCYDE with *m*T improved the total CK pool in both roots and shoots of the tissue-cultured 'Williams' banana regenerants (Aremu et al. 2012b). A similar pattern was determined in the shoots when BAP was supplemented with roscovitine and INCYDE; however, both compounds reduced the total CK pool in the roots as well as the sum total in the plantlets. It is noteworthy to highlight that the reduction in total CK pool was mainly due to the decrease in the level of 9-glucosides, which are generally detrimental to plant growth (Aremu et al. 2012b).

6.1 Future Perspectives: Application of Fluorinated Compounds

Fluorination has a long tradition in nucleoside chemistry. It was demonstrated that replacement of the 2′ or 3′ hydroxyl groups of a nucleoside with a fluorine atom causes only a minor change in the total structure, but significantly affects the stereoelectronic properties of the sugar moiety (Thibaudeau et al. 1998). Fluorine substitution has been extensively investigated in drug research and biochemistry as a means of increasing biological activity and enhancing chemical or metabolic stability. However, to date only a few fluorinated CK derivatives have been prepared and their biological activity tested (Clemenceau et al. 1996; Doležal et al. 2006, 2007). Later, Murvanidze et al. (2019) evaluated the impact of 6-(3-fluorobenzylamino) purine (*m*F-BAP) and its 9-β-D-riboside (mF-BAPR) (Fig. 6.5) on in vitro cloning of *Phalaenopsis* hybrids, which are usually characterized by slow growth and low multiplication rates. The plantlets formed significantly more but smaller new shoots when treated with *m*F-BAPR (25.3) compared to *m*F-BAP (14.6) and BAP (7.0). The results suggest the following strategy: massive micropropagation of small shoots for a number of cycles on *m*F-BAPR in closed containers, followed by a

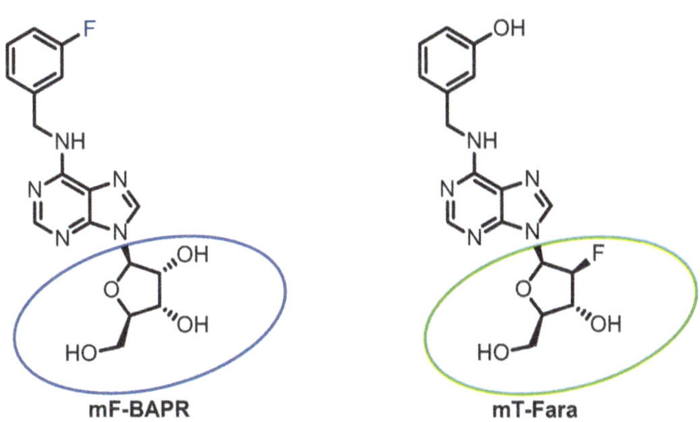

Fig. 6.5 Chemical structure of 6-(3-fluorobenzylamino)purine-9-β-D-ribofuranoside (*m*F-BAPR) and 6-(3-hydroxybenzylamino)purine-9-(2′-fluoro-2′-deoxy-β-D-arabinofuranoside) (*m*T-Fara)

final step with BAP in filter vessels to produce large shoots with roots. For this reason, the use of fluorinated topolins might present a breakthrough in the in vitro micropropagation of *Phalaenopsis* (Murvanidze et al. 2019). Another class of N^6-substituted-2′-deoxy-2′-fluoro-9-β-D-arabinofuranosylpurine derivatives (Fig. 6.5) was very recently prepared, and their biological activity is currently being evaluated (Bryksová et al. 2020).

Acknowledgements We would like to thank Alex Oulton, PhD, for the careful and critical reading of the chapter. This work was funded by the project "Plants as a tool for sustainable global development" (No. CZ.02.1.01/0.0/0.0/16_019/0000827) within the program Research, Development and Education (OP RDE) and the Internal Grant Agency of Palacký University (IGA_PrF_2021_011).

References

Amoo SO, Aremu AO, Moyo M, Szüčová L, Doležal K, Van Staden J (2014) Physiological effects of a novel aromatic cytokinin analogue in micropropagated *Aloe arborescens* and *Harpagophytum procumbens*. Plant Cell Tiss Org Cult 116:17–26

Amoo SO, Aremu AO, Moyo M, Sunmonu TO, Plíhalová L, Doležal K, Van Staden J (2015) Physiological and biochemical effects of a tetrahydropyranyl-substituted meta-topolin in micropropagated *Merwilla plumbea*. Plant Cell Tiss Org Cult 121:579–590

Aremu AO, Bairu MW, Doležal K, Finnie JF, Van Staden J (2012a) Topolins: a panacea to plant tissue culture challenges? Plant Cell Tiss Org Cult 108:1–16

Aremu AO, Bairu MW, Novák O, Plačková L, Zatloukal M, Doležal K, Finnie JF, Strnad M, Van Staden J (2012b) Physiological responses and endogenous cytokinin profiles of tissue-cultured 'Williams' bananas in relation to roscovitine and an inhibitor of cytokinin oxidase/dehydrogenase (INCYDE) treatments. Planta 236:1775–1790

Aremu AO, Plačková L, Bairu MW, Novák O, Plíhalová L, Doležal K, Finnie JF, Van Staden J (2014) How does exogenously applied cytokinin type affect growth and endogenous cytokinins in micropropagated *Merwilla plumbea*? Plant Cell Tiss Org Cult 118:245–256

Aremu AO, Doležal K, Van Staden J (2017a) New cytokinin-like compounds as a tool to improve rooting and establishment of micropropagated plantlets. Acta Hortic 1155:497–503

Aremu AO, Plačková L, Masondo NA, Amoo SO, Moyo M, Novák O, Doležal K, Van Staden J (2017b) Regulating the regulators: responses of four plant growth regulators during clonal propagation of *Lachenalia montana*. Plant Growth Regul 82:305–315

Bairu MW, Stirk WA, Doležal K, Van Staden J (2007) Optimizing the micropropagation protocol for the endangered *Aloe polyphylla*: can *meta*-topolin and its derivatives serve as replacement for benzyladenine and zeatin? Plant Cell Tiss Org Cult 90(1):15–23

Bairu MW, Novák O, Doležal K, Van Staden J (2011) Changes in endogenous cytokinin profiles in micropropagated *Harpagophytum procumbens* in relation to shoot-tip necrosis and cytokinin treatments. Plant Growth Regul 63:105–114

Blagoeva E, Dobrev PI, Malbeck J, Motyka V, Strnad M, Hanuš J, Vaňková R (2004) Cytokinin N-glucosylation inhibitors suppress deactivation of exogenous cytokinins in radish, but their effect on active endogenous cytokinins is counteracted by other regulatory mechanisms. Physiol Plant 121:215–222

Bryksová M, Hybenová A, Hernándiz AE, Novák O, Pěnčík A, Spíchal L, De Diego N, Doležal K (2020) Hormopriming for mitigating abiotic stress effects: a case study of N9-substituted cytokinin derivatives with fluorinated carbohydrate moiety. Frontiers in Plant Sciences, submitted.

Clemenceau D, Cousseau J, Martin V, Molines H, Wakselman C, Mornet R, Nogue F, Laloue M (1996) Synthesis and cytokinin activity of two fluoro derivatives of N^6-isopentenyladenine. J Agric Food Chem 44:320–323

Doležal K, Popa I, Kryštof V et al (2006) Preparation and biological activity of 6 benzylaminopurine derivatives in plants and human cancer cells. Bioorg Med Chem 14:875–884

Doležal K, Popa I, Hauserová E et al (2007) Preparation, biological activity and endogenous occurrence of N^6-benzyladenosines. Bioorg Med Chem 15:3737–3747

Doležal K, Plíhalová L, Vylíčilová H, Zatloukal M, Plíhal O, Voller J, Strnad M, Van Staden J, Baskaran P, Kumari A, Aremu AO (2017) 6-aryl-9-glycosylpurines and use thereof. US 0326168

Dwivedi S, Vanková R, Motyka V, Herrera C, Zizkova E, Auer C (2010) Characterization of *Arabidopsis thaliana* mutant ror-1 (roscovitine-resistant) and its utilization in understanding of the role of cytokinin N-glucosylation pathway in plants. Plant Growth Regul 61:231–242

Holub J, Hanuš J, Hanke DE, Strnad M (1998) Biological activity of cytokinins derived from *ortho*- and *meta*-hydroxybenzyladenine. Plant Growth Regul 26(2):109–115

Kaminek M, Vaněk T, Motyka V (1987) Cytokinin activities of N6-Benzyladenosine derivatives hydroxylated on the side-chain phenyl ring. J Plant Growth Regul 6:113–120

Letham DS, Parker CW, Duke CC, Summons RE, MaCleod JK (1977) O-glucosylzeatin and related compounds - a new group of cytokinin metabolites. Ann Bot 41:261–263

Matušková V, Zatloukal M, Voller J, Grúz J, Pěkná Z, Briestenská K, Mistríková J, Spíchal L, Doležal K, Strnad M (2020) New aromatic 6-substituted 2′-deoxy-9-(β)-D-ribofuranosylpurine derivatives as potential plant growth regulators. Bioorg Med Chem 28(2):115230

Motte H, Galuszka P, Spíchal L, Tarkowski P, Plíhal O, Šmehilová M, Jaworek P, Vereecke D, Werbrouck S, Geelen D (2013) Phenyl-adenine, identified in a LIGHT-DEPENDENT SHORT HYPOCOTYLS4-assisted chemical screen, is a potent compound for shoot regeneration through the inhibition of CYTOKININ OXIDASE/DEHYDROGENASE activity. Plant Physiol 161:1229–1241

Murvanidze N, Dolezal K, Werbrouck SPO (2019) Fluorine containing topolin cytokinins for *Phalaenopsis amabilis* (L.) Blume micropropagation. Propag Ornam Plants 19(2):48–51

Podlešáková K, Zalabák D, Čudejková M, Plíhal O, Szűčová L, Doležal K, Spíchal L, Strnad M, Galuszka P (2012) Novel cytokinin derivatives do not show negative effects on root growth and proliferation in submicromolar range. PLOS One 7, e39293

Spichal L, Gemrotova M, Zatloukal M, Frebortova J, Galuszka P, Werner T, Schmulling T, Dolezal K, Strnad M (2012) Substituted 6-anilinopurine derivatives as inhibitors of cytokinin oxidase/dehydrogenase and preparations containing these derivatives. US 08222260

Strnad M (1996) Enzyme immunoassays of N^6-benzyladenine and N^6-(meta-hydroxybenzyl)adenine cytokinins. J Plant Growth Regul 15:179–188

Strnad M (1997) The aromatic cytokinins. Physiol Plant 101:674–688

Strnad M, Hanuš J, Vaněk T et al (1997) Meta-Topolin, a highly active aromatic cytokinin from poplar leaves (*Populus x canadensis* Moencho cv. Robusta). Phytochemistry 42:213–218

Szüčová L, Spíchal L, Doležal K, Zatloukal M, Greplová J, Galuszka P, Kryštof V, Voller J, Popa I, Massino FJ, Jørgensen JE, Strnad M (2009) Synthesis, characterization and biological activity of ring-substituted 6-benzylamino-9-tetrahydropyran-2-yl and 9-tetrahydrofuran-2-ylpurine derivatives. Bioorg Med Chem 17:1938–1947

Thibaudeau C, Plavec J, Chattopadhyaya J (1998) A new generalized karplus-type equation relating vicinal proton-fluorine coupling constants to H-C-C-F torsion angles. J Org Chem 63 (15):4967–4984

Valero-Aracama C, Kane M, Wilson S, Philman N (2010) Substitution of benzyladenine with *meta*-topolin during shoot multiplication increases acclimatization of difficult- and easy-to acclimatize sea oats (*Uniola paniculata* L.) genotypes. Plant Growth Regul 60:43–49

Werbrouck SPO (2010) Merits and drawbacks of new aromatic cytokinins in plant tissue culture. Acta Hortic 865:103–108

Werbrouck SP, van der Jeugt B, Dewitte W, Prinsen E, Van Onckelen HA, Debergh PC (1995) The metabolism of benzyladenine in *Spathiphyllum floribundum* 'Schott Petite' in relation to acclimatisation problems. Plant Cell Rep 14:662–665
Werbrouck SP, Strnad M, Van Onckelen HA, Debergh PC (1996) Meta-topolin, an alternative to benzyladenine in tissue culture? Physiol Plant 98(2):291–297
Zatloukal M, Gemrotová M, Doležal K, Havlíček L, Spíchal L, Strnad M (2008) Novel potent inhibitors of *A. thaliana* cytokinin oxidase/dehydrogenase. Bioorg Med Chem 16:9268–9275

Effect of *Meta*-Topolins on Senescence 7

Cemil Işlek

Contents

Abstract

The phenomenon by which the death of cells, tissues, and organs occurs in plants is called "senescence." It is characterized by a decrease in photosynthesis, gradual decomposition of chloroplast pigment, and changes in protein and nucleic acid levels. It has long been recognized that plant hormones affect leaf senescence. Some plant hormones like ethylene, jasmonic acid, abscisic acid, and salicylic acid promote senescence. Others such as cytokinin and auxin delay leaf senescence. *Meta*-topolin (*m*T) is a more recently discovered aromatic cytokinin isolated from poplar leaves. *m*T prevents the rapid decomposition of chlorophyll and protein molecules during senescence in various plants. Delay in senescence and decrease in loss of total chlorophyll amount in plants that were treated with *m*T have been reported by different researchers. It is thought that *m*T can be a potential alternative to other cytokinins controlling the senescence process.

Keywords

Senescence · Cytokinins · Plant growth regulators · Topolins · Biotic and abiotic stress

C. Işlek (✉)
Biotechnology Department, Art and Sciences Faculty, Nigde Omer Halisdemir University, Nigde, Turkey
e-mail: cislek@ohu.edu.tr

N. Ahmad, M. Strnad (eds.), *Meta-topolin: A Growth Regulator for Plant Biotechnology and Agriculture*, https://doi.org/10.1007/978-981-15-9046-7_7

Senescence is a natural developmental process in plants and concludes with the death of plant organ, tissue, and cell (Borrás et al. 2003; Chandlee 2001; Sağlam 2015; Thomas et al. 2003; Thompson et al. 1998). Concluding with death at the level of cells, tissues, organs, and the organism or ending of the life cycle, senescence is the last stage in the development of a plant (Gregersen et al. 2013; Gully et al. 2015). The observations about senescence were made by Hildebrand (1882) and Molisch (1929). The studies in this matter have shown a great increase since the 1960s.

Senescence is one of the most important causes of agricultural product loss. Many important features from the agricultural aspect like seed number and quality, fruit maturation, seed germination time, etc. are affected by senescence (Iqbal et al. 2017). The development of fungus and other microorganisms on senescent tissues can affect the nutrition quality at an important level, and delaying senescence can allow expanding the length of life. In terms of carrying some important nutrients like nitrogen, sulfur, phosphor, and potassium and their recycling, senescence plays a vital importance in the development of a plant (Thomas 2013). These nutrients are transported from the leaves with senescence to actively growing tissues, so the growth and proliferation of the plant are supported (Gregersen et al. 2013; Gully et al. 2015).

Despite the importance of the senescence process, our knowledge of the mechanisms that regulate the senescence process is still insufficient. In fact, senescence is the loss of normal cellular functions in an orderly manner under the control of nucleus genes (Khan et al. 2014).

In senescence leaves, the degradation products of cell components are transported as nutrients to younger organs or reproductive structures. This shows that senescence is an active process necessary for plant life (Himelblau and Amasino 2001). Senescence is an event which is controlled by internal and external signals, and it can be delayed or accelerated by changing these signals (Kaplan 2005; Majid and Abbas 2019; Nooden et al. 1997).

Genetic and epigenetic mechanisms regulating phase change from juvenility to maturity influence directly the capacity for responding to senescence signals and factors determining reproduction-related patterns of disruptive aging and death. Senescence is responsive to communication between sources and sinks in which sugar signalling and hormonal regulation play central roles (Thomas 2013). Srivastava (2002) defines senescence as a normal stage of plant development like growth and reproduction and states that it is determined by a genetic program and seen in all parts of the plant. Plants respond to disadvantageous environments that include temperature, drought, nitrogen deficiency, insufficient light, disease, and pathogens with such changes as senescence and abscission, early seed development, and reduced life span (He et al. 2001; Srivastava 2002). The difference between these changes and naturally programmed senescence is that they can be reversed when the growth conditions are suitable. However, natural senescence, which is part of the developmental program, occurs even under the most favorable environmental conditions. Senescence process is associated with a specific genetic program with the control mechanism regulated by various internal and environmental factors (Smart 1994; Gan and Amasino 1997).

Since the genetically regulated senescence includes decreasing chlorophyll and proteins and cessation the photosynthetic activity in a programmed manner (Buchanan-Wollaston and Ainsworth 1997), it is considered to be the phenomenon of programmed cell death (Gan and Amasino 1997; Pennell and Lamb 1997; Quirino et al. 2000; Srivalli and Khanna-Chopra 2001; Thomas et al. 2003; Zentgraf et al. 2004). The senescence process generally occurs slower than the other types of programmed cell death which are generally limited to in particular site and showing rapid cell death. Cell death, which occurs within this process, is observed in plant organs like leaves, petals, fruit, and entire plant (Lim et al. 2003). The formation of abscission zone, xylogenesis, embryogenesis, and extreme sensitivity to pathogen infections are the other events that are controlled with programmed cell death (Greenberg 1996; Jones and Dangl 1996).

It was found that many genes related to the senescence process (Buchanan-Wollaston 1997; Nam 1997). Named as "senescence-associated genes (SAGs)," these genes are isolated from various plant species (Ansari et al. 2005; Biswal and Biswal 1999; Buchanan-Wollaston 1997).

A group of researchers states that senescence consists of three stages as beginning, breakdown, and terminal (Yoshida 2003; Springer et al. 2016; Kalra and Bhatla 2018). The beginning mechanisms can change under different conditions. As soon as the senescence program is triggered, the breakdown of the cellular components begins. It is known that the proteolytic process is activated during leaf senescence (Belknap and Garbarino 1996; Nooden et al. 1997). Rapid catabolic reactions occur during the breakdown stage. In the terminal stage, it is defined to be cell death. Senescence involves the process of cell death similar to apoptosis seen in animals, taking into account the changes that occur at this stage.

Another group of researchers reached the conclusion that senescence consists of two stages (Buchanan-Wollaston et al. 2003). According to these researchers, all cells can sustain their existence in this first stage that is recoverable. However, the second stage is concluded with cell death. While *Arabidopsis* and many other monocarpic plants are in the second stage, their leaves are brown and dried in appearance. And other species (e.g., deciduous trees) can shed their leaves before reaching this stage. Senescence process continues with sudden metabolic and physiologic changes that cause the expression of many genes (Gan and Amasino 1997; Yoshida 2003). As the mechanism that regulates the inducement and advancement of leaf senescence is very complex, it could not be fully clarified (Yoshida 2003).

The initiation of senescence has been understood as a result of biochemical examinations caused by various internal and external factors. The most important ones among the internal factors that control senescence are genetic factors. Thimann (1978) found two genes that delay senescence in *Podospora*. Ambler et al. (1987) reported in *Sorghum* that some genes prevented senescence from occurring by changing the cytokine content of some genes. In addition to genetic factors, the role of plant growth regulators is very important in the process of senescence.

Plant hormones affect the formation process of leaf senescence (Jibran et al. 2013). Senescence process can be reversibly controlled with hormone application

(Bleecker and Patterson 1997). While some plant hormones like ethylene, jasmonic acid, abscisic acid, and salicylic acid stimulate leaf senescence, some hormones like cytokinins and auxin delay it (Karataş et al. 2016; Li et al. 2012).

Such growth hormones like ethylene and cytokinins are the most significant ones of the senescence regulating factors (Dangl 2000; Gan and Amasino 1997; Nam 1997). In contrast to the delaying effects of cytokinins, ethylene and abscisic acid generally stimulate the early beginning of senescence (Grbić and Bleecker 1995; Kaplan 2005; Nooden and Leopold 1988).

It was seen that senescence started early in ethylene applied plants, and it was found that senescence accelerated particularly in older leaves (Buchanan-Wollaston et al. 2003). Similarly, it was observed that leaf senescence delayed in the sensitive tomatoes plant that synthesizes ethylene at very low levels (Picton et al. 1993). It is known that abscisic acid (ABA) induces senescence by initiating chlorophyll breakdown and inhibiting chlorophyll biosynthesis (Nooden and Leopold 1988). Although ABA's role in senescence is not understood very well, it was observed that it induced senescence in the leaves of *Arabidopsis* (He et al. 2001; Park et al. 1998).

Even though it was observed that salicylic acid accelerated senescence (Morris et al. 2000), the observation of the occurrence of senescence normally in the plants deficient of salicylic acid indicates that the genes, controlled by salicylic acid, are not necessary for senescence (Buchanan-Wollaston et al. 2003). The chlorophyll loss and decreasing RUBISCO level that occur with JA (jasmonic acid) or MeJA's exogenous application to barley leaves revealed that these molecules induced senescence (Parthier 1990). JA's *Arabidopsis* application led to early senescence, and this situation did not occur in a mutant insensitive to jasmonic acid (Xie et al. 1998). It is thought that brassinosteroids also play an important role in the occurrence of senescence (Hayat et al. 2010; Kaplan 2005).

The gibberellin levels decrease during or before senescence. Gibberellins have the capacity of preventing the chlorophyll loss in leaves, fruit, and flower stalks (Arteca 1996). The gibberellin exogenously applied to *Tropaeolum majus* (nasturtium) and *Alstroemeria hybrida* (Peruvian lily) delayed leaf senescence (Beevers and Guernsey 1967; Kappers et al. 1998). Auxins can delay senescence or stay without making any effect in some plant tissues (Arteca 1996). With a study, made in *Pisum*, an increase occurred in the production of auxins during senescence, and this increase induced ethylene synthesis (Belimov et al. 2009; Galston and Davies 1970).

Cytokinins are plant hormones that delay senescence. It is known that cytokinins increase such metabolic events as meiosis and the syntheses of proteins, DNA, RNA and also protects RNA from fragmentation (Paranjothy and Wareing 1971; Zwack and Rashotte 2013).

As it is known that cytokinins that delay senescence are synthesized in the root and indole-3-acetic acid (IAA) also induces cytokinin oxidation (Hare and Van Staden 1994; Lindoo and Noodén 1977; Moore 1979; Palni et al. 1988), it is thought that cytokinin distribution to xylem components at different concentrations can happen as the result of cytokinin oxidation depending on IAA, occurring in the root meristem in the occurrence of ordinal leaf senescence. It was concluded that

IAA controls the cytokinin concentration and distribution in the plant in the hormonal order of ordinal leaf senescence, and the cytokinin deficiency also causes the occurrence of senescence in organs (Saglam Cag 1997; Zwack and Rashotte 2013).

The decrease in the endogenous cytokinin level in the leaves that had senescence is seen, and senescence is generally delayed with external cytokinin application (Gan and Amasino 1997; Nooden and Leopold 1988; Woolhouse 1983). On the other hand, the application of cytokinin on leaves was not very effective in the occurrence of leaf senescence in *Arabidopsis* (Bleecker and Patterson 1997). Cytokinins that delay leaf senescence cannot prevent the occurrence of senescence (Kaplan 2005; Srivastava 2002).

However, data which explain cytokinins' delaying effect on senescence very commonly come across in general (James et al. 1965; Osborne 1967; Osborne and Hallaway 1960). For example, some researchers (Even-Chen et al. 1978) examined the concentrations of the hormones at the beginning and end of senescence in leaves and saw that the cytokinin activity and auxin amount were low in ripe leaves, plucked off at the beginning of senescence, and the cytokinin amount was low while the auxin amount was high at the moment of senescence. On the other hand, some researchers stated that the cytokinin synthesis decreased in roots in the course of the ripening of seeds (Heindl et al. 1982; Sitton et al. 1967). Some other researchers also emphasized that the decrease occurring in the cytokinin amount can cause senescence (Nooden et al. 1997; Saglam Cag 1997).

The role of cytokinins in delaying senescence was initially studied by externally applying them to plant tissues, and it was seen that it delays senescence of the cut leaves (Richmond and Lang 1957). Later on, it was observed that cytokinin internal production regulated all of the senescence programs of the entire plant (Gan and Amasino 1997). The cytokinin amount decreases in the leaves that have had senescence (Singh et al. 1992). Increasing cytokinin production prevents the occurrence of senescence in leaves (Guarente et al. 1998). Cytokinins delay leaf senescence but cannot prevent it indefinitely (Brault and Maldiney 1999; Downs et al. 1997; Srivastava 2002). However, after senescence starts, the normal process continues until the cytokinin application (Goren 2004).

Along with cytokinins' senescence delaying effect, the conducted studies showed that N^6-benzyladenine (BA), applied at high concentrations, induced senescence (Carimi et al. 2003, 2004; Iqbal et al. 2017). There are studies about cytokinins and their effects on senescence (Gan and Amasino 1997; Goren 2004; Smart et al. 1991; Srivastava 2002; Wingler et al. 1998). The external cytokinin application decreases the lipoxygenase activity on *Pisum sativum* L. leaves and delays leaf senescence (Goren 2004; Liu and Huang 2002).

Strnad (1997) discovered a new endogenous aromatic cytokinin family, named *meta*-topolin (*m*T, 6-(3-hydroxybenzylamine)purine). Firstly, it was isolated from poplar tree leaves. The name of *meta*-topolin was derived from "topol" which means poplar tree in the Czech language. *Meta*-topolin's metabolism is similar to other cytokinins. *m*T is generally an adenine-based cytokinin that causes less root inhibition in comparison to BA (Holub et al. 1998; Werbrouck et al. 1996).

*m*T prevented the rapid breakdown of chlorophyll and protein in the course of senescence in the leaf segments of *Triticum vulgare*. It was reported that in the *m*T applied leaf segments, the peroxidase level increased, senescence was delayed, the total chlorophyll loss was decreased, and the total nitrogen content also showed a similar tendency, compared to the control leaves (Palavan-Ünsal et al. 2004). Nevertheless, a group of the researchers proved that auxins delayed senescence (James et al. 1965; Osborne 1967); another group of them proved the same for gibberellins.

It was stated that the use of *m*T increased shoot regeneration, protected tissue stability, increased rooting efficiency, and then decreased production costs of in vitro systems (Bogaert et al. 2004; Werbrouck et al. 1996). *m*T was advocated to be a convenient alternative for BAP in microshoot induction (Bairu et al. 2007; Bogaert et al. 2004; Werbrouck et al. 1996). The reason for this is that *m*T has the best anti-senescence effect for in vitro propagated roses when they were compared with other used cytokinins (Mutui et al. 2012).

Thidiazuron has shown both auxin- and cytokinin-like effects, although, chemically, it is totally different from commonly used auxins and cytokinins (Guo et al. 2011; Turkyilmaz Unal 2018). Since *m*T is very active in delaying senescence and the ability to induce rooting of *m*T-treated portions, it is a suitable alternative for thidiazuron in delaying the initial leaf chlorosis (Mutui et al. 2012).

Palavan-Ünsal et al. (2002) reported that the protease activity decreased in the aged wheat leaf segments and the protein content increased on the tenth day of the *m*T application. Exogenous application of *m*T was effective in increasing chlorophyll content. Fresh weight and soluble protein content exhibited linear stimulation by the application of increasing *m*T concentration. The same researchers stated that *m*T inhibited DNA degradation during senescence but did not depend on the change in concentration (Palavan-Ünsal et al. 2002).

Palavan-Ünsal et al. (2004) found that there is an increase in the total nitrogen content with the application of *m*T. In the aged wheat leaf segments, the total nitrogen content was found to be lower than those applied *m*T. Storey and Beevers (1977) displayed the decrease in the total nitrogen content during senescence in leaves or cotyledons (Palavan-Ünsal et al. 2004).

Cytokinins are the strongest anti-senescence hormones (Thimann 1980) and exist as isoprenoid (the compounds related to zeatin and N^6-isopentenyladenine) and aromatic (N^6-benzyladenine and its derivatives) cytokinins (Holub et al. 1998). It was shown in various studies that methoxy and hydroxyl BA derivatives have a strong cytokinin activity (Holub et al. 1998; Mutui et al. 2014).

In delaying chlorophyll deterioration in excised wheat leaves, *m*T activity was found to be two times more than BA (Tarkowská et al. 2003). Additionally, the chlorophyll contents increased in the wheat leaf segments when *m*T was applied (Palavan-Ünsal et al. 2002, 2004). Çag et al. (2003) stated that *m*T increased the chlorophyll amount two or three times in low concentrations in contrast to another cytokinins. Furthermore, they observed a significant increase in the chlorophyll and protein levels during the greening of cotyledons. They remarked that *m*T can be a potential alternative to other well-known cytokinins controlling the chlorophyll level (Çag et al. 2003).

The conducted studies show that cytokinins displayed parallel effects in maintaining the protein levels while inhibiting senescence. For this reason, it is thought that it is not possible to generalize that cytokinins delay senescence by protecting or initiating the protein synthesis. Çag et al. (2003) found an increase in the dissoluble proteins during the greening of the cotyledon, treated with different cytokinins. When the *m*T applied samples were compared with the zeatin-, BA-, and kinetin-treated samples, the protein amount increased two times (Çag et al. 2003).

Vlčková et al. (2006) examined the effect of the external *m*T application on *Triticum aestivum* L. cv. Hereward leaves' artificial senescence and compared the leaves that are constantly kept under light to those kept in darkness. While *m*T effectively slowed down all senescence-originated changes in all measured parameters in the darkened leaf segments, the *m*T's effect was reversed a few days after separation in the light applied senescence segments.

Wheat leaf senescence assay (WLSA) is commonly used to evaluate the anti-senescence features of cytokinins. Hönig et al. (2018) compared the anti-senescence activities of different cytokinin free bases in the WLSA, made according to the protocol recommended by Holub et al. (1998). They found that *trans*-zeatin had the highest activity and the *m*T activity is very close to this cytokinin. They stated that the other examined cytokinins have much lower anti-senescence activity.

7.1 Conclusion

In this review, the effect of *m*T, which is an alternative to N^6-benzyladenine cytokinin, an important plant growth regulator, on senescence was evaluated. Senescence is the genetically programmed aging process of plants. It can be stimulated by dark, abiotic, and biotic stresses and some hormones or delayed by some hormones or minerals. Controlling senescence process is of economic importance because of its effect on parameters such as quality of plant products and their shelf life. The *m*T from natural aromatic cytokinins has been found to be very active in delaying senescence. The results of the studies reveal that *m*T is a very promising plant growth regulator for physiological research. It can be concluded that *m*T is a plant growth regulator that plays a directive role in anti-senescence activity.

References

Ambler JR, Morgan PW, Jordan WR (1987) Genetic regulation of senescence in a tropical grass. In: Thomson WW, Nothnagel EA, Huffaker RC (eds) Plant senescence: its biochemistry and physiology. American Society of Plant Physiologists, USA, Rockville, MD, pp 43–53

Ansari MI, Lee RH, Chen SCG (2005) A novel senescence-associated gene encoding γ-aminobutyric acid (GABA): pyruvate transaminase is upregulated during rice leaf senescence. Physiol Plant 123(1):1–8

Arteca RN (1996) Plant growth substances: principles and applications. Chapman and Hall, New York, NY

Bairu MW, Stirk WA, Dolezal K, Van Staden J (2007) Optimizing the micropropagation protocol for the endangered Aloe polyphylla: can meta-topolin and its derivatives serve as replacement for benzyladenine and zeatin? Plant cell. Tiss Organ Cult 90(1):15–23

Beevers L, Guernsey FS (1967) Interaction of growth regulators in the senescence of Nasturtium leaf disks. Nature 214(5091):941–942

Belimov AA, Dodd IC, Hontzeas N, Theobald JC, Safronova VI, Davies WJ (2009) Rhizosphere bacteria containing 1-aminocyclopropane-1-carboxylate deaminase increase yield of plants grown in drying soil via both local and systemic hormone signalling. New Phytol 181 (2):413–423

Belknap WR, Garbarino JE (1996) The role of ubiquitin in plant senescence and stress responses. Trends Plant Sci 1(10):331–335

Biswal B, Biswal UC (1999) Leaf senescence: physiology and molecular biology. Curr Sci 77:775–782

Bleecker AB, Patterson SE (1997) Last exit: senescence, abscission, and meristem arrest in Arabidopsis. Plant Cell 9(7):1169

Bogaert I, Van Cauter S, Werbrouck SPO, Dolezal K (2004) New aromatic cytokinins can make the difference. ISHS Acta Hortic 725:265–270

Borrás L, Maddonni GA, Otegui ME (2003) Leaf senescence in maize hybrids: plant population, row spacing and kernel set effects. Field Crop Res 82(1):13–26

Brault M, Maldiney R (1999) Mechanisms of cytokinin action. Plant Physiol Biochem 37 (5):403–412

Buchanan-Wollaston V (1997) The molecular biology of leaf senescence. J Exp Bot 48(2):181–199

Buchanan-Wollaston V, Ainsworth C (1997) Leaf senescence in Brassica napus: cloning of senescence related genes by subtractive hybridisation. Plant Mol Biol 33(5):821–834

Buchanan-Wollaston V, Earl S, Harrison E, Mathas E, Navabpour S, Page T, Pink D (2003) The molecular analysis of leaf senescence—a genomics approach. Plant Biotechnol J 1(1):3–22

Çag S, Palavan-Ünsal N, Büyüktuncer D (2003) Comparison of the effects of meta-topolin and other cytokinins on chlorophyll and protein contents and peroxidase activity in cucumber cotyledons. Israel J Plant Sci 51(4):261–265

Carimi F, Zottini M, Formentin E, Terzi M, Schiavo FL (2003) Cytokinins: new apoptotic inducers in plants. Planta 216(3):413–421

Carimi F, Terzi M, De Michele R, Zottini M, Schiavo FL (2004) High levels of the cytokinin BAP induce PCD by accelerating senescence. Plant Sci 166(4):963–969

Chandlee JM (2001) Current molecular understanding of the genetically programmed process of leaf senescence. Physiol Plant 113(1):1–8

Dangl JL (2000) Senescence and programmed cell death. In: Biochemistry and molecular biology of plants, pp 1044–1100

Del Duca S, Serafini-Fracassini D, Cai G (2014) Senescence and programmed cell death in plants: polyamine action mediated by transglutaminase. Front Plant Sci 5:1–17

Downs CG, Somerfield SD, Davey MC (1997) Cytokinin treatment delays senescence but not sucrose loss in harvested broccoli. Postharvest Biol Technol 11(2):93–100

Even-Chen Z, Atsmon D, Itai C (1978) Hormonal aspects of senescence in detached tobacco leaves. Physiol Plant 44(4):377–382

Galston AW, Davies PJ (1970) Control mechanisms in plant development. Prentice-Hall, Upper Saddle River, NJ

Gan S, Amasino RM (1997) Making sense of senescence (molecular genetic regulation and manipulation of leaf senescence). Plant Physiol 113(2):313–319

Goren N (2004) The effect of auxin and cytokinins on sequential leaf senescence on *Helianthus annuus L.* seedlings. Master's Thesis, Istanbul University, Istanbul, Turkey

Grbić V, Bleecker AB (1995) Ethylene regulates the timing of leaf senescence in Arabidopsis. Plant J 8(4):595–602

Greenberg JT (1996) Programmed cell death: a way of life for plants. Proc Natl Acad Sci 93 (22):12094–12097

Gregersen PL, Culetic A, Boschian L, Krupinska K (2013) Plant senescence and crop productivity. Plant Mol Biol 82(6):603–622

Guarente L, Ruvkun G, Amasino R (1998) Aging, life span, and senescence. Proc Natl Acad Sci 95 (19):11034–11036

Gully K, Hander T, Boller T, Bartels S (2015) Perception of Arabidopsis AtPep peptides, but not bacterial elicitors, accelerates starvation-induced senescence. Front Plant Sci 6:1–10

Guo B, Abbasi BH, Zeb A, Xu LL, Wei YH (2011) Thidiazuron: a multi-dimensional plant growth regulator. Afr J Biotechnol 10(45):8984–9000

Hare PD, Van Staden J (1994) Cytokinin oxidase: biochemical features and physiological significance. Physiol Plant 91(1):128–136

Hayat S, Mori M, Fariduddin Q, Bajguz A, Ahmad A (2010) Physiological role of brassinosteroids: an update. Indian J Plant Physiol 15:99–109

He Y, Tang W, Swain JD, Green AL, Jack TP, Gan S (2001) Networking senescence-regulating pathways by using Arabidopsis enhancer trap lines. Plant Physiol 126(2):707–716

Heindl JC, Carlson DR, Brun WA, Brenner ML (1982) Ontogenetic variation of four cytokinins in soybean root pressure exudate. Plant Physiol 70(6):1619–1625

Hildebrand F (1882) Die Lebensdauer und Vegetationsweise der Pflanzen, ihre Ursachen und Entwicklung. Botanisches Jahrbuch 2:51–135

Himelblau E, Amasino RM (2001) Nutrients mobilized from leaves of Arabidopsis thaliana during leaf senescence. J Plant Physiol 158(10):1317–1323

Holub J, Hanuš J, Hanke DE, Strnad M (1998) Biological activity of cytokinins derived from ortho- and meta-hydroxybenzyladenine. Plant Growth Regul 26(2):109–115

Hönig M, Plíhalová L, Husičková A, Nisler J, Doležal K (2018) Role of cytokinins in senescence, antioxidant defence and photosynthesis. Int J Mol Sci 19(12):1–23

Iqbal N, Khan NA, Ferrante A, Trivellini A, Francini A, Khan M (2017) Ethylene role in plant growth, development and senescence: interaction with other phytohormones. Front Plant Sci 8:1–19

James AL, Anderson IC, Greer HAL (1965) Effects of naphthaleneacetic acid on field-grown soybeans 1. Crop Sci 5(5):472–474

Jibran R, Hunter DA, Dijkwel PP (2013) Hormonal regulation of leaf senescence through integration of developmental and stress signals. Plant Mol Biol 82(6):547–561

Jones AM, Dangl JL (1996) Logjam at the Styx: programmed cell death in plants. Trends Plant Sci 1(4):114–119

Kalra G, Bhatla SC (2018) Senescence and programmed cell death. In: Plant physiology, development and metabolism. Springer, Singapore, pp 937–966

Kaplan E (2005) Investigation of relationship between senescence and gravitropism with Brassinosteroids in *Helianthus annuus L.* seedlings. Master's Thesis, Istanbul University, Istanbul, Turkey

Kappers IF, Jordi W, Maas FM, Stoopen GM, Van Der Plas LHW (1998) Gibberellin and phytochrome control senescence in alstroemeria leaves independently. Physiol Plant 103 (1):91–98

Karataş İ, Öztürk L, Okatan Y (2016) Effects of gibberellic acid and 6-benzylaminopurine on some biochemical parameters during dark-induced leaf senescence. Gaziosmanpaşa Üniversitesi Ziraat Fakültesi Dergisi 33(1):17–24

Khan M, Rozhon W, Poppenberger B (2014) The role of hormones in the aging of plants—a mini-review. Gerontology 60(1):49–55

Li Z, Peng J, Wen X, Guo H (2012) Gene network analysis and functional studies of senescence-associated genes reveal novel regulators of arabidopsis leaf senescence f. J Integr Plant Biol 54 (8):526–539

Lim PO, Woo HR, Nam HG (2003) Molecular genetics of leaf senescence in Arabidopsis. Trends Plant Sci 8(6):272–278

Lindoo SJ, Noodén LD (1977) Studies on the behavior of the senescence signal in Anoka soybeans. Plant Physiol 59(6):1136–1140
Liu X, Huang B (2002) Cytokinin effects on creeping bentgrass response to heat stress. Crop Sci 42 (2):466–472
Majid I, Abbas N (2019) Signal transduction in leaf senescence: an overview. In: Sarwat M, Tuteja N (eds) Senescence signalling and control in plants. Elsevier, Amsterdam, pp 41–59
Molisch H (1929) Die Lebensdauer der Pflanzen. Gustav Fischer, Jena, 168 pp
Moore TC (1979) Ethylene. In: Moore TC (ed) Biochemistry and physiology of plant hormones. Springer, New York, NY, pp 208–229
Morris K, Mackerness SAH, Page T, John CF, Murphy AM, Carr JP, Buchanan-Wollaston V (2000) Salicylic acid has a role in regulating gene expression during leaf senescence. Plant J 23 (5):677–685
Mutui TM, Mibus H, Serek M (2012) Effect of meta-topolin on leaf senescence and rooting in Pelargonium×hortorum cuttings. Postharvest Biol Technol 63(1):107–110
Mutui TM, Mibus H, Serek M (2014) Cytokinins inhibit leaf senescence in Pelargonium cuttings. ISHS Acta Hortic 1079:325–330
Nam HG (1997) The molecular genetic analysis of leaf senescence. Curr Opin Biotechnol 8 (2):200–207
Nooden LD, Leopold AC (1988) Senescence and aging in plants. Academic Press, San Diego, CA
Nooden LD, Guiamet JJ, John I (1997) Senescence mechanisms. Physiol Plant 101:746–753
Osborne DJ (1967) Hormonal regulation of leaf senescence. Symp Soc Exp Biol 21:305–321
Osborne DJ, Hallaway M (1960) Auxin control of protein-levels in detached autumn leaves. Nature 188(4746):240–241
Palavan-Ünsal N, Çağ S, Çetin E, Büyüktunçer D (2002) Retardation of senescence by meta-topolin in wheat leaves. J Cell Mol Biol 1:101–108
Palavan-Ünsal N, Cag S, Çetin E (2004) The role of meta-topolin in senescence of wheat leaf segments. J Mol Cell Biol 3:23–31
Palni LMS, Burch L, Horgan R (1988) The effect of auxin concentration on cytokinin stability and metabolism. Planta 174(2):231–234
Paranjothy K, Wareing PF (1971) The effects of abscisic acid, kinetin and 5-fluorouracil on ribonucleic acid and protein synthesis in senescing radish leaf disks. Planta 99(2):112–119
Park JH, Oh SA, Kim YH, Woo HR, Nam HG (1998) Differential expression of senescence-associated mRNAs during leaf senescence induced by different senescence-inducing factors in Arabidopsis. Plant Mol Biol 37(3):445–454
Parthier B (1990) Jasmonates: hormonal regulators or stress factors in leaf senescence? J Plant Growth Regul 9(1–4):57–63
Pennell RI, Lamb C (1997) Programmed cell death in plants. Plant Cell 9(7):1157–1168
Picton S, Barton SL, Bouzayen M, Hamilton AJ, Grierson D (1993) Altered fruit ripening and leaf senescence in tomatoes expressing an antisense ethylene-forming enzyme transgene. Plant J 3 (3):469–481
Quirino BF, Noh Y-S, Himelblau E, Amasino RM (2000) Molecular aspects of leaf senescence. Trends Plant Sci 5(7):278–282
Richmond AE, Lang A (1957) Effect of kinetin on protein content and survival of detached Xanthium leaves. Science 125(3249):650–651
Sağlam NG (2015) Leaf senescence: a view of its physiological and molecular regulation. Marmara Fen Bilimleri Dergisi 27(3):83–92
Saglam Cag S (1997) The effects of mineral nutrients and growth regulators on the mechanism of sequential leaf senescence in *Helianthus annuus L.* Doctoral Thesis, Istanbul University, Istanbul, Turkey
Singh S, Letham DS, Palni LMS (1992) Cytokinin biochemistry in relation to leaf senescence. VIII. Translocation, metabolism and biosynthesis of cytokinins in relation to sequential leaf senescence of tobacco. Physiol Plant 86(3):398–406

Sitton D, Itai C, Kende H (1967) Decreased cytokinin production in the roots as a factor in shoot senescence. Planta 73(3):296–300

Smart CM (1994) Gene expression during leaf senescence. New Phytol 126(3):419–448

Smart CM, Scofield SR, Bevan MW, Dyer TA (1991) Delayed leaf senescence in tobacco plants transformed with tmr, a gene for cytokinin production in agrobacterium. Plant Cell 3 (7):647–656

Springer A, Kang C, Rustgi S, Von Wettstein D, Reinbothe C, Pollmann S, Reinbothe S (2016) Programmed chloroplast destruction during leaf senescence involves 13-lipoxygenase (13-LOX). Proc Natl Acad Sci 113(12):3383–3388

Srivalli B, Khanna-Chopra R (2001) Induction of new isoforms of superoxide dismutase and catalase enzymes in the flag leaf of wheat during monocarpic senescence. Biochem Biophys Res Commun 288(4):1037–1042

Srivastava LM (2002) Plant growth and development hormones and environment. Academic Press, San Diego, CA

Storey R, Beevers L (1977) Proteolytic activity in relationship to senescence and cotyledonary development in Pisum sativum L. Planta 137(1):37–44

Strnad M (1997) The aromatic cytokinins. Physiol Plant 101(4):674–688

Tarkowská D, Doležal K, Tarkowski P, Åstot C, Holub J, Fuksová K, Schmülling T, Sandberg G, Strnad M (2003) Identification of new aromatic cytokinins in Arabidopsis thaliana and Populus× canadensis leaves by LC-(+) ESI-MS and capillary liquid chromatography/frit–fast atom bombardment mass spectrometry. Physiol Plant 117(4):579–590

Thimann KV (1978) Senescence. Bot Mag Tokyo 1:19–43

Thimann KV (1980) The senescence of leaves. In: Thimann KV (ed) Senescence in plants. CRC, Boca Raton, FL, pp 85–115

Thomas H (2013) Senescence, aging and death of the whole plant. New Phytol 197(3):696–711

Thomas H, Ougham HJ, Wagstaff C, Stead AD (2003) Defining senescence and death. J Exp Bot 54 (385):1127–1132

Thompson JE, Froese CD, Madey E, Smith MD, YuWen H (1998) Lipid metabolism during plant senescence. Progr Lipid Res (UK) 37(2–3):119–141

Turkyilmaz Unal B (2018) Thidiazuron as an elicitor in the production of secondary metabolite. In: Ahmad N, Faisal M (eds) Thidiazuron: from urea derivative to plant growth regulator. Springer, Singapore, pp 463–470

Vlčková A, Špundová M, Kotabová E, Novotný R, Doležal K, Nauš J (2006) Protective cytokinin action switches to damaging during senescence of detached wheat leaves in continuous light. Physiol Plant 126(2):257–267

Werbrouck SP, Strnad M, Van Onckelen HA, Debergh PC (1996) Meta-topolin, an alternative to benzyladenine in tissue culture? Physiol Plant 98(2):291–297

Wingler A, von Schaewen A, Leegood RC, Lea PJ, Quick WP (1998) Regulation of leaf senescence by cytokinin, sugars, and light: effects on NADH-dependent hydroxypyruvate reductase. Plant Physiol 116(1):329–335

Woolhouse HW (1983) Hormonal control of senescence allied toreproduction in plants. In: Meudt WJ (ed) Strategies of plant reproduction. Allanheld, Osmun and Co, Totowa, NJ, pp 201–236

Xie DX, Feys BF, James S, Nieto-Rostro M, Turner JG (1998) COI1: an Arabidopsis gene required for jasmonate-regulated defense and fertility. Science 280(5366):1091–1094

Yoshida S (2003) Molecular regulation of leaf senescence. Curr Opin Plant Biol 6(1):79–84

Zentgraf U, Jobst J, Kolb D, Rentsch D (2004) Senescence-related gene expression profiles of rosette leaves of Arabidopsis thaliana: leaf age versus plant age. Plant Biol 6(2):178–183

Zwack PJ, Rashotte AM (2013) Cytokinin inhibition of leaf senescence. Plant Signal Behav 8(7): e24737

Effects of Aromatic Cytokinins on Senescence-Induced Alterations in Photosynthesis

8

Helena Janečková

Contents

Abstract

Leaf senescence constitutes the final stage in leaf development involving multiple alterations in cell physiology and biochemistry. There are pronounced changes in chloroplasts, the structures of the photosynthetic apparatus are degraded, and photosynthetic activity declines. The progress of leaf senescence as well as the senescence-induced alterations in photosynthesis can be retarded or mitigated by plant hormone cytokinin. Today, cytokinins are well-known decelerators of leaf senescence. To study the role of cytokinins in this process, exogenous application on senescing leaves is often used. For this purpose most researchers use aromatic cytokinins, such as 6-benzylaminopurine, kinetin and *meta*-topolin. This chapter provides an overview of the alterations in photosynthesis during leaf senescence and how they can be affected by exogenous application of aromatic cytokinins.

Keywords

Aromatic cytokinin · Senescence · Photosynthesis · *meta*-Topolin

H. Janečková (✉)
Department of Biophysics, Centre of the Region Haná for Biotechnological and Agricultural Research, Faculty of Science, Palacký University, Olomouc, Czech Republic
e-mail: helena.janeckova@upol.cz

N. Ahmad, M. Strnad (eds.), *Meta-topolin: A Growth Regulator for Plant Biotechnology and Agriculture*, https://doi.org/10.1007/978-981-15-9046-7_8

8.1 Changes in Chloroplast During Leaf Senescence

Leaf senescence is a very complex developmental process characterized by numerous alterations of cell structure, metabolism and gene expression. During senescence, cellular components are gradually dismantled, and valuable compounds are remobilized.

Chloroplasts are the first organelles to manifest the symptoms of senescence. The number and size are reduced, and the shape changes from nearly oval to more spherical. The ultrastructural modification of membrane system in chloroplasts is sequential and includes the unstacking of granal thylakoids and consequently the formation of loose and elongated lamellae with concomitant formation of plastoglobules (for a review, see, e.g. Biswal 1997; Khanna-Chopra 2012). The inner components are degraded, thylakoid membranes are disintegrated, and chloroplasts gradually transform into gerontoplasts (for a review, see, e.g. Biswal et al. 2012). Senescent chloroplast is characterized by pronounced loss of chlorophyll (Chl) molecules and degradation of pigment-protein complexes, and essential macromolecules including proteins, nucleic acids and lipids are decomposed (Mae et al. 1993; Humbeck et al. 1996; Miersch et al. 2000; Tang et al. 2005; Żelisko and Jackowski 2004; Nath et al. 2013).

The loss of thylakoid membrane proteins has been reported in various plant species during both developmental (Humbeck et al. 1996; Miersch et al. 2000; Tang et al. 2005; Nath et al. 2013) and induced senescence (Oh et al. 2003; Żelisko and Jackowski 2004; Oh et al. 2005; Kusaba et al. 2007; Park et al. 2007; Zavaleta-Mancera et al. 2007; Sato et al. 2009; Sakuraba et al. 2013). For example, Kusaba et al. (2007) have observed that senescence of detached rice (*Oryza sativa* L. cv Koshihikari) leaves kept in the dark is accompanied by a loss of proteins of both light-harvesting complexes (LHCI and LHCII) and components of core complexes of both photosystems. A pronounced decrease in the amount of CP47, a photosystem II (PSII) core antenna, and PsaF, a component of reaction centre I (RCI), has been found. Moreover, the results also showed that the major trimeric LHCII proteins, Lhcb1 and Lhcb2, are more stable during induced senescence than the monomeric proteins (Lhcb4 and Lhcb6) (Kusaba et al. 2007).

Degradation of pigment-protein complexes during senescence includes the action of chloroplastic proteases. For example, chloroplast-localized DegP and FtsH proteases are known to be involved in the degradation of the D1 protein of PSII (Adam and Clarke 2002; Khanna-Chopra 2012). Despite some known proteases, there is still a largely unknown field of proteases participating in this process.

The most obvious symptom of leaf senescence is the loss of green colour of leaves caused by degradation of Chl molecules. Chl degradation is achieved by the action of specific enzymes and occurs in a multistep pathway leading to colourless non-fluorescent catabolite products that are stored in vacuoles (for reviews, see, e.g. Hörtensteiner 2013; Kuai et al. 2018). Chl degradation is tightly linked to the degradation of pigment-binding proteins. The removal of Chl *b* is required for LHCII degradation, while the degradation of LHCII appears to be necessary for the breakdown of Chl *a* and some carotenoids contained within LHCII (Kusaba et al. 2007; Sato et al. 2009). Moreover, it seems that the retention of Chl *b* together with stabilization of LHCII complexes has substantial impact on the progress of leaf senescence (Kusaba et al. 2007; Sakuraba et al. 2012; Voitsekhovskaja and Tyutereva 2015; Yang et al. 2016). While retained Chl *b* content associated with higher stability of LHCII was found to be accompanied by slower progress of senescence-associated changes (Kusaba et al. 2007; Sakuraba et al. 2012; Voitsekhovskaja and Tyutereva 2015), lack of Chl *b* appears to accelerate leaf senescence (Kusaba et al. 2007; Yang et al. 2016; Janečková et al. 2019). In our study (Janečková et al. 2019), we observed that the leaf senescence of a Chl *b*-deficient barley mutant of chlorina *f2*^{f2}(*chlorina*) was accompanied by a substantial acceleration of the inhibition of PSII photochemistry which appeared to be related to more pronounced deterioration of reaction centres of PSII (RCII) (Janečková et al. 2019).

The gradual loss of Chl, as well as the structural disorganization of photosynthetic complexes, is accompanied by deterioration of the electron transport and a drop in photosynthetic activity. Alternative electron transport pathways, such as cyclic electron flow or the chlororespiratory pathway, may be employed (for a review, see, e.g. Krieger-Liszkay et al. 2019). The photochemical efficiency of PSII (Oh et al. 1996; Špundová et al. 2003, 2005b; Vlčková et al. 2006; Nath et al. 2013; Krieger-Liszkay et al. 2015; Janečková et al. 2018, 2019) as well as PSI (Nath et al. 2013; Krieger-Liszkay et al. 2015; Janečková et al. 2019) decreases during leaf senescence. However, which of the two photosystems is first affected remains unclear as the data are inconsistent. In our study, the efficiency of PSII photochemistry in the dark-senescing leaves of barley was more impaired than that of PSI (Janečková et al. 2019). Krieger-Liszkay et al. (2015) however found that PSI activity declined earlier than PSII activity during the senescence of flag barley leaves of cv. Carina. Similar trends were observed by Tang et al. (2005), who showed that PSII appears to remain more functional than PSI during flag leaf senescence in rice (Tang et al. 2005).

In contrast to the functionality of PSII versus PSI, the data for the primary versus secondary phase of photosynthesis are consistent. The electron transport via the PSII is usually inhibited later than the activity of the Calvin-Benson cycle (Camp et al. 1982; Grover 1993; Špundová et al. 2005a; Vlčková et al. 2006).

As a consequence of the impaired photosynthetic apparatus, the light energy cannot be properly utilized by photochemistry, and the efficiency of non-photochemical processes increases (e.g. Lu and Zhang 1998; Lu et al. 2001; Wingler et al. 2004; Janečková et al. 2019). The excess light energy is dissipated via regulatory non-photochemical quenching processes which protect the photosynthetic apparatus against photooxidative damage and/or non-regulatory quenching processes (for a review, see, e.g. Lazár 2015). The regulatory non-photochemical quenching processes are related to the activation of the xanthophyll cycle which dissipates the excessive excitation energy as heat in the antenna complexes before it reaches the reaction centres. Violaxanthin is converted into zeaxanthin by de-epoxidation catalysed by violaxanthin de-epoxidase, through the intermediate antheraxanthin, and zeaxanthin then stimulates non-photochemical quenching within LHC (e.g. Lu et al. 2001; Demmig-Adams et al. 2014).

If the generation of electrons in the electron transport chain exceeds demand, there may be undesirable formation of reactive oxygen species (ROS). ROS may then oxidize PSII proteins and lipids and increase the extent of lipid peroxidation (Dhindsa et al. 1982; Chang and Kao 1998; Pospíšil 2016). These changes then pronouncedly contribute to the damage of the photosynthetic apparatus and cell membranes during leaf senescence.

Leaf senescence is regulated at the level of the cell nucleus, and its management involves highly complex genetic programmes, many regulatory molecules as well as a series of signalling pathways (for a review, see, e.g. Kim et al. 2018). Its progress can be affected by diverse external or endogenous factors. Various abiotic (e.g. drought, extreme temperatures, nutritional deficiency or limitation, shading) and biotic (e.g. pathogen infection) stresses are external (environmental) factors (Lim et al. 2007; Khanna-Chopra 2012). The essential environmental factor which regulates leaf senescence is light. Light can slow down senescence-induced changes, while light deprivation, severe shade or darkness usually accelerates senescence (Weaver and Amasino 2001; Špundová et al. 2005a; Liebsch and Keech 2016; Janečková et al. 2018). On the other hand, high doses of light can have effects on senescence that are directly opposed to those induced by less intense light. Whereas less intense light usually retards the onset of senescence, high light intensities have been reported to accelerate decreases in Chl levels (Hidema et al. 1992; Kar et al. 1993; Wilhelmová et al. 1997). Likewise, continuous light can induce premature senescence by causing photodamage under conditions where the application of a day/night cycle would be more benign (Procházková and Wilhelmová 2004).

Endogenous factors which influence leaf senescence include phytohormones and different transcription factors (Lim et al. 2007). To date, there is evidence for numerous transcription factors (TFs) which have been described as important players in modulating senescence, for example, PIFs (phytochrome-interacting factors), ORE1 (ORESARA1) or WRKY TFs (Kim et al. 2009; Balazadeh et al. 2010; Hornitschek et al. 2012; Sakuraba et al. 2014; for a review, see, e.g. Liebsch and Keech 2016). One well-known group of phytohormones linked to leaf senescence is the cytokinins. Cytokinins play key roles in numerous processes of plant growth and development (for a review, see, e.g. Kieber and Schaller 2014) and are known for their ability to decelerate or postpone the progress of leaf senescence.

8.2 Effects of Aromatic Cytokinins on Photosynthesis During Induced Senescence

Cytokinins (CKs) are today well-known as negative regulators of leaf senescence. They are able to mitigate, slow down or even reverse some senescence-induced processes (for a review, see, e.g. Zwack and Rashotte 2013; Hönig et al. 2018). During leaf senescence, the content of active CK forms decreases (Tao et al. 1983; Singh et al. 1992; Gan and Amasino 1995, 1996, 1997; Ananieva et al. 2008; Hönig et al. 2018). The decrease may be a result of inhibited CK biosynthesis as expression of CK biosynthetic genes has been shown to decline during senescence (Buchanan-Wollaston et al. 2005; Breeze et al. 2011) or it may be a result of faster CK degradation by cytokinin oxidase/dehydrogenase (Werner et al. 2006). Irreversible inactivation of CKs by uridine diphosphate glycosyltransferases may also play a role (Šmehilová et al. 2016). It should be pointed out that a low CK content is not the trigger of senescence but one of the factors that allow senescence to proceed (Werner et al. 2003). Leaf senescence delay can be achieved by exogenous application of active CKs as well as by increase in their endogenous content (Smart et al. 1991; Gan and Amasino 1995; Cortleven and Valcke 2012; for a review, see, e.g. Zwack and Rashotte 2013; Hönig et al. 2018).

Aromatic CKs (ARCKs) are largely used in senescence assays. The first record of the antisenescent effect of CKs was reported by Richmond and Lang (1957), who used kinetin (6-furfurylaminopurine; K) in the exogenous treatment of detached leaves of cocklebur plants (*Xanthium pennsylvanicum* L.). They observed that leaves treated with K retained higher Chl and protein content. Since then, other CKs with an aromatic side chain were found and their antisenescence activity repeatedly shown.

Treatment with 6-benzylaminopurine (BAP) or *meta*-topolin (*m*T) leads to retention of Chl content in detached senescing leaves of various plant species such as wheat (Holub et al. 1998; Pavalan-Ünsal et al. 2002; Vlčková et al. 2006; Zavaleta-Mancera et al. 2007; Huang et al. 2011), barley (Selivankina et al. 2001; Janečková et al. 2019), *Arabidopsis* (Zacarias and Reid 1990; Kudryakova et al. 2001; Oh et al. 2005; Vylíčilová et al. 2016) and rice (Kao 1980; Talla et al. 2016). Re-greening of yellowing leaves has even been reported in the presence of BAP (Zavaleta-Mancera et al. 1999). Talla et al. (2016) showed that the retained Chl content in BAP-treated rice leaves senescing in dark was associated with the accumulation of 7-hydroxymethyl Chl *a*, an intermediate of the Chl cycle. This finding suggests that CKs could delay senescence-induced Chl loss by regulating the conversion of Chl *b* to Chl *a* (Talla et al. 2016).

The CK-mediated retention of Chl during leaf senescence could also be related to the CK-induced changes in the expression of genes involved in Chl synthesis and degradation. In senescing rice leaves treated with BAP, Talla et al. (2016) reported an increase in the gene expression of chlorophyllide *a* oxygenase (*CAO*) which is involved in the biosynthesis of Chl *b* and of non-yellow colouring1-like (*NOL*) and 7-hydroxymethyl Chl *a* reductase (*HCAR*), involved in the conversion of Chl *b* to Chl *a* via 7-hydroxymethyl Chl *a*. The CK-mediated retention of Chl in senescent leaves may also be associated with the reduction in activity of enzymes that participate in Chl catabolism, as BAP was shown to reduce the activity of Mg-dechelatase and Chl degrading peroxidase during post-harvest senescence of broccoli florets (Costa et al. 2005). Furthermore, Vylíčilová et al. (2016) reported the downregulation of genes involved in Chl degradation including genes for a non-yellow colouring protein 1 (NYC1), non-yellowing proteins 1 and 2 (NYE1 and NYE2), pheophorbide *a* oxygenase/accelerated cell death 1 (PAO/ACD1) or NOL.

During leaf senescence, treatment with ARCKs also led to the retention of higher content of carotenoids and xanthophylls (Vlčková et al. 2006; Zavaleta-Mancera et al. 2007; Vylíčilová et al. 2016; Janečková et al. 2019), inhibition of protease activity (Pavalan-Ünsal et al. 2002), prevention of formation of plastoglobules and modification of chloroplast shape (Hudák et al. 1996; Vlčková et al. 2006; Zavaleta-Mancera et al. 2007). In the presence of ARCKs, a senescence-induced increase in lipid peroxidation is mitigated (Dhindsa et al. 1982; Huang et al. 1997; Vlčková et al. 2006; Huang et al. 2011; Mik et al. 2011; Liu et al. 2016). This CK action could be linked to their ability to activate the xanthophyll cycle (e.g. Vlčková et al. 2006) or antioxidant enzymes, such as catalase, superoxide dismutase, peroxidase and ascorbate peroxidase (Dhindsa et al. 1982; Zavaleta-Mancera et al. 2007; Wu et al. 2012; Ren et al. 2018; Yang et al. 2018), all of which contribute considerably to the protection of the photosynthetic apparatus and cell membranes against oxidative damage during senescence.

Treatment of senescing leaves with ARCKs led to preserved photosynthetic activity (Vlčková et al. 2006; Zubo et al. 2008; Talla et al. 2016; Vylíčilová et al. 2016; Janečková et al. 2019). In senescing leaves treated with ARCKs (BAP or *m*T), maintained content and activity of Rubisco (Weidhase et al. 1987; Zavaleta-Mancera et al. 2007) and CO_2 assimilation (Vlčková et al. 2006) have been observed. The CK-mediated retention of Rubisco activity and stability during leaf senescence could also be associated with maintained expression of genes related to the Calvin cycle. It has been shown that BAP-treated leaves senescing in darkness had maintained expression of genes for a small subunit of Rubisco (*RBCS*), Rubisco methyltransferase-like protein or fructose-bisphosphate aldolase (Talla et al. 2016; Vylíčilová et al. 2016). Interestingly, Zavaleta-Mancera et al. (2007) suggested that the preventive effect of BAP on Rubisco activity and stability could be related to BAP-induced accumulation of the antioxidant enzyme, ascorbate peroxidase which would protect Rubisco against its degradation (Zavaleta-Mancera et al. 2007).

With regard to the primary phase of photosynthesis, exogenous treatment by ARCKs delays the inhibition of PSII photochemistry in dark-senescing leaves (Oh et al. 2005; Vlčková et al. 2006; Vylíčilová et al. 2016; Janečková et al. 2019). The function of PSII photochemistry, estimated from Chl fluorescence parameters (F_V/F_M and Φ_P, maximal and effective quantum yield of PSII photochemistry in the dark- and in the light-adapted state, respectively), was shown to be highly maintained in dark-senescing BAP-treated leaves of *Arabidopsis* (Oh et al. 2005), rice (Talla et al. 2016), barley (Janečková et al. 2019) and *m*T-treated leaves of wheat (Vlčková et al. 2006). A well-preserved photosynthetic apparatus was also shown by Vylíčilová et al. (2016) in senescing *Arabidopsis* leaves treated with newly prepared 2-chloro-6-(halogenobenzylamino)purine-9-ribosides. These ARCK derivatives were even more effective than BAP treatment. Besides maintained PSII photochemistry, delayed loss of oxygen evolution activity has been also shown in BAP-treated rice leaves senescing in the dark (Talla et al. 2016).

In the presence of ARCKs, the senescence-induced alterations in the structure of pigment-protein complexes are suppressed. It has been shown that treatment with ARCKs (K, BAP) can lead to greater stabilization of LHCII as well as RCII in dark-senescing leaves (Jackowski 1996; Oh et al. 2005; Zavaleta-Mancera et al. 2007; Talla et al. 2016; Janečková et al. 2019). The stabilizing effect of CKs on photosynthetic complexes has also been reported under stress conditions (e.g. Monakhova and Chernyad'ev 2004).

The effect of CKs on LHCII stability could be related to the CK-dependent retention of Chl *b*, as this pigment is preferentially bound by LHCII. The preferential retention of Chl *b* accompanied by higher stability of LHCII proteins was reported, for example, by Zavaleta-Mancera et al. (2007) in dark-senescing detached leaves of wheat (*Triticum aestivum* L.) treated with BAP. Chl *b* is important for the stability of LHCII, and, as described previously, its presence has significant impact on the progress of leaf senescence (Kusaba et al. 2007; Sakuraba et al. 2012; Voitsekhovskaja and Tyutereva 2015; Yang et al. 2016; Janečková et al. 2019). Deficiency in Chl *b* has been shown to result in reduced amount of LHCs (Ghirardi et al. 1986; Bossmann et al. 1997), increased sensitivity of PSII to stress conditions due to the decreased amount of LHCs (Havaux and Tardy 1997) and faster progress of senescence (Kusaba et al. 2007; Yang et al. 2016; Janečková et al. 2019). In our study with the Chl *b*-deficient barley mutant *chlorina*, we observed that BAP treatment led to delayed degradation of photosynthetic pigments and suppressed senescence-induced impairment of PSII photochemistry in both *wild-type* and *chlorina* (Janečková et al. 2019). However the effect of BAP was more pronounced in *chlorina*. The more marked BAP effect in *chlorina* implies that exogenous application of BAP was able to compensate for the destabilizing effect of Chl *b*/LHC deficiency on PSII. Further, we suggested that the protective effect of BAP on PSII function during leaf senescence was based mainly on attenuated deterioration of RCII, as the impaired PSII function in non-CK-treated leaves appeared to be associated with more pronounced damage of RCII. In addition, our results allow us to conclude that the presence of Chl *b* is not decisive for the protective CK effect on PSII photochemistry in dark-senescing leaves (Janečková et al. 2019).

The CK-mediated protection of photosynthetic structures during senescence may be the result of maintained expression of various genes related to both PSII and PSI or cytochrome b_6/f complex (Talla et al. 2016; Vylíčilová et al. 2016). It has been found that senescing *Arabidopsis* leaves treated with halogenated ARCK ribosides have upregulated the expression of genes related to RC and LHC of both PSII and PSI, including *Lhcb2.2*, *Lhcb3*, *Lhcb5*, *Lhcb6*, *Lhca1* or *PsaK* (subunit K of PSI) (Vylíčilová et al. 2016). In dark-senescing rice leaves, CK-dependent retention of higher expression was observed for *Lhcb4* and *Lhcb6* genes, encoding Chl *a/b*-binding proteins of the LHC, and for oxygen-evolving enhancer genes *PsbO* and *PsbP* and genes coding the subunits of the cytochrome b_6/f (Talla et al. 2016).

Taken together, treatment with ARCKs can led to preserved photosynthetic machinery during leaf senescence. That aside, our understanding of the precise mechanisms by which CKs exert their protection of photosynthesis during leaf senescence is still limited. Regarding the signalling components involved in CK-regulated senescence, the cytokinin receptor *Arabidopsis* histidine kinase 3 (AHK3) is involved in mediating CK signalling which is coupled with the phosphorylation/activation of the B-type *Arabidopsis* response regulator (ARR),

ARR2 (Kim et al. 2006; Riefler et al. 2006; Hwang et al. 2012). Besides AHK3, the AHK2 receptor is considered to be redundant in senescence regulation, whereas the AHK4 receptor is thought to have only minor relevance (Riefler et al. 2006). On the other hand, AHK4 appears to have the main responsibility for CK-induced inhibition of lipid peroxidation during dark-induced senescence (Janečková et al. 2018). Further, CK response factors (CRFs), downstream components of CK receptors, mediate the CK signal for regulating senescence (Zwack and Rashotte 2013; Raines et al. 2016). Interestingly, in one study (Oh et al. 2005), the authors observed that only the degradation of Chl was delayed in the *Arabidopsis* mutant of the *ore9* gene in the presence of BAP, while PSII photochemistry was preserved. On the other hand, in *wild-type*, both processes were well preserved by BAP. This observation may implicate the involvement of ORE9 in the mediation of CK signals for maintaining PSII function during dark-induced senescence. The signalling components of the CK-mediated protection of photosynthesis still remain largely unknown.

It should be highlighted that the effect of CK on the progress of leaf senescence is dependent on their structural form. Different CK forms possess various biological activities (see, e.g. Holub et al. 1998; Mik et al. 2011; Vylíčilová et al. 2016). For example, Holub et al. (1998) compared the structure-activity relationships of various ARCK forms. In the senescence bioassay, they showed that *m*T was more active than BAP, whereas *ortho*-topolin was less active than BAP (Holub et al. 1998). The final CK effect on leaf senescence is also influenced by light conditions (see, e.g. Zacarias and Reid 1990; Vlčková et al. 2006; Prokopová et al. 2010). And finally, the CK effect depends on applied concentration (see, e.g. Zacarias and Reid 1990; Kudryakova et al. 2001). There are studies showing application of high CK concentrations leads to accelerated leaf senescence. Thus, the CKs were acting as positive regulators of senescence (see, e.g. Pospíšilová et al. 1993; Genkov et al. 1997; Carimi et al. 2004; Vescovi et al. 2012). Although CK-induced acceleration of senescence is described in the literature, the exact mechanism is unknown. One possible explanation could be the CK-dependent induction of oxidative stress observed by Mlejnek et al. (2005) in a suspension of tobacco cells.

8.3 Conclusion

ARCKs are often used in exogenous treatments in various senescence experiments to reveal new roles for CKs. During leaf senescence, the application of ARCKs can lead to maintained content of photosynthetic pigments and preserved structure and function of the photosynthetic apparatus. More detailed understanding of the underlying molecular mechanisms as well as identification of signalling molecules and pathways involved will still need to be clarified. The accelerated CK-mediated leaf senescence, which usually occurs when high concentrations are applied, also remains to be elucidated along with the exact molecular mechanisms.

References

Adam Z, Clarke AK (2002) Cutting edge of chloroplast proteolysis. Trends Plant Sci 7:451–456

Ananieva KT, Ananiev ED, Doncheva S, Georgieva KM, Tzvetkova N, Kamínek M, Motyka V, Dobrev PI, Gajdošová S, Malbeck (2008) Senescence progression in a single darkened cotyledon depends on the light status of the other cotyledon in *Cucurbita pepo* (zucchini) seedlings: potential involvement of cytokinins and cytokinin oxidase/dehydrogenase activity. Physiol Plant 134:609–623

Balazadeh S, Siddiqui H, Allu AD, Matallana-Ramirez LP, Caldana C, Mehrnia M, Zanor MI, Kohler B, Mueller-Roeber B (2010) A gene regulatory network controlled by the NAC transcription factor ANAC092/AtNAC2/ORE1 during salt-promoted senescence. Plant J 62:250–264

Biswal B (1997) Chloroplast metabolism during leaf greening and degreening. In: Pessarakali M (ed) Handbook of photosynthesis. Marcel Dekker, New York, pp 71–81

Biswal B, Mohapatra PK, Biswal UC, Raval MK (2012) Leaf senescence and transformation of chloroplasts to gerontoplasts. In: Eaton-Rye J, Tripathy B, Sharkey T (eds) Photosynthesis. Advances in photosynthesis and respiration, vol 34. Springer, Dordrecht, pp 217–230

Bossmann B, Knoetzel J, Jansson S (1997) Screening of *chlorina* mutants of barley (*Hordeum vulgare* L.) with antibodies against light-harvesting proteins of PSI and PSII: absence of specific antenna proteins. Photosynth Res 52:127–136

Breeze E, Harrison E, McHattie S, Hughes L, Hickman R, Hill C, Kiddle S, Kim YS, Penfold CA, Jenkins D, Zhang C, Morris K, Jenner C, Jackson S, Thomas B, Tabrett A, Legaie R, Moore JD, Wild DL, Ott S, Rand D, Beynon J, Denby K, Mead A, Buchanan-Wollaston V (2011) High-resolution temporal profiling of transcripts during Arabidopsis leaf senescence reveals a distinct chronology of processes and regulation. Plant Cell 23:873–894

Buchanan-Wollaston V, Page T, Harrison E, Breeze E, Lim PO, Nam HG, Lin JF, Wu SH, Swidzinski J, Ishizaki K, Leaver CJ (2005) Comparative transcriptome analysis reveals significant differences in gene expression and signalling pathways between developmental and dark/starvation-induced senescence in *Arabidopsis*. Plant J 42:567–585

Camp PJ, Huber SC, Burke JJ, Moreland DE (1982) Biochemical changes that occur during senescence of wheat leaves. Plant Physiol 70:1641–1646

Carimi F, Terzi M, De Michele R, Zottini M, Lo Schiavo F (2004) High levels of the cytokinin BAP induce PCD by accelerating senescence. Plant Sci 166:963–969

Chang CJ, Kao CH (1998) H_2O_2 metabolism during senescence of rice leaves: changes in enzyme activities in light and darkness. Plant Growth Regul 25:11–15

Cortleven A, Valcke R (2012) Evaluation of the photosynthetic activity in transgenic tobacco plants with altered endogenous cytokinin content: lessons from cytokinins. Plant Physiol 144:394–408

Costa ML, Civello PM, Chaves AR, Martínez GA (2005) Effect of ethephon and 6-benzylaminopurine on chlorophyll degrading enzyme and a peroxidase-linked chlorophyll bleaching during post-harvest senescence of broccoli (*Brassica oleracea* L.) at 20 degrees C. Postharvest Biol Technol 35:191–199

Demmig-Adams B, Stewart JJ, Adams WW III (2014) Chloroplast photoprotection and the trade-off between abiotic and biotic defense. In: Demmig-Adams B, Garab G, Adams WW III, Govindjee (eds) Non-photochemical quenching and energy dissipation in plants, algae and cyanobacteria. Springer, Dordrecht, pp 631–643

Dhindsa RS, Plumb-Dhindsa P, Reid DM (1982) Leaf senescence and lipid peroxidation: effect of some phytohormones, and scavengers of free radicals and singlet oxygen. Physiol Plant 56:453–457

Gan S, Amasino RM (1995) Inhibition of leaf senescence by autoregulated production of cytokinin. Science 270:1986–1988

Gan S, Amasino RM (1996) Cytokinins in plant senescence: from spray and pray to clone and play. BioEssays 18:557–565

Gan S, Amasino RM (1997) Making sense of senescence (molecular genetic regulation and manipulation of leaf senescence). Plant Physiol 113:313–319

Genkov T, Tsoneva P, Ivanova I (1997) Effects of cytokinins on photosynthetic pigments and chlorophyllase activity *in vitro* cultures of axillary buds of *Dianthus caryophyllus* L. J Plant Growth Regul 16:169–172

Ghirardi ML, McCauley SW, Melis A (1986) Photochemical apparatus organization in the thylakoid membrane of *Hordeum vulgare* wild type and chlorophyll *b*-less chlorina f2 mutant. Biochim Biophys Acta 851:331–339

Grover A (1993) How do senescing leaves lose photosynthetic activity? Curr Sci 64:22–234

Havaux M, Tardy F (1997) Thermostability and photostability of photosystem II in leaves of the *chlorina-f2* barley mutant deficient in light-harvesting chlorophyll *a/b* protein complexes. Plant Physiol 113:913–923

Hidema J, Makino A, Kurita Y, Mae T, Ojima K (1992) Changes in the levels of chlorophyll and light-harvesting chlorophyll *a/b* protein of PS II in rice leaves aged under different irradiances from full expansion through senescence. Plant Cell Physiol 33:1209–1214

Holub J, Hanuš J, Hanke DE, Strnad M (1998) Biological activity of cytokinins derived from *ortho*- and *meta*-hydroxybenzyladenine. Plant Growth Regul 26:109–115

Hönig M, Plíhalová L, Husičková A, Nisler J, Doležal K (2018) Role of cytokinins in senescence, antioxidant defence and photosynthesis. Int J Mol Sci 19:4045–4068

Hornitschek P, Kohnen MV, Lorrain S, Rougemont J, Ljung K, López-Vidriero I, Franco-Zorrilla JM, Solano R, Trevisan M, Pradervand S, Xenarios I, Fankhauser C (2012) Phytochrome interacting factors 4 and 5 control seedling growth in changing light conditions by directly controlling auxin signaling. Plant J 71:699–711

Hörtensteiner S (2013) Update on the biochemistry of chlorophyll breakdown. Plant Mol Biol 82:505–517

Huang FY, Philosoph-Hadas S, Meir S, Callaham DA, Sabato R, Zelcer A, Hepler PK (1997) Increases in cytosolic Ca^{2+} in parsley mesophyll cells correlate with leaf senescence. Plant Physiol 115:51–60

Huang J, Han B, Xu S, Zhou M, Shen W (2011) Heme oxygenase-1 is involved in the cytokinin-induced alleviation of senescence in detached wheat leaves during dark incubation. J Plant Physiol 168:768–775

Hudák J, Vizárová G, Šikulová J, Ovečková O (1996) Effect of cytokinins produced by strains of Agrobacterium tumefaciens with binary vectors on plastids in senescent barley leaves. Acta Physiol Plant 18:205–210

Humbeck K, Quast S, Krupinska K (1996) Functional and molecular changes in the photosynthetic apparatus during senescence of flag leaves from field-grown barley plants. Plant Cell Environ 19:337–344

Hwang I, Sheen J, Müller B (2012) Cytokinin signaling networks. Annu Rev Plant Biol 63:353–380

Jackowski G (1996) Senescence-related changes in the subcomplex organization of the major light-harvesting chlorophyll *a/b*-protein complex of photosystem II (LHCII) as influenced by cytokinin. Z Naturforsch C 51:464–472

Janečková H, Husičková A, Ferretti U, Prčina M, Pilařová E, Plačková L, Pospíšil P, Doležal K, Špundová M (2018) The interplay between cytokinins and light during senescence in detached *Arabidopsis* leaves. Plant Cell Environ 41:1870–1885

Janečková H, Husičková A, Lazár D, Ferretti U, Pospíšil P, Špundová M (2019) Exogenous application of cytokinin during dark senescence eliminates the acceleration of photosystem II impairment caused by chlorophyll *b* deficiency in barley. Plant Physiol Biochem 136:43–51

Kao CH (1980) Senescence of rice leaves. IV. Influence of benzyladenine on chlorophyll degradation. Plant Cell Physiol 21:1255–1262

Kar M, Streb P, Hertwig B, Feierabend J (1993) Sensitivity to photodamage increases during senescence in excised leaves. J Plant Physiol 141:538–544

Khanna-Chopra R (2012) Leaf senescence and abiotic stresses share reactive oxygen species-mediated chloroplast degradation. Protoplasma 249:469–481

Kieber JJ, Schaller GE (2014) Cytokinins. Arabidopsis Book 12:e0168
Kim HJ, Ryu H, Hong SH, Woo HR, Lim PO, Lee IC, Sheen J, Nam HG, Hwang I (2006) Cytokinin-mediated control of leaf longevity by AHK3 through phosphorylation of ARR2 in Arabidopsis. Proc Natl Acad Sci U S A 103:14–819
Kim JH, Woo HR, Kim J, Lim PO, Lee IC, Choi SH, Hwang D, Nam HG (2009) Trifurcate feed-forward regulation of age-dependent cell death involving miR164 in *Arabidopsis*. Science 323:1053–1057
Kim J, Kim JH, Lyu JI, Woo HR, Lim PO (2018) New insights into the regulation of leaf senescence in *Arabidopsis*. J Exp Bot 69:787–799
Krieger-Liszkay A, Trösch M, Krupinska K (2015) Generation of reactive oxygen species in thylakoids from senescing flag leaves of the barley varieties Lomerit and Carina. Planta 241:1497–1508
Krieger-Liszkay A, Krupinska K, Shimakawa G (2019) The impact of photosynthesis on initiation of leaf senescence. Physiol Plant 166:148–164
Kuai B, Chen J, Hörtensteiner S (2018) The biochemistry and molecular biology of chlorophyll breakdown. J Exp Bot 69:751–767
Kudryakova NV, Burkhanova EA, Rakitin VY, Yakovleva LA, Smith A, Hall MA, Kulaeva ON (2001) Russ J Plant Physiol 48:624–627
Kusaba M, Ito H, Morita R, Iida S, Sato Y, Fujimoto M, Kawasaki S, Tanaka R, Hirochika H, Nishimura M, Tanaka A (2007) Rice NON-YELLOW COLORING1 is involved in light-harvesting complex II and grana degradation during leaf senescence. Plant Cell 19:1362–1375
Lazár D (2015) Parameters of photosynthetic partitioning. J Plant Physiol 175:131–147
Liebsch D, Keech O (2016) Dark-induced leaf senescence: new insights into a complex light-dependent regulatory pathway. New Phytol 212:563–570
Lim PO, Kim HJ, Nam HG (2007) Leaf senescence. Annu Rev Plant Biol 58:115–136
Liu L, Li HR, Zeng H, Cai Q, Zhou X, Yin C (2016) Exogenous jasmonic acid and cytokinin antagonistically regulate rice flag leaf senescence by mediating chlorophyll degradation, membrane deterioration, and senescence-associated genes expression. J Plant Growth Regul 35:366–376
Lu CM, Zhang JA (1998) Changes in photosystem II function during senescence of wheat leaves. Physiol Plant 104:239–247
Lu C, Lu Q, Zhang J, Kuang T (2001) Characterization of photosynthetic pigment composition, photosystem II photochemistry and thermal energy dissipation during leaf senescence of wheat plants grown in the field. J Exp Bot 52:1805–1810
Mae T, Thomas H, Gay AP, Makino A, Hidema J (1993) Leaf development in *Lolium temulentum*: photosynthesis and photosynthetic proteins in leaves senescing under different irradiances. Plant Cell Physiol 34:391–399
Miersch I, Heise J, Zelmer I, Humbeck K (2000) Differential degradation of the photosynthetic apparatus during leaf senescence in barley (*Hordeum vulgare* L.). Plant Biol 2:618–623
Mik V, Szüčová L, Šmehilová M, Zatloukal M, Doležal K, Nisler J, Grúz J, Galuszka P, Strnad M, Spíchal L (2011) N9-substituted derivatives of kinetin: effective anti-senescence agents. Phytochemistry 72:821–831
Mlejnek P, Doležal P, Procházka S (2005) Intracellular conversion of cytokinin bases into corresponding mononucleotides is related to cell death induction in tobacco BY-2 cells. Plant Sci 168:389–395
Monakhova OF, Chernyad'ev II (2004) Effects of cytokinin preparations on the stability of the photosynthetic apparatus of two wheat cultivars experiencing water deficiency. Appl Biochem Microbiol 40:573–580
Nath K, Phee B-K, Jeong S, Lee SY, Tateno Y, Allakhverdiev SI, Lee C-H, Nam HG (2013) Age-dependent changes in the functions and compositions of photosynthetic complexes in the thylakoid membranes of *Arabidopsis thaliana*. Photosynth Res 117:547–556

Oh S, Lee S, Chung I, Lee C, Nam H (1996) A senescence-associated gene of Arabidopsis thaliana is distinctively regulated during natural and artificially induced leaf senescence. Plant Mol Biol 30:739–754

Oh M, Moon Y, Lee C (2003) Increased stability of LHCII by aggregate formation during dark-induced leaf senescence in the Arabidopsis mutant, *ore10*. Plant Cell Physiol 44:1368–1377

Oh MH, Kim JH, Zulfugarov IS, Moon Y-H, Rhew T-H, Lee C-H (2005) Effects of benzyladenine and abscisic acid on the disassembly process of photosystems in an Arabidopsis delayed-senescence mutant, *ore9*. J Plant Biol 48:170–177

Park SY, Yu JW, Park JS, Li J, Yoo SC, Lee NY, Lee SK, Jeong SW, Seo HS, Koh HJ, Jeon JS, Park YI, Paek NC (2007) The senescence-induced staygreen protein regulates chlorophyll degradation. Plant Cell 19:1649–1664

Pavalan-Ünsal N, Çağ S, Çetin E, BüyüktunÇer D (2002) Retardation of senescence by *meta*-topolin in wheat leaves. J Cell Mol Biol 1:101–108

Pospíšil P (2016) Production of reactive oxygen species by photosystem II as a response to light and temperature stress. Front Plant Sci 7:1950

Pospíšilová J, Čatský J, Synková H, Macháčková I, Solárová J (1993) Gas exchange and *in vivo* chlorophyll fluorescence in potato and tobacco plantlets *in vitro* as affected by various concentrations of 6-benzylaminopurine. Photosynthetica 29:1–12

Procházková D, Wilhelmová N (2004) Changes in antioxidative protection in bean cotyledons during natural and continuous irradiation-accelerated senescence. Biol Plant 48:33–39

Prokopová J, Špundová M, Sedlářová M, Husičková A, Novotný R, Doležal K, Nauš J, Lebeda A (2010) Photosynthetic responses of lettuce to downy mildew infection and cytokinin treatment. Plant Physiol Biochem 48:716–723

Raines T, Shanks C, Cheng C-Y, McPherson D, Argueso CT, Kim HJ, Franco-Zorrilla JM, López-Vidriero I, Solano R, Vaňková R, Schaller GE, Kieber JJ (2016) The cytokinin response factors modulate root and shoot growth and promote leaf senescence in Arabidopsis. Plant J 85:134–147

Ren B, Zhang J, Dong S, Liu P, Zhao B (2018) Exogenous 6-benzyladenine improves antioxidative system and carbon metabolism of summer maize waterlogged in the field. J Agric Crop Sci 204:175–184

Richmond AE, Lang A (1957) Effect of kinetin on protein content and survival of detached Xanthium leaves. Science 125:650–651

Riefler M, Novák O, Strnad M, Schmülling T (2006) Arabidopsis cytokinin receptor mutants reveal functions in shoot growth, leaf senescence, seed size, germination, root development, and cytokinin metabolism. Plant Cell 18:40–54

Sakuraba Y, Balazadeh S, Tanaka R, Mueller-Roeber B, Tanaka A (2012) Overproduction of chl *b* retards senescence through transcriptional reprogramming in Arabidopsis. Plant Cell Physiol 53:505–517

Sakuraba Y, Kim Y, Yoo S, Hörtensteiner S, Paek N (2013) 7-Hydroxymethyl chlorophyll *a* reductase functions in metabolic channeling of chlorophyll breakdown intermediates during leaf senescence. Biochem Biophys Res Commun 430:32–37

Sakuraba Y, Jeong J, Kang MY, Kim J, Paek NC, Choi G (2014) Phytochrome interacting transcription factors PIF4 and PIF5 induce leaf senescence in Arabidopsis. Nat Commun 5:4636–4649

Sato Y, Morita R, Katsuma S, Nishimura M, Tanaka A, Kusaba M (2009) Two short-chain dehydrogenase/reductases, NON-YELLOW COLORING 1 and NYC1-LIKE, are required for chlorophyll *b* and light-harvesting complex II degradation during senescence in rice. Plant J 57:120–131

Selivankina S, Karavaiko N, Kuiper D, Novikova GV, Kulaeva ON (2001) Cytokinin activity of zeatin allylic phosphate, a natural compound. Plant Growth Regul 33:157–164

Singh S, Letham DS, Palni LMS (1992) Cytokinin biochemistry in relation to leaf senescence. VIII. Translocation, metabolism and biosynthesis of cytokinins in relation to sequential leaf senescence of tobacco. Physiol Plant 86:398–406

Smart CM, Scofield SR, Bevan MW, Dyer TA (1991) Delayed leaf senescence in tobacco plants transformed with *tmr*, a gene for cytokinin production in *Agrobacterium*. Plant Cell 3:647–656

Šmehilová M, Dobrůšková J, Novák O, Takáč T, Galuszka P (2016) Cytokinin-specific glycosyltransferases possess different roles in cytokinin homeostasis maintenance. Front Plant Sci 7:1264–1283

Špundová M, Popelková H, Ilík P, Skotnica J, Novotný R, Nauš J (2003) Ultra-structural and functional changes in the chloroplasts of detached barley leaves senescing under dark and light conditions. J Plant Physiol 160:1051–1058

Špundová M, Slouková K, Hunková M, Nauš J (2005a) Plant shading increases lipid peroxidation and intensifies senescence-induced changes in photosynthesis and activities of ascorbate peroxidase and glutathione reductase in wheat. Photosynthetica 43:403–409

Špundová M, Strzałka K, Nauš J (2005b) Xanthophyll cycle activity in detached barley leaves senescing under dark and light. Photosynthetica 43:117–124

Talla SK, Panigrahy M, Kappara S, Nirosha P, Neelamraju S, Ramanan R (2016) Cytokinin delays dark-induced senescence in rice by maintaining the chlorophyll cycle and photosynthetic complexes. J Exp Bot 67:1839–1851

Tang Y, Wen X, Lu C (2005) Differential changes in degradation of chlorophyll-protein complexes of photosystem I and photosystem II during flag leaf senescence of rice. Plant Physiol Biochem 43:193–201

Tao G-Q, Letham DS, Palni LMS, Summons RE (1983) Cytokinin biochemistry in relation to leaf senescence I. The metabolism of 6-benzylaminopurine and zeatin in oat leaf segments. J Plant Growth Regul 2:89–102

Vescovi M, Riefler M, Gessuti M, Novák O, Schmülling T, Lo Schiavo F (2012) Programmed cell death induced by high levels of cytokinin in *Arabidopsis* cultured cells is mediated by the cytokinin receptor CRE1/AHK4. J Exp Bot 63:2825–2832

Vlčková A, Špundová M, Kotabová E, Novotný R, Doležal K, Nauš J (2006) Protective cytokinin action switches to damaging during senescence of detached wheat leaves in continuous light. Physiol Plant 126:257–267

Voitsekhovskaja OV, Tyutereva EV (2015) Chlorophyll *b* in angiosperms: functions in photosynthesis, signaling and ontogenetic regulation. J Plant Physiol 189:51–64

Vylíčilová H, Husičková A, Spíchal L, Srovnal J, Doležal K, Plíhal O, Plíhalová L (2016) C2-substituted aromatic cytokinin sugar conjugates delay the onset of senescence by maintaining the activity of the photosynthetic apparatus. Phytochemistry 122:22–33

Weaver LM, Amasino RM (2001) Senescence is induced in individually darkened *Arabidopsis* leaves, but inhibited in whole darkened plants. Plant Physiol 127:876–886

Weidhase RA, Lehmann J, Kramell H, Sembdner G, Parthier B (1987) Degradation of ribulose-1,5-bisphosphate carboxylase and chlorophyll in senescing barley leaf segments triggered by jasmonic acid methylester, and counteraction by cytokinin. Physiol Plant 69:161–166

Werner T, Motyka V, Laucou V, Smets R, Van Onckelen H, Schmülling T (2003) Cytokinin-deficient transgenic Arabidopsis plants show multiple developmental alterations indicating opposite functions of cytokinins in the regulation of shoot and root meristem activity. Plant Cell 15:2532–2550

Werner T, Köllmer I, Bartrina I, Holst K, Schmülling T (2006) New insights into the biology of cytokinin degradation. Plant Biol 8:371–381

Wilhelmová N, Wilhelm J, Kutík J, Haisel D (1997) Changes in French bean cotyledon composition associated with modulated life-span. Photosynthetica 34:377–391

Wingler A, Marès M, Pourtau N (2004) Spatial patterns and metabolic regulation of photosynthetic parameters during leaf senescence. New Phytol 161:781–789

Wu X, Zhu Z, Li X, Zha D (2012) Effects of cytokinin on photosynthetic gas exchange, chlorophyll fluorescence parameters and antioxidative system in seedlings of eggplant (*Solanum melongena* L.) under salinity stress. Acta Physiol Plant 34:2105–2114

Yang Y, Xu J, Huang L, Leng Y, Dai L, Rao Y, Chen L, Wang Y, Tu Z, Hu J, Ren D, Zhang G, Zhu L, Guo L, Qian Q, Zeng D (2016) *PGL*, encoding chlorophyllide *a* oxygenase 1, impacts leaf senescence and indirectly affects grain yield and quality in rice. J Exp Bot 67:1297–1310

Yang DQ, Luo YL, Dong WH, Yin YP, Li Y, Wang ZL (2018) Response of photosystem II performance and antioxidant enzyme activities in stay-green wheat to cytokinin. Photosynthetica 56:567–577

Zacarias L, Reid SM (1990) Role of growth regulators in the senescence of *Arabidopsis thaliana* leaves. Physiol Plant 80:549–554

Zavaleta-Mancera HA, Franklin KA, Ougham HJ, Thomas H, Scott IM (1999) Regreening of senescent *Nicotiana* leaves. I. Reappearance of NADPH-protochlorophyllide oxidoreductase and light-harvesting chlorophyll *a/b* binding protein. J Exp Bot 50:1677–1682

Zavaleta-Mancera HA, López-Delgadob H, Loza-Taverac H, Mora-Herrerab M, Trevilla-García C, Vargas-Suárezc M, Oughame H (2007) Cytokinin promotes catalase and ascorbate peroxidase activities and preserves the chloroplast integrity during dark-senescence. J Plant Physiol 164:1572–1582

Żelisko A, Jackowski GR (2004) Senescence-dependent degradation of Lhcb3 is mediated by a thylakoid membrane-bound protease. J Plant Physiol 161:1157–1170

Zubo YO, Yamburenko MV, Selivankina SY, Shakirova FM, Avalbaev AM, Kudryakova NV, Zubkova NK, Liere K, Kulaeva ON, Kusnetsov VV, Börner T (2008) Cytokinin stimulates chloroplast transcription in detached barley leaves. Plant Physiol 148:1082–1093

Zwack PJ, Rashotte AM (2013) Cytokinin inhibition of leaf senescence. Plant Signal Behav 8: e24737

Meta-topolin and Related Cytokinins as a Solution to Some In Vitro Problems

9

Stefaan P. O. Werbrouck

Contents

Abstract

Plant micropropagation does not always run smoothly. With a number of plants, problems such as hyperhydricity, shoot-tip necrosis, early senescence, and poor rooting are bottlenecks. These difficulties prevent the large-scale exploitation of elite genotypes of food crops, ornamental plants and fruit, forest, and timber trees. In this chapter the current evidence is summarized that shows that important in vitro problems can be alleviated using the right type and concentration of *meta*-topolin or a member of its expanding group of derived analogues.

Keywords

Cytokinin metabolism · Hyperhydricity · Rooting · Micropropagation · Aromatic cytokinin

S. P. O. Werbrouck (✉)
Laboratory for Applied In Vitro Plant Biotechnology, Faculty of Bioscience Engineering, Ghent University, Ghent, Belgium
e-mail: stefaan.werbrouck@ugent.be

N. Ahmad, M. Strnad (eds.), *Meta-topolin: A Growth Regulator for Plant Biotechnology and Agriculture*, https://doi.org/10.1007/978-981-15-9046-7_9

9.1 Introduction

Topolins are derivatives of N^6-benzyladenine (BA). After the surprising discovery of *ortho*-topolin (*o*T) and *meta*-topolin (*m*T) in poplar and other plant species (Horgan et al. 1973; Strnad et al. 1994, 1997), the first successful in vitro protocol was developed (Werbrouck et al. 1996). It was shown that, in contrast to BA, *m*T does not inhibit root formation in *Spathiphyllum* and combines this feature with a good in vitro multiplication rate. A large family of *m*T derivatives was then developed, by replacing the *m*T hydroxyl on the aromatic ring by a methoxy (CH_3O-), F-, or Cl-group. The corresponding 9-riboside derivatives were also synthesized in order to improve topolin uptake. A number of bioassays, such as tobacco callus, amaranthus, and wheat senescence bioassay, revealed strong cytokinin bioactivities such as induction of cell division and anti-senescence (Doležal et al. 2006). It is worth mentioning that *meta*-methoxy-topolin (Me*m*T) and its riboside (Me*m*TR) are naturally occurring cytokinins (Tarkowská et al. 2003). In 2009, Szüčová et al. developed 9-tetrahydopyran-2-yl derivatives that also showed beneficial effects. There are major metabolic differences between BA and topolins which have practical consequences. As the N^6-benzyl substituents of topolins give them more resistance to cleavage by cytokinin oxidase/dehydrogenase, their glucosylation appears to be favored more than their degradation. However, *m*T is not stored as a stable 9-glucoside in plants, unlike BA. Instead, it is reversibly glycosylated to form 6-(3-*O*-β-D-glucopyranosyl)benzyladenine-9-riboside (*m*TROG), a non-toxic O-glucoside that is an easily convertible storage form which does not accumulate during subsequent cultures (Werbrouck et al. 1996). Developing efficient micropropagation protocols faces challenges such as hyperhydricity, shoot-tip necrosis, poor rooting ability, early senescence, and decomposing chimeras. This chapter aims at summarizing the available evidence that in a number of plant species these in vitro problems can be alleviated using the right type and concentration of topolins.

9.2 Hyperhydricity

Hyperhydricity is a syndrome that recurrently affects micropropagated shoots. Characteristic symptoms are swollen, curled leaves that appear fragile and translucent. Hyperhydric shoots are difficult to acclimatize and root poorly. The phenomenon is often irreversible and may cause complete loss of in vitro cultures (Rojas-Martínez et al. 2010). A high relative humidity in the culture vessels, the use of Gelrite, and presence of NH_4^+ are principal causative factors. Under these conditions, cytokinins can stimulate hyperhydricity in a concentration-dependent manner (Ivanova and van Staden 2008). Quite a number of reports have shown that the cytokinin type plays a role (Kadota and Niimi 2003). Using *m*T instead of BA can solve the problem in a number of herbaceous and woody plant species. Adding *m*T in the place of BA to the micropropagation medium of apple shoots caused a significant decrease in hyperhydricity. As a result, the subsequent adventitious shoot regeneration per leaf segment increased (Dobránszki et al. 2002). Bairu et al. (2007)

reported that *m*T was the preferred cytokinin in terms of both multiplication rate and rooting of *Aloe polyphylla*. It completely resolved the hyperhydricity problem in this species. Van der Westhuizen (2014), while micropropagating *Eucalyptus*, noted less vitrified shoots with 0.2 mg/l *m*T compared to BA. The *Corylus colurna* hyperhydricity observed in in vitro shoots that developed on BA was absent when applying *m*T, regardless of the concentration used (Gentile et al. 2017). In order to increase the efficiency of a cryopreservation protocol, Poisson et al. (2016) had to resolve the problem of hyperhydricity of shoots that regenerated from isolated apices of dihaploid apple. By replacing BA with *m*T, the shoots were significantly less hyperhydric. Hyperhydricity was also alleviated with the application of *m*T in *Scadoxus puniceus* regenerated from bulb scales (Naidoo et al. 2017). Cultures of *Syzygium cumini*, an Indian medicinal tree, that were raised on *m*T appeared green and healthy, while BA induced hyperhydricity (Naaz et al. 2019).

9.3 Shoot-Tip Necrosis

Shoot-tip necrosis is a physiological disorder caused by a number of factors which, in addition, are genotype dependent. Typically, the apical shoot parts turn brown and die which confounds propagating in vitro cultures. There is probably a distinctive role of cytokinins in this phenomenon. By analyzing the cytokinin pool in *Harpagophytum procumbens*, Bairu et al. (2009) provided evidence that shoot-tip necrosis is associated with a lack of active cytokinins. In this particular plant, STN is associated with large callus formation at the plant base. While passing the callus, BA is irreversibly converted in and stored as its inactive 9-glucoside form. Therefore the apical meristem is depleted of active cytokinins which are essential for maintaining the cell cycle and other developmental processes. Thanks to their hydroxyl group, *m*T and *m*TR are able to reversibly form O-glucosides. They can readily access the continuously dividing cells of the apical shoot meristem and thus avoid hyperhydricity.

9.4 Decomposing Chimeras

Chimeras are plants with one or more mutated histogenic layers. It is a challenge to micropropagate them in a stable manner. A commercially acceptable multiplication rate is often linked to the production of undesired non-variegated shoots of adventitious or axillary origin. Kinetin was the only cytokinin to safely micropropagate these valuable ornamentals, but the slow multiplication rate had to be taken into account. Bogaert et al. (2006) used a leaf-variegated GWG petunia to demonstrate how a true-to-type micropropagation could be achieved with *meta*-methoxy-topolin riboside (Me*m*TR) instead of BA. On a medium with 1–2 μM Me*m*TR, the petunia explants produced a significant number of new variegated shoots, with only a small number of unwanted non-variegated green or white shoots. The same concentration of BA induced almost exclusively green or white non-variegated shoots. These results suggested that Me*m*TR is less active in inducing adventitious shoots and that Me*m*TR prevents incorrect cell divisions in the distinct histogenic layers of the axillary meristems.

9.5 Rooting Inhibition During Acclimatization

The aforementioned difference in metabolism between BA and *m*T has marked consequences. In *Spathiphyllum* it was shown that BA accumulates as a rather stable N-glucoside at the plant base and callus that inhibits post vitro (Werbrouck et al. 1995). *m*T, on the other hand, is mainly transformed into *m*TROG, which is broken down so fast that it can no longer inhibit rooting during acclimatization (Werbrouck et al. 1996). Likewise, Ahmad and Anis (2019) reported that, compared to BA, *m*T improved in vitro shoot multiplication and rhizogenesis in *Pterocarpus marsupium*, a medicinal tree. Saeiahagh et al. (2019) cultured red-fleshed *Actinidia chinensis* on media with *m*T, BA, or zeatin. Not only shoot number and weight were significantly higher with *m*T, but plantlets that were propagated in *m*T-supplemented media readily produced roots once transferred to rooting media. This was not the case for plantlets propagated in BA or zeatin containing media. Acclimatized banana plantlets that were regenerated using Me*m*T, Me*m*TR, and Me*m*TTHP (Szüčová et al. 2009) had significantly longer roots and better shoot/root ratios compared to the control and BA-treated plants (Aremu et al. 2012). Amoo et al. (2014), studying the micropropagation of *Aloe* and *Harpagophytum*, reported that *m*TTHP had a better rooting stimulatory effect than other (Bairu et al. 2011). Valero-Aracama et al. (2010) noticed that 10 μM *m*T or more was inhibitory to in vitro rooting of *Uniola paniculata* during its multiplication phase. This aside, survival during acclimatization was significantly better when the shoots were propagated in the presence of 2.2–30 μM *m*T, compared to 2.2 μM BA.

In fact, roots may already appear during the acclimatization phase. Werbrouck et al. (1996) reported that *m*T allows rooting during the micropropagation phase of *Spathiphyllum*, although raising its concentration from 5 to 40 μM clearly reduced the number of in vitro roots and their total length. *Cannabis sativa* grew vigorously in MS plus 1–4 μM *m*T and started rooting within 3–4 weeks. A separate medium for root initiation was not needed (Lata et al. 2016). Also Behera et al. (2019) demonstrated that in vitro propagation and rooting of *Hedychium coronarium* could be combined using 12.4 μM *m*T.

9.6 Early Senescence

Leaf senescence is a developmental process actively initiated as part of an age-dependent genetic program or as a response to environmental stress. Cytokinins are important endogenous negative regulators of senescence in plants (Zwack and Rashotte 2013, Van Staden et al. 1988). It should come as no surprise that one of the three bioassays by which the activity of topolin derivatives was tested during their development was the wheat senescence bioassay (Doležal et al. 2006). Although Me*m*TR, Me*o*TR, and 3F-BA popped up as very active anti-senescence compounds in this assay, they remain to be fully exploited in this regard. This has already been achieved for roses. Roses that are micropropagated with BA often senesce too early in the vessel. This process is generally initiated from 4 to 6 weeks on. Some leaves and shoots start yellowing, and a few weeks later, all the plants in the jar are lifeless. Bogaert et al. (2006) demonstrated that 2.5 μM Me*m*TR had much better anti-senescing effects than BA, *m*T, Me*m*T, 3F-BA, or 3F-BAR.

In *Pelargonium* cuttings, *m*T was very active, delaying the onset of leaf yellowing and senescence, without affecting the rooting percentage (Mutui et al. 2012). Palavan-Ünsal et al. (2002) revealed that *m*T retarded senescence in excised wheat leaf segments and increased chlorophyll a and b content. Also Wojtania (2010) demonstrated that *m*T was superior to BA in inhibiting early senescence and retaining chlorophyll in *Pelargonium* cultivars. In acclimatized banana, the chlorophyll a/b ratio was significantly improved when produced on medium with Me*m*T, *m*TR, and *m*T compared to control (Aremu et al. 2012). In *Corylus colurna*, the chlorophyll a content and chlorophyll a/b ratio were higher with 8.2 μM *m*T application, compared to BA (Gentile et al. 2017).

9.7 Conclusion

Advanced commercial in vitro labs have been using topolins for more than 20 years to improve the quality of their plants. But micropropagation is a big business, and as a consequence, a lot of industrial protocols remain confidential. Luckily, academic reports revealing the use and beneficial effects of topolins are now increasing in number. They may not be miracle compounds, but for a number of difficult crops, they can make the difference between success and failure. Therefore, these cytokinins deserve a place in the toolbox of every commercial and academic tissue culture lab.

References

Ahmad A, Anis M (2019) Meta-topolin improves in vitro morphogenesis, rhizogenesis and biochemical analysis in *Pterocarpus marsupium* Roxb.: a potential drug-yielding tree. J Plant Growth Regul 38(3):1007–1016

Amoo SO, Aremu AO, Moyo M et al (2014) Physiological effects of a novel aromatic cytokinin analogue in micropropagated *Aloe arborescens* and *Harpagophytum procumbens*. Plant Cell Tissue Organ Cult 116(1):17–26

Aremu AO, Bairu MW, Szüčová L et al (2012) Assessment of the role of meta-topolins on in vitro produced phenolics and acclimatization competence of micropropagated 'Williams' banana. Acta Physiol Plant 34(6):2265–2273

Bairu MW, Stirk WA, Dolezal K, Van Staden J (2007) Optimizing the micropropagation protocol for the endangered *Aloe polyphylla*: can meta-topolin and its derivatives serve as replacement for benzyladenine and zeatin? Plant Cell Tissue Organ Cult 90(1):15–23

Bairu MW, Stirk WA, Van Staden J (2009) Factors contributing to in vitro shoot-tip necrosis and their physiological interactions. Plant Cell Tissue Organ Cult 98(3):239–248

Bairu MW, Novák O, Doležal K, Van Staden J. (2011) Changes in endogenous cytokinin profiles in micropropagated *Harpagophytum* procumbens in relation to shoot-tip necrosis and cytokinin treatments. Plant Growth Regul 63(2):105–114

Behera S, Kar SK, Rout KK et al (2019) Assessment of genetic and biochemical fidelity of field-established *Hedychium coronarium* J. Koenig regenerated from axenic cotyledonary node on meta-topolin supplemented medium. Indus Crop Prod 134:206–215

Bogaert I, Van Cauter S, Prinsen E et al (2006) New aromatic cytokinin can make the difference. Acta Hortic 725:265–270

Dobránszki J, Magyar-Tábori K, Jámbor-Benczúr E, et al. (2002) Effect of conditioning apple shoots with *meta*-topolin on the morphogenic activity of in vitro leaves. Acta Agron Hung 50 (2):117–126

Doležal K, Popa I, Kryštof V et al (2006) Preparation and biological activity of 6-benzylaminopurine derivatives in plants and human cancer cells. Bioorg Med Chem 14:875–884

Gentile A, Frattarelli A, Nota P et al (2017) The aromatic cytokinin *meta*-topolin promotes in vitro propagation, shoot quality and micrografting in *Corylus colurna* L. Plant Cell Tissue Organ Cult 128(3):693–703

Horgan R, Hewett EW, Purse JG, Wareing PF (1973) A new cytokinin from *Populus x robusta*. Tetrahedron 30:2827–2828

Ivanova M, van Staden J (2008) Effect of ammonium ions and cytokinins on hyperhydricity and multiplication rate of in vitro regenerated shoots of *Aloe polyphylla*. Plant Cell Tissue Organ Cult 92(2):227–231

Kadota M, Niimi Y (2003) Effects of cytokinin types and their concentrations on shoot proliferation and hyperhydricity in in vitro pear cultivar shoots. Plant Cell Tissue Organ Cult 72(3):261–265

Lata H, Chandra S, Techen N et al (2016) In vitro mass propagation of *Cannabis sativa* L.: a protocol refinement using novel aromatic cytokinin meta-topolin and the assessment of eco-physiological, biochemical and genetic fidelity of micropropagated plants. J Appl Res Med Arom Plants 3(1):18–26

Mutui TM, Mibus H, Serek M (2012) Effect of *meta*-topolin on leaf senescence and rooting in *Pelargonium*× *hortorum* cuttings. Posthar Biol Technol 63(1):107–110

Naaz A, Hussain SA, Anis M, Alatar AA (2019) Meta-topolin improved micropropagation in *Syzygium cumini* and acclimatization to *ex vitro* conditions. Biol Plant 63(1):174–182

Naidoo D, Aremu AO, Van Staden J, Finnie JF (2017) In vitro plant regeneration and alleviation of physiological disorders in *Scadoxus puniceus*. S Afr J Bot 109:316–322

Palavan-Ünsal N, Çağ S, Çetin E et al (2002) Retardation of senescence by *meta*-topolin in wheat leaves. J Cell Mol Biol 1:101–108

Poisson AS, Berthelot P, Le BC a (2016) A droplet-vitrification protocol enabled cryopreservation of doubled haploid explants of *Malus x domestica Borkh*. 'Golden delicious'. Sci Hortic 209:187–191

Rojas-Martínez L, Visser RG, De Klerk GJ (2010) The hyperhydricity syndrome: waterlogging of plant tissues as a major cause. Propag Ornam Plants 10(4):169–175

Saeiahagh H, Mousavi M, Wiedow C et al (2019) Effect of cytokinins and sucrose concentration on the efficiency of micropropagation of 'Zes006' *Actinidia chinensis* var. chinensis, a red-fleshed kiwifruit cultivar. Plant Cell Tissue Organ Cult 138:1–10

Strnad M, Peters W, Hanuš J, Beck E (1994) Ortho-topolin-9-glucoside, an aromatic cytokinin from *Populusxcanadensis* cv Robusta leaves. Phytochemistry 37(4):1059–1062

Strnad M, Hanuš J, Vaněk T et al (1997) Meta-topolin, a highly active aromatic cytokinin from poplar leaves (*Populus*×*canadensis* Moench., cv. Robusta). Phytochemistry 45(2):213–218

Szüčová L, Spíchal L, Doležal K et al (2009) Synthesis, characterization and biological activity of ring-substituted 6-benzylamino-9-tetrahydropyran-2-yl purine derivatives. Bioorg Med Chem 17(5):1938–1947

Tarkowská D, Doležal K, Tarkowski P et al (2003) Identification of new aromatic cytokinins in *Arabidopsis thaliana* and *Populus× canadensis* leaves by LC (+) ESI MS and capillary liquid chromatography/frit–fast atom bombardment mass spectrometry. Physiol Plant 117(4):579–590

Valero-Aracama C, Kane ME, Wilson SB et al (2010) Substitution of benzyladenine with *meta*-topolin during shoot multiplication increases acclimatization of difficult-and easy-to-acclimatize sea oats (*Uniola paniculata* L.) genotypes. Plant Growth Regul 60(1):43

Van der Westhuizen A (2014) The use of *meta*-topolin as an alternative cytokinin in the tissue culture of *Eucalyptus* species. Acta Hortic 1055(1055):25–28

Van Staden J, Cook EL, Noodén LD (1988) Cytokinins and senescence. In: Noodén LD, Leopold AC (eds) Senescence and aging in plants. Academic, San Diego, pp 281–328

Werbrouck SP, van der Jeugt B, Dewitte W et al (1995) The metabolism of benzyladenine in *Spathiphyllum floribundum* 'Schott Petite' in relation to acclimatisation problems. Plant Cell Rep 14(10):662–665

Werbrouck SP, Strnad M, Van Onckelen HA et al (1996) Meta-topolin, an alternative to benzyladenine in tissue culture? Physiol Plant 98(2):291–297

Wojtania A (2010) Effect of *meta*-topolin on in vitro propagation of *Pelargonium x hortorum* and *Pelargonium x hederaefolium* cultivars. Acta Soc Bot Pol 79:101–106

Zwack PJ, Rashotte AM (2013) Cytokinin inhibition of leaf senescence. Plant Signal Behav 8(7): e24737

The Role of *Meta*-topolin in Plant Morphogenesis In Vitro

10

C. P. Krishna Vrundha, N. V. Aswathi, and T. Dennis Thomas

Contents

C. P. Krishna Vrundha · N. V. Aswathi · T. D. Thomas (✉)
Department of Plant Science, Central University of Kerala, Periye (PO), Kasaragod, Kerala, India

N. Ahmad, M. Strnad (eds.), *Meta-topolin: A Growth Regulator for Plant Biotechnology and Agriculture*, https://doi.org/10.1007/978-981-15-9046-7_10

Abstract

Over 80% of world population relies on plant-based traditional medicine for health care. The genetic diversity of medicinal plants is declining due to excessive and unsustainable harvesting. Hence there is strong need to develop protocols for effective and reproducible micropropagation of these plants. Micropropagation is undoubtedly the most reliable method for the production of a large number of plants in a short time. The technique also has immense value in plant genetic engineering. The efficiency of micropropagation depends on the composition of culture medium, especially the type and concentration of plant growth regulators (PGRs) added to the medium. N^6-(3-hydroxybenzyl)adenine (*meta*-topolin, *m*T) is a natural aromatic cytokinin first isolated from the leaves of *Populus robusta*. Since its discovery, especially in the past two decades, there are many studies conducted worldwide using *m*T and its derivatives to develop new protocols for the micropropagation of a variety of plant species, resolve defects due to in vitro propagation, and compare its properties and efficiency with other cytokinins. Various studies show that *m*T is more effective in morphogenesis than other cytokinins. It has superior regeneration capacity, reduced hyperhydricity, and no root inhibition in culture compared with commonly used cytokinins, thus gaining global interest, even though its efficacy is not universally applicable to all plant species. This chapter focuses on the use of *m*T in the micropropagation of plants with special emphases on its effects on the physiology of the plant culture and morphogenesis in vitro.

Keywords

Cytokinin · Medicinal plants · Micropropagation · Plant growth regulators · Plant tissue culture

10.1 Introduction

Micropropagation is considered as more reliable and attractive protocol over traditional propagation methods due to its high propagation rate to obtain a large number of plants in a short period of time and in a limited space. The various factors influencing micropropagation include composition of the medium, type and concentration of plant growth regulators (PGRs), explant type, age of the explants, varietal differences of plant, etc. Various protocols are available for micropropagation of a variety of plant species using different explant types, medium, and PGRs especially cytokinins (Cheruvathur and Thomas 2011, 2014; Howell et al. 2003). Cytokinins are N^6-substituted adenine derivatives that were first identified as cell division promoter factors (Miller et al. 1955, 1956). There are more than 200 natural and synthetic cytokinins which are used for various purposes (Huyluoğlu et al. 2008). The most commonly used cytokinin in plant tissue culture is N^6-benzyladenine (BA). Apart from cell division, cytokinins also influence seed germination, shoot initiation, apical dominance, and senescence (Noodeen and Leopold 1988; Mok

1994; Dewitte et al. 1999; Werner et al. 2001). The type and concentration of cytokinins play a critical role in micropropagation (Aremu et al. 2012a). The major goal of micropropagation is the large-scale production of uniform and healthy plants. However, in many plant species, in vitro propagation causes abnormalities such as stunted growth, deformities, somaclonal variation, epigenetic changes, etc. which limit its industrial and research applications. For this reason, there is urgent need for more research in search of superior and new cytokinins (Bairu et al. 2011a; Smulders and De Klerk 2011).

A highly active hydroxylated BA analogue was found in poplar (*P. robusta*) leaves, and this group of compounds was termed topolins by Strnad et al. (1997). Topolins including *m*T and its derivatives have been demonstrated as alternative cytokinins in PTC (Werbrouck et al. 1996; Strnad et al. 1997; Werbrouck 2010). Topolins have been found to be more active than *trans*-zeatin (*t*Z) and BA in promoting shoot formation in some PTCs (Kubalakova and Strnad 1992; Werbrouck et al. 1996). Its structure was confirmed to be N^6-(3-hydroxylbenzyl)adenine, and it was labeled *meta*-topolin (*m*T) by Strnad et al. (1997). In the recent past, there has been an array of promising results with the use of *m*T and its derivatives in various PTC systems (Table 10.1). Positive reports on important tissue culture parameters like higher shoot proliferation, reduction in commonly observed physiological disorders, improved rooting, and better acclimatization have increased the popularity of topolins in PTCs (Aremu et al. 2012b).

10.2 *Meta*-topolin Properties and Discovery

Cytokinins like BA are glucosylated on the purine scaffold into biologically inactive storage forms either as the glucoside itself or by the β-glucosidase (McGaw et al. 1985). These storage forms accumulate in plants and later slowly release and cause rooting inhibition and abnormal growth patterns (Werbrouck et al. 1996). To overcome these problems, it is necessary to identify alternative cytokinins that are conjugated at positions more prone to glucosylation than the nitrogen atoms of the purine nucleus. Horgan et al. (1975) isolated the first representative of these aromatic cytokinins (ARCKs), a hydroxyl analogue 6-(2-hydroxylbenzyl)adenine riboside (*ortho*-topolin riboside; *o*TR) from poplar leaves. They termed this particular class of compounds "populins". The free ARCK bases, 6-(2-hydroxylbenzyl)adenine (*ortho*-topolin; *o*T) and 6-(3-hydroxylbenzyl)adenine (*meta*-topolin; *m*T), were also identified in poplar leaves. Later Strnad et al. (1997) proposed alternative term "topolin" for this class of compounds as the term populins was already in use for representing salicin benzoate. Among the natural aromatic cytokinins, *m*T is one of the most active and thus was proposed to be a better alternative to BA for a number of different biotechnological applications (Werbrouck et al. 1996).

The effectiveness of *m*T over other cytokinins is due to its chemical structure. The basic structure of topolin is differentiated from BA by the presence of an extra hydroxyl group on the benzyl ring (Fig. 10.1). The hydroxyl group from the side chain of *m*T allows the formation of *O*-glucoside metabolites, a convertible

Table 10.1 Some recent reports on *meta*-topolin-induced morphogenesis in various systems

	Species	Explant	PGR tested	Preferred PGR and its concentration	Effect of *m*T	References
1.	*Pterocarpus marsupium*	CN	BA, *m*T	7.5 μM *m*T	Shoot multiplication Rhizogenesis	Ahmad and Anis (2019)
2.	*Syzygium cumini*	N	BA, iP, *m*T, KIN, IBA, IAA, NAA	5.0 μM *m*T + 2.0 μM NAA	Shoot multiplication Shoot elongation	Naaz et al. (2019)
3.	*Daphne mezereum*	N	*t*Z, *m*T, BA, NAA	4.4 μM BA + 0.5 μM NAA	Shoot elongation Vividly colored and better quality shoots	Nowakowska et al. (2019)
4.	*Mammillaria hernandezii*	SC	BA, KIN, *m*T, TDZ	2.2 μM *m*T	Shoot multiplication Shoot elongation Callus production	Lázaro-Castellanos et al. (2018)
5.	*M. dixanthocentron*	SC	BA, KIN, *m*T, TDZ	1.1 μM *m*T	Shoot multiplication Shoot elongation Callus production	Lázaro-Castellanos et al. (2018)
6.	*M. lanata*	SC	BA, KIN, *m*T, TDZ	1.1 μM *m*T (lateral explants) 8.9 μM TDZ (apical explants)	Shoot multiplication Shoot elongation Callus production	Lázaro-Castellanos et al. (2018)
7.	*Juglans nigra*	N	BA, *m*T, *t*Z	8.9 μM BA + 4.1 μM *m*T	Shoot elongation	Stevens and Pijut (2018)
8.	*Dendrobium aphyllum*	*t*-TCL	*m*T, TDZ, IBA, NAA	15.0 μM *mT* + 10.0 μM TDZ + 10.0 μM $AgNO_3$	Shoot proliferation Genetic stability Phytochemical yield	Bhattacharyya et al. (2018)
9.	*Carthamus tinctorius*	ST	*m*T, BA, TDZ, KIN, iP, *t*Z, Diuron, CPPU, Monuron	12.3 μM *m*T + 2.0 μM CPPU	Shoot proliferation Shoot multiplication	Vijayakumar et al. (2017)
10.	*C. tinctorius*	C	*m*T, TDZ, KIN, BA, TDZ, iP, Diuron, Monuron, *t*Z, CPPU	16.4 μM *m*T + 6.1 μM CPPU	Organogenesis	Vijayakumar et al. (2017)

11.	*Malaxis wallichii*	TSPS tTCL	BA, KIN, iP, *m*T, TDZ, NAA	4.1 µM *m*T + 2.7 µM NAA	Shoot proliferation Plant height and growth	Bose et al. (2017)
12.	*Corylus colurna*	AB	*m*T, BA, IBA, GA_3	8.2 µM *m*T	Shoot multiplication Photosynthetic pigments	Gentile et al. (2017)
13.	*Eriocephalus africanus*	ST	BA, *m*T, iP, *m*TTHP, IBA, *m*TR	5 µM *m*T	Shoot multiplication Shoot elongation	Madzikane-Mlungwana et al. (2017)
14.	*Scadoxus puniceus*	L	BA, *m*T, PG, GCA	GCA (1.0, 5.0, 10.0 µM) + 5.0 µM BA	Shoot multiplication Hyperhydricity	Naidoo et al. (2017)
15.	*Dendrocalamus asper*	N	*m*T, BA	15.0 µM BA Or 20.0 µM *m*T	Culm proliferation Culm height No phytotoxicity	Ornellas et al. (2015)
16.	*Ceratonia siliqua*	CN	BA, *m*T, KIN, TDZ, NAA, IBA, IAA	10.0 µM BA+ 0.5 µM NAA Or 10.0 µM *m*T	Shoot multiplication	Shahzad et al. (2017)
17.	*Nardostachys jatamansi*	L P	BA, *m*T, iP	4.1 µM *m*T	Callusing	Bose et al. (2016)
18.	*N. jatamansi*	C	BA, TDZ, *m*T, KIN, iP	4.1 µM *m*T	Shoot induction Multiplication	Bose et al. (2016)
19.	*Cannabis sativa*	N	TDZ, *m*T	2.0 µM *m*T	Shoot multiplication Shoot elongation	Lata et al. (2016)
20.	*Pyrus communis*	S	*m*T	6.0–8.0 µM *m*T	Shoot multiplication Shoot quality Improve leaf gas exchange and antioxidant activity	Dimitrova et al. (2016)
21.	*Dendrobium nobile*	N	*m*T, BA, spermine, spermidine, TDZ, putrescine, NAA	4.1 µM *m*T + 9.0 µM putrescine	Shoot proliferation	Bhattacharyya et al. (2016)

(continued)

Table 10.1 (continued)

	Species	Explant	PGR tested	Preferred PGR and its concentration	Effect of *m*T	References
22.	*Ginkgo biloba*	ST	BA, iP, *m*T	5.5 µM *m*T	Shoot multiplication Shoot length Least deformities	Nacheva et al. (2017)
23.	*Persea americana*	N	*m*T, GA_3	0.4 µM *m*T + 0.3 µM GA_3	Shoot multiplication Shoot elongation Eliminates shoot tip dieback	Bandaralage et al. (2015)
24.	*Coleonema album*	ST	BA, KIN, TDZ, *m*T	5.0 µM *m*T	Shoot multiplication Shoot length	Fajinmi et al. (2014)
25.	*Paulownia elongata*	AB	BA, *m*T	4.1 µM *m*T	Shoot proliferation Shoot multiplication	Clapa et al. (2014)
26.	*P. fortunei*	AB	BA, *m*T	4.1 µM *m*T	Shoot proliferation Shoot multiplication	Clapa et al. (2014)
27.	*P. elongata* × *P. fortunei*	AB	BA, *m*T	4.1 µM *m*T	Shoot proliferation Shoot multiplication	Clapa et al. (2014)
28.	*Merwilla plumbea*	L	iP, BA, *m*TR, *m*T, Me*m*TTHP	0.1 µM *m*TR	Shoot proliferation	Aremu et al. (2014)
29.	*Prunus domestica*	MS	BA, *m*T	2.1 µM BA	Growth and quality	Gentile et al. (2014)
30.	*P. insititia* × *domestica*	MS	BA, *m*T	2.1 µM BA	Growth and quality	Gentile et al. (2014)
31.	*Hypoxis hemerocallidea*	CM	BA, iP, *m*T	10.0 µM *m*T	Shoot multiplication	Moyo et al. (2014)
32.	*H. hemerocallidea*	C	BA, iP, *m*T	2.5 µM BA	Not mentioned	Moyo et al. (2014)

33.	*Ocimum basilicum*	SL	BA, *m*T	4.4 μM BA Or 2.1 μM *m*T	Shoot multiplication Shoot quality	Kőszeghi et al. (2014)
34.	*Huernia hystrix*	S	BA, BAR, *m*T, *m*TR, Me*m*TR, Me*m*TTHP	20.0 μM *m*T Or 25.0 μM BA	Shoot multiplication	Amoo and Van Staden (2013)
35.	*Agapanthus praecox*	ST	AdeS, KIN, *m*T, BA, iP, *m*TR, TDZ	4.4 μM BA	Shoot proliferation	Baskaran and Van Staden (2013)
36.	*Aloe arborescens*	ST	*m*T, *m*TR, Me*m*T	5.0 μM *m*T 5.0 μM BAR	Shoot proliferation	Amoo et al. (2012a)
37.	*M. plumbea*	L	AdeS, BA, iP, *m*T, *m*TR, KIN, TDZ, NAA	0.5 μM TDZ Or 0.5 μM *m*T Or 0.28 μM *m*TR	Moderate shooting Shoot length	Baskaran et al. (2012)
38.	*Aloe polyphylla*	ST	*m*T, Me*m*T, Me*m*TR, BA, *t*Z	5.0 μM *m*T	Shoot multiplication Reduced hyperhydricity Acclimatization	Bairu et al. (2007)

AB axillary buds, *C* callus, *CN* cotyledonary nodes, *CM* corm, *L* leaf, MS micro shoots, *N* nodes, *P* petiole, *S* shoot, *SC* seedling cuttings, *SL* seedlings, *ST* shoot tip, *t-TCL* thin cell layer, *TSPS* transversely sliced pseudo-stem segments, *iP* N^6-isopentyladenine, *Ads* adenine sulfate, *BA* N^6-benzyladenine, *BAR* N^6-benzyladenine riboside, *CPPU* *N*-(2-chloro-4-pyridyl)-*N*′-phenylurea, GA_3 gibberellic acid, *GCA* gallic acid, *IAA* indole-3-acetic acid, *KIN* kinetin, Me*m*T *meta*-methoxytopolin, Me*m*TR *meta*-methoxytopolin riboside, Me*m*TTHP *meta*-methoxytopolin tetrahydropyran-2-yl, *m*T *meta*-topolin, *m*TR *meta*-topolin riboside, *NAA* 1-naphthaleneacetic acid, *PG* phloroglucinol, *TDZ* thidiazuron, *tZ* *trans*-zeatin

Fig. 10.1 Chemical structure of BA and *m*T

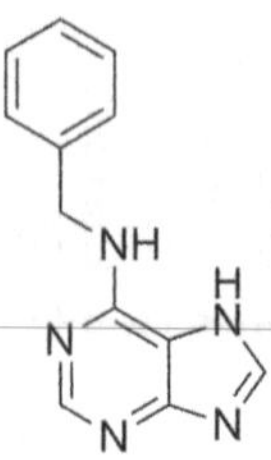

BA

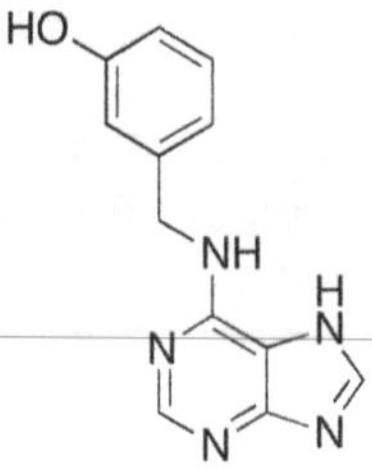

***m*T**

cytokinin storage form which is nontoxic. This also accounts for the rapid translocation of *m*T in plant tissues (Kamínek et al. 1987; Werbrouck et al. 1996). Hydroxylation of the side chain is known to have remarkable influence on the biological activity of cytokinins (Aremu et al. 2012a). Position-specific hydroxylation can also influence the biological activity of cytokinins by affecting the dihedral angle and the formation of hydrogen bond with nitrogen at the N1-position (Trávníček et al. 1997; Nirmalram et al. 2011). A decrease in activity of BA was observed after hydroxylation at the *ortho*- and even more so at the *para*- position of the phenyl ring (Holub et al. 1998). Similarly, in the case of thidiazuron (TDZ), the greatest activity was observed for *meta*-substituents followed by *ortho*- and *para*-, respectively (Mok and Mok 1985). Thus, the comparatively greater activity of *m*T may be attributed to hydroxylation at the *meta*-position (Aremu et al. 2012a).

10.3 Interaction of *Meta*-topolin with Auxins for Shoot Multiplication

Cytokinins can interact with auxins in a synergistic, additive, or antagonistic manner to produce a physiological response in cultured explants. The action of these phytohormones depends on the concentrations at which they are applied (Coenen and Lomax 1997). In the case of *Coleonema album*, *m*T with α-naphthaleneacetic acid (NAA) has additive effects as it increased the number of shoots and shoot length when combined (Fajinmi et al. 2014). In *Huernia hystrix* (family Asclepiadaceae), use of *m*T with high concentration of NAA resulted in three times higher shoot production compared with medium supplemented with *m*T only. This shows the additive or synergetic effect of auxins and cytokinins in the Asclepiadaceae family (Amoo and Van Staden 2013). A two-step protocol has been standardized for shoot organogenesis process in *Manihot esculenta* by Chauhan and Taylor (2018) where *m*T in combination with 2,4-dichlorophenoxyacetic acid (2,4-D) in the first step followed by high concentration of *m*T (6.0 μM) in the second step was found to be greatly superior to medium containing *m*T alone. *m*T and indole-3-butyric acid (IBA) act synergistically or additively in *Coleonema album* resulting in highest shoot induction and growth (Fajinmi et al. 2014). In *Eucomis autumnalis* subspecies *autumnalis*, a 1.5-fold increase in shoot production was observed in the medium with *m*T in combination with NAA compared to *m*T alone (Masondo et al. 2015). Shoot tip necrosis of in vitro culture is affected by cytokinin-auxin interaction. The excessive accumulation of cytokinin 9-glucosides can cause shoot tip necrosis of in vitro grown plants (Bairu et al. 2011b).The presence of indole-3-acetic acid (IAA) increased the level of 9-glucosides in BA-treated cultures but decreased it in *m*T-treated cultures of *Harpagophytum procumbens* (Bairu et al. 2011b).

10.4 *Meta*-topolin as a Solution to Common Micropropagation Problems

Hyperhydricity, yellowing of shoots and excessive shooting with abnormalities, failure to root, reduced survival rate during acclimatization, and genetic instability are abnormalities and problems associated with BA, the most commonly used cytokinin in PTC (Bairu et al. 2007; Werbrouck et al. 1995).

Hyperhydricity is a physiological disorder caused by active absorption of water because of metabolic abnormalities or passive diffusion of water into the formed shoots during in vitro propagation (Pâques 1991). The type and concentration of gelling agent, humidity in the culture vessel, and level of mineral ions such as NH^{+} and Cl^{-} are among the major factors that cause hyperhydricity along with type and concentration of PGR added to the medium (Gaspar et al. 1987; Kevers et al. 2004). Hyperhydricity increases with the cytokinin concentration. The problem is more severe in the case of BA, the most frequently used cytokinin (Bairu et al. 2007). To reduce the problem, the cytokinin, especially BA, could be replaced by *m*T and related analogues. *m*T in the PTC medium has been shown to eliminate the hyperhydricity problem compared to other cytokinins (Bairu et al. 2007; Naidoo et al. 2017; Bose et al. 2016).

The better acclimatization ability of *m*T-treated plants may be partly due to its influence on photosynthetic parameters. A well-developed photosynthetic apparatus greatly improves successful acclimatization. Prolonged culture and high concentrations of cytokinin can cause reduction in photosynthetic pigment concentration. In William's banana, compared to other commonly used cytokinins, *m*T and *meta*-topolin riboside (*m*TR)-treated plants remained visually greener and had a higher chlorophyll/carotenoid ratio after 10 days of culture (Aremu et al. 2012b). Gentile et al. (2014) reported that in *Prunus* species, *m*T-treated plantlets had significantly high chlorophyll content than control plants.

Shoot tip necrosis is the browning and death of young leaves and buds that eventually leads to the death of the plant. This makes micropropagation of many plants extremely difficult. The role of cytokinins in shoot tip necrosis is a controversial issue as different species react differently to the necrosis (Piagnani et al. 1996; Pérez-Tornero and Burgos 2000; Bairu et al. 2009).Studies in *Harpagophytum procumbens* showed an increase in accumulation of 9-glucosides (a toxic cytokinin form) and a decrease in *O*-glucosides (reusable, nontoxic compound) in the treatment with BA compared to *m*T-treated plants (Bairu et al. 2011b). Bandaralage et al.

(2015) reported successful elimination of shoot tip dieback, defoliation, and necrosis in *Persea americana* with the application of *m*T to the medium.

One of the major problems in PTC is abnormal or inadequate rooting or no rooting from the micropropagated shoots. These problems affect the survival of the multiplied plants during the acclimatization period. In many plant species, *m*T-treated plants are found to be more adapted to rooting and have a better survival rate during acclimatization. Lata et al. (2016) reported that in *Cannabis sativa*, using MS (Murashige and Skoog 1962) medium supplemented with 2.0 μM *m*T led to production of more roots than on 1/2 MS medium supplemented with 3.0 μM IBA. The *m*T-induced roots were healthier and longer than IBA. These plants successfully acclimatized with 100% survival rate. In *Prunus domestica* and *P. insititia* × *domestica*, *m*T-treated shoots produced higher rooting percentage, root number, and length compared to BA (Gentile et al. 2014). Werbrouck et al. (1996) reported that *m*T stimulated better in vitro rooting activity in *Spathiphyllum floribundum* compared to BA. In *Aloe polyphylla, m*T was observed to promote rooting of the micropropagated shoots in the multiplication medium. Compared to BA-treated plants (65%), a significant increase in acclimatization rate (90%) was observed for *m*T-treated plants (Bairu et al. 2007).

Along with propagation efficiency, the genetic stability of the regenerated plants has a high priority in large-scale micropropagation (Haisel et al. 2001). Passing through different micropropagation stages is reported to cause genetic variations in the regenerated plants. Somaclonal variation limits the applications of in vitro propagation. The type of PGRs and explant play major role in inducing somaclonal variation (Devi et al. 2014). Most traditional cytokinins induce genetic variations as a result of their low translocation rate and residual toxicity causing oxidative stress that leads to DNA damage (Jackson et al. 1998). In various experiments, *m*T-derived plants showed less clonal variability compared to plants grown in other commonly used cytokinins. Bhattacharyya et al. (2017) observed better clonal fidelity in *Ansellia africana* in *m*T-treated plants than plants multiplied in medium supplemented with BA. Similar results were also observed in *Dendrobium nobile* (Bhattacharyya et al. 2016).

The following section discusses the most effective use of *m*T in various genera.

10.4.1 *Nardostachys jatamansi* DC.

N. jatamansi is a critically endangered rhizomatous medicinal herb restricted to specialized regions of the alpine Himalayas (Chauhan and Nautiyal 2005). The roots and rhizomes of *N. jatamansi* are a rich source of biologically active compounds like coumarins and sesquiterpenes and are thus overexploited for medicinal purposes (Rekha et al. 2013).

Bose et al. (2016) optimized efficient protocols for both direct and indirect shoot organogenesis in *N. jatamansi.* For callus induction, MS medium supplemented with various auxins (2, 4-D, NAA, and picloram) either alone or in combination with cytokinins [BA, *m*T, and N^6-isopentyladenine (iP)] in a range of concentrations was employed. MS medium supplemented with a combination of 8.0 μM NAA and 4.1 μM *m*T resulted in optimum callusing from the cut ends of leaf and petiole explants after 15 days of culture. MS medium supplemented with cytokinins [BA, TDZ, *m*T, kinetin (KIN), and iP] in various ranges of concentration either alone or in combination with 2.7 μM NAA was used to study the regeneration from callus. Of these, medium supplemented with a combination of 4.1 μM *m*T and 2.7 μM NAA gave optimum results. In this medium, 92.5% cultures responded and produced a maximum number of 25.2 shoots with 6.1 leaves per shoot. Shoot tip and petiole were used as explants for direct shoot induction. Of the various concentrations of different cytokinins (KIN, iP, BA, and *m*T) used, *m*T was found to be the most effective in inducing multiple direct shoots. At 4.1 μM *m*T, shoot tip explants produced 14.5 direct shoots with a maximum length of 8.0 cm after 6 weeks of culture. Similarly, petioles produced the highest number of shoots (12.4) and maximum shoot length (7.3 cm) on the same medium. *m*T was found to be twice as effective as other cytokinins used in this experiment for shoot generation. Another advantage of using *m*T in the medium was that it does not produce hyperhydric shoots at high concentrations unlike other cytokinins.

10.4.2 *Coleonema album* (Thunb.) Bartl.

C. album is an ornamental and aromatic medicinal plant belonging to the family Rutaceae (Esterhuizen et al. 2006). The plant is in high demand in traditional medicine as it has antimicrobial, anti-inflammatory, and antioxidant activities (Eldeen and Van Staden 2008).

Fajinmi et al. (2014) compared various conventional cytokinins (BA, KIN, and TDZ) with *m*T for the direct shoot proliferation from shoot tip explants of *C. album*. Of the various concentrations (2.5, 5.0, 7.5, and 10.0 μM) of cytokinins (BA, TDZ, KIN, and *m*T) tested, 5.0 μM *m*T produced maximum number of shoots (14.5) in MS medium. An overall increase in shoot length, shoot number, and longest shoot was observed in the medium fortified with *m*T along with auxins other than *m*T alone. The maximum number of shoots (18.0 per explant) was found in the medium with 5.0 μM *m*T and 2.0 μM IBA. In the case of *C. album*, the combination of auxin and cytokinin had a synergistic effect on shoot proliferation.

10.4.3 *Huernia hystrix* (Hook.f.) N.E.Br.

H. hystrix is a heavily exploited medicinal and ornamental succulent plant belonging to the family Asclepiadaceae. It is endemic to the eastern region of South Africa and is used in traditional medicine by the Zulu among other cultures. It has antimicrobial, antioxidant, and anti-inflammatory activities (Amoo et al. 2012b). *H. hystrix* is endangered due to overharvesting as whole plant for medicinal purposes (Amoo et al. 2009).

Amoo and Van Staden (2013) developed a micropropagation protocol for this highly valued medicinal plant. According to these researchers, in comparison with BA, *m*T was found to be more efficient in shoot proliferation and secondary metabolite production in the *H. hystrix* species. Stem explants from in vitro regenerated shoots were used in their studies. MS medium supplemented with six different cytokinins BA, BA riboside (BAR), *m*T, *m*TR, *meta*-methoxytopolin riboside (Me*m*TR), and *meta*-methoxytopolin-9-tetrahydropyran-2-yl (Me*m*TTHP) in four different concentrations (10.0, 15.0, 20.0, and 25.0 μM) was tested. Of these, 20.0 μM *m*T was found to be the most effective in shoot proliferation even though the results did not differ greatly from BA at 25.0 μM concentration. In the next step, 20.0 and 25.0 μM BA or *m*T along with different concentrations of NAA (0.0, 2.5, 5.0, and 10.0 μM) were used to study the effect of auxin along with cytokinin. The maximum number of shoots (12.2 shoots per explant) was recorded on MS medium containing 20.0 μM *m*T and 10.0 μM NAA.

10.4.4 *Ginkgo biloba* L.

G. biloba is a medicinal and ornamental plant which is used all around the world. *G. biloba* extracts are widely sold herbal supplements and have antiaging, anti-inflammatory, antioxidant, antitumor, and memory-improving properties (Chan et al. 2007). Despite its high demand, not many studies have been conducted on its rapid propagation.

Nacheva et al. (2017) developed a micropropagation protocol for the lateral bud proliferation from shoot apices of *G. biloba* using *m*T. In this study, two different media, viz., MS and Driver and Kuniyuki Walnut (DKW; Driver and Kuniyuki 1984) media, with different concentrations of three cytokinins BA (2.5 or 4.44 μM), iP (50.0 or 75.0 μM), or *m*T (2.5, 4.0, 5.5, 7.0, or 8.0 μM) were employed for micropropagation. The DKW medium was found to be more suitable for *G. biloba*. The best result was obtained from DKW medium supplemented with 5.5 μM *m*T which had the highest mean number of shoots per explant (4.2) with good physiological parameters. Other cytokinins produced similar number of shoots, but they were very small; some shoots failed to elongate, and the leaves were abnormal in shape. Some plants exhibited hyperhydricity when supplied with high concentration of cytokinin.

10.4.5 *Ocimum basilicum* L.

O. basilicum or sweet basil is an aromatic medicinal plant from the family Lamiaceae which is distributed all around the world (Asghari et al. 2012). The common sweet basil is a rich source of volatile essential oils which are used for pharmaceutical, aromatic, and culinary applications (Saha et al. 2010; Sahoo et al. 1997; Gopi and Ponmurugan 2006).

The *m*T application on micropropagation of *O. basilicum* was reported by Kőszeghi et al. (2014). In their experiments, 15-day-old in vitro germinated plantlets were cultured in multiplication medium with BA and *m*T. Of the various concentrations (1.1–8.3 μM) of BA and *m*T, MS medium supplemented with 4.4 μM BA and 2.1 μM *m*T had the best results for shoot morphogenesis in *O. basilicum*. *m*T-derived plants showed better growth parameters such as healthy growth and fewer abnormalities than BA. No defects were noticed in the roots formed from *m*T-treated plants. In sum, for the micropropagation of sweet basil, 2.1 μM *m*T was found to be the most effective.

10.4.6 *Paulownia elongata* Thunb., *P. fortunei* (Seem.) Hemsl., and *P. elongata* × *P. fortunei*

Paulownia species belonging to the family Paulowniaceae is one of the most important forest crops because of its high growth rate, light weight, strength, and density. It is used to make furniture, containers, and high-quality paper (Rafighi and Tabarsa 2011; Feria et al. 2013).

Clapa et al. (2014) conducted a micropropagation study on *P. elongata*, *P. fortunei*, and *P. elongata* × *P. fortunei* using *m*T. Axillary buds and seeds were used for the initiation of in vitro culture. For multiplication, MS medium supplemented with BA (2.2 μM) and *m*T (2.0, 2.9, and 4.1 μM) was used. In *P. elongata* × *P. fortunei* and *P. elongata*, 4.1 μM *m*T gave the highest shoot proliferation. However, healthier shoots were observed in 2.0 and 2.9 μM *m*T. MS medium supplemented with 4.1 μM *m*T produced an average proliferation rate of 9.3 and average multiplication rate of 12.9 in *P. fortunei*. MS medium supplemented with 2.2 μM BA resulted in intense and irregular proliferation as well as callus formation from the base of the shoot.

10.4.7 *Agapanthus praecox* Willd.

A. praecox is an evergreen plant belonging to the family Amaryllidaceae and native to South Africa. It is a highly valuable medicinal plant which is used for heart disease, coughs, colds, chest discomfort, and paralysis by the Zulu culture (Varga and Veale 1997). Xhosa women use the roots to make an antenatal medicine and necklace as a charm for producing healthy babies (Sharaibi and Afolayan 2017). It is taken with other plants to prevent complications during pregnancy and to ensure the delivery of a healthy child (Ndhlala et al. 2013). The plant has anti-inflammatory, antiedemic, antitumor, immunoregulatory, antifungal, and antibacterial properties (Ndhlala et al. 2013). Anticancer property of this plant has also been reported (Koduru et al. 2007).

Baskaran and Van Staden (2013) reported the culture of shoot tip explants excised from 30-day-old seedlings of the plant, on MS medium supplemented with different cytokinins [adenine sulfate (AdeS), KIN, iP, *m*T, *m*TR, TDZ, and BA] at low concentrations (2.7–5.4 μM). From the study it is clear that 4.4 μM BA supported better shoot regeneration (10.5 shoots per explant) than other cytokinins. MS supplemented with 4.1 μM *m*T produced 3.7 shoots per explant, and MS with 2.7 μM *m*TR produced 3.4 shoots per explant.

10.4.8 *Aloe arborescens* Mill.

Aloe species (family Xanthorrhoeaceae) is highly valued for their medicinal and ornamental value and is overused for various purposes which pose a threat to the genus. Plants belonging to the genus *Aloe* have been used for centuries in traditional medicine for the treatment of gastrointestinal disorders, burns, insect bites, and other skin injuries (Grace et al. 2008). This genus has more than 200 different biologically active phytochemicals (Radha and Laxmipriya 2015). It is also used commercially as a food in certain countries (Grace 2011). *A. arborescens* is one of the most important medicinal plants in *Aloe* genus (Amoo et al. 2012a). Amoo et al. (2012a) optimized a protocol for the efficient regeneration of *A. arborescens*. Shoot tips collected from in vitro grown seedlings were cultured on MS medium supplemented with five different concentrations (2.5, 5.0, 7.5, 10.0, and 15.0 μM) of aromatic cytokinins [*m*T, *m*TR, *meta*-methoxytopolin (Me*m*T), Me*m*TR, Me*m*TTHP, BA, and BAR] and a control without any PGR. On MS medium supplemented with 5.0 or 7.5 μM *m*T, *m*TR, Me*m*T, and BAR, the adventitious shoot production and shoot length (above 1.0 cm) was satisfactory. Overall, the highest number of shoots (6.2) with an optimum length of 1.0 cm was observed on MS medium supplemented with 5.0 μM *m*T. Here, *m*T in the medium significantly increased the response both in terms of shoot number and length compared to other cytokinins. Additional shooting was observed during rooting of the proliferated *m*T-treated plants. This protocol with *m*T is simple and cost-effective due to its high shoot proliferation rate at low concentration. The production of additional shoots during rooting is also favorable, along with the high survival frequency on acclimatization.

10.4.9 *Aloe polyphylla* Pillans

A. polyphylla is an endangered plant belonging to the family Asphodelaceae, native to South Africa (Chukwujekwu et al. 2002). The plant has high medicinal and ornamental value. Unsustainable harvesting along with its specific habitat requirement is reducing the wild population of the plant at an alarming rate, and as a result, this plant is currently registered in the Red Data List of endangered species by SABONET (Southern African Botanical Diversity Network). A micropropagation protocol for *A. polyphylla* has been developed by Abrie and Van Staden (2001) and Chukwujekwu et al. (2002).

The effect of topolin derivatives, viz., *m*T, Me*m*T, Me*m*TR, BA, and *t*Z, on micropropagation was investigated with the objective of finding an alternative to BA and *t*Z (Bairu et al. 2007). Shoot tip explants with 2–4 young leaves were cultured on

MS medium supplemented with different concentrations (0.0, 0.5, 2.5, 5.0, 7.5, and 15.0 μM) of *m*T, Mc*m*T, Mc*m*TR, BA, and *t*Z. At lower concentrations, BA produced more shoots, but at higher concentrations, large numbers of shoots were induced by *m*T. The multiplication rate increased with an increase in concentration up to 15.0 μM with all PGRs. A decline in multiplication rate and excessive abnormal growth were observed at 15.0 μM. However, a concentration of 5.0 μM cytokinin was observed to be the optimum level for shoot induction. The highest rate of shoot multiplication (8.0 shoots per explant) was seen for *m*T at this optimum level. The response of *m*T at 5.0 μM was found to be significantly greater than other PGRs. *m*T-treated plants showed significantly superior quality and quantity than plants treated with BA. Bairu et al. (2007) suggested that *m*T would be a good replacement for *t*Z in the tissue culture of *A. polyphylla* and the better multiplication rate makes *m*T the preferred cytokinin. In *A. polyphylla* all hormone treatments caused hyperhydricity at higher concentrations; of these BA treatment was most severe. But at the optimum concentration of *m*T (5.0 μM), no hyperhydric shoots were observed. The results were similar in Me*m*T-treated plants. In plants treated with BA, failure to rooting, yellowing, and excessive abnormal growth of shoot were observed. However, the presence of *m*T in the medium significantly overcomes these problems. Compared to BA-treated plants, about 25% more plants treated with *m*T acclimatized successfully.

10.4.10 *Cannabis sativa* L.

C. sativa is a member of the family Cannabaceae. This plant is a rich source of fiber, and its medicinal potential is well-known. The major biologically active compounds isolated from *C. sativa* are cannabinoids. These include 9-tetrahydrocannabinol (9-THC), cannabidiol (CBD), tetrahydrocannabivarin (THCV), cannabichromene (CBC), cannabigerol (CBG), and cannabinol (CBN). The plant has hallucinogenic and medicinal properties and is used to treat various diseases such as glaucoma, nausea, depression, asthma, and insomnia (Mechoulam et al. 1976; Andre et al. 2016; Corroon and Phillips 2018).

Different concentrations of TDZ and *m*T were tested for shoot formation in *C. sativa* by Lata et al. (2016). The authors describe an efficient *m*T regeneration protocol for *C. sativa* using nodal explants. The effect of MS medium supplemented with different concentrations (0.05, 0.5, 1.0, 2.0, 3.0, 4.0, and 5.0 μM) of TDZ and IBA with 1/2 MS was compared with similar concentrations of *m*T. MS medium supplemented with 2.0 μM *m*T gave the highest number of shoots (13.4) and maximum shoot length (11.4 cm). However, of the different concentrations of TDZ, 0.5 μM gave highest number of shoots (11.8) with a mean shoot length of 7.3 cm. Comparatively, *m*T gave superior results in terms of number of shoots and mean shoot length. This study reported that separate root initiation medium was not necessary for *C. sativa*. MS medium supplemented with 2.0 μM *m*T was found to be the best concentration for rooting. Of the regenerated shoots, about 96% developed roots on the medium. An average of 14.0 roots per shoots was observed. In this study *m*T was found to be superior to IBA for parallel rooting, but at higher concentrations, *m*T was inhibitory. Of the different concentrations of IBA tested, the best rooting was observed on MS medium supplemented with 3.0 μM IBA. Here, an average number of 4.6 roots were produced per explant with a mean root length of 5.3 cm which was significantly lower than with *m*T (19.0 cm).

10.4.11 *Carthamus tinctorius* L.

C. tinctorius or safflower is an important oil plant belongs to the family Asteraceae. Its seeds and leaves are edible, and the flowers are used to color cosmetics and dye fabrics. PTC protocols using BA and TDZ along with auxins are found to be less effective in *C. tinctorius* as it produces fewer shoots and causes hyperhydricity. For this reason, attempts were made to develop new protocols using other cytokinins.

Vijayakumar et al. (2017) standardized a micropropagation protocol for *C. tinctorius* with shoot tip and nodal explants. They used MS medium supplemented with various cytokinins [*m*T, TDZ, BA, KIN, *t*Z, Monuron, Diuron, and *N*-(2-chloro-4-pyridyl)-*N′*-phenyl urea (CPPU)] in different concentrations. In the next step, CPPU in the range 0.4–10.0 μM in combination with optimum concentration of *m*T (12.3 μM), TDZ (22.7 μM), or BA (31.1 μM) was tested. MS medium supplemented with 12.3 μM *m*T + 2.0 μM CPPU produced optimum result of 97.7% response with 42.4 shoots from shoot tip explants. A combination of 16.4 μM *m*T with 6.1 μM CPPU was found to be optimum for organogenesis from leaf-derived callus.

10.4.12 *Merwilla plumbea* Lindl.

M. plumbea is a vulnerable South African plant with numerous medicinal uses. It plays a crucial role in traditional medicine. Overexploitation coupled with other anthropogenic activities has become a major reason behind the decline of the *M. plumbea* population (McCartan and Van Staden 1998). The plant is used for treating stomachaches, nausea, intestinal worms, diarrhea, constipation, indigestion, and paralysis. It is also a remedy for sprains, fractures, internal tumors, and cancers. *Merwilla* is a rich source of bioactive compounds which shows anti-inflammatory, anthelmintic, antibacterial, antischistosomal, antimutagenic, angioprotective, and phosphodiesterase inhibitory properties (Ncube et al. 2011). In vitro propagation protocols have been developed in this species by some workers (Baskaran et al. 2012; Baskaran and Van Staden 2012; McCartan and Van Staden 1998).

Baskaran et al. (2012) standardized a protocol for in vitro propagation of *M. plumbea* using leaf explants excised from in vitro grown plantlets on MS medium supplemented with various PGRs such as AdeS, BA, iP, *m*T, *m*TR, TDZ, KIN, and NAA. Even though TDZ (0.5 μM) was found most effective (20.4 shoots per explant), both *m*T (0.5 μM) and *m*TR (0.3 μM) showed moderate shoot induction in *M. plumbea* with 10.6 and 10.0 shoots per explants, respectively. BA (0.4 μM) produced 7.8 shoots per explant. Shoot length was significantly higher on MS medium supplemented with *m*T (5.9 cm) and *m*TR (6.3 cm) compared to other PGRs (BA, 3.5 cm; TDZ, 3.1 cm; iP, 5.1 cm; and KIN, 5.3 cm). In another study conducted by Aremu et al. (2014), leaf explants were cultured on MS medium supplemented with 1.0 μM of different cytokinins (iP, BA, *m*T, *m*TR, and Me*m*TTHP). Of the various cytokinins used, *m*TR was the best and generated approximately 8.0 shoots per explant, followed by *m*T (7.7) compared to the control (4.4) and other cytokinins.

10.4.13 *Prunus domestica* L. and *P. insititia* × *domestica*

P. domestica or European plum is an important fruit tree grown throughout the world (Okie and Ramming 1999). Plum fruits are known for their antioxidant, anti-inflammatory, and memory-improving qualities (Igwe and Charlton 2016).

Gentile et al. (2014) compared the effect of *m*T and BA in the micropropagation of rootstock of *P. domestica* and *P. insititia* × *domestica*. Micro shoots from in vitro grown plantlets were cultured on MS medium supplemented with 2.1 μM BA or 2.1, 4.2, or 6.3 μM *m*T. BA showed better shoot multiplication rate than *m*T; however the presence of *m*T in the medium significantly increased the growth and quality of the shoots in both *P. domestica* and *P. insititia* × *domestica*. The BA and *m*T gave similar response in rooting of *P. domestica*, but in *P. insititia* × *domestica*, *m*T-treated plants showed higher root length and rooting percentage (92%) than BA. For adventitious shoot regeneration, apical leaves from multiplied plants were transferred to MS medium containing BA, *t*Z, or TDZ. The optimum result of 42% and 65% regeneration for *P. insititia* × *domestica* and *P. domestica*, respectively, was obtained in TDZ containing medium, only from shoots previously grown in *m*T.

10.4.14 *Hypoxis hemerocallidea* Fisch. & C.A. Mey.

H. hemerocallidea, commonly known as African star grass or African potato, is a medicinal plant from the family Hypoxidaceae. It is used to boost immunity and treat impotency, cardiac problems, dizziness, and inflammation (Owira and Ojewole 2009). The antidiabetic properties of this plant are well documented in many studies (Ojewole 2006; Owira and Ojewole 2009; Mahomed and Ojewole 2003).

Micropropagation of *H. hemerocallidea* using MS medium supplemented with different cytokinins was conducted by Moyo et al. (2014). Cytokinins (BA, iP, and *m*T) at different concentrations (2.5, 5.0, 7.5, and 10.0 μM) were tested for shoot multiplication from surface sterilized corms. MS medium supplemented with 10.0 μM *m*T produced the highest number of shoots per explant (4.7 shoots). The greatest number of transferable healthy shoots (89%) was also obtained from *m*T containing medium compared to BA and iP. Both *m*T and BA were found to be effective in callus proliferation and shoot regeneration. BA at 2.5 μM concentration produced the highest number of shoot buds (18.1) per jar during shoot regeneration from callus. Plants grown on iP produced significantly more roots and root growth than other cytokinins tested. The longest root (11.3 cm) was observed in MS basal medium, and these plants were found to be easier to acclimatize than hormone-treated plantlets.

10.4.15 *Daphne mezereum* L.

D. mezereum, commonly known as February daphne, is a shrub of the family Thymelaeaceae. This is a commercially important plant due to its ornamental value. Daphne is traditionally used to treat cancer, and an antileukemic principle called mezerein is isolated from the plant (Kupchan and Baxter 1975).

Nowakowska et al. (2019) conducted an experiment aimed at finding out the effect of PGRs on the micropropagation of *D. mezereum.* They tested different cytokinins (*t*Z, *m*T, and BA) and other PGRs [gibberellic acid (GA_3) and NAA] at different concentrations (0.5, 1.0, 2.0, and 3.0 μM) in both MS and Woody Plant Medium (WPM; McCown and Lloyd 1981). Nodal segments were used for culture initiation. MS medium supplemented with 4.4 μM BA and 0.5 μM NAA was found optimum for shoot induction. MS medium with 0.2 μM GA_3 and 4.1 μM *m*T produced the longest shoots. Most vividly colored and better quality shoots were obtained in WPM supplemented with *m*T.

10.4.16 *Mammillaria hernandezii* Glass and R.A. Foster, *M. dixanthocentron* Backeb., and *M. lanata* (Britton and Rose)

M. hernandezii, *M. dixanthocentron*, and *M. lanata* belong to the family Cactaceae. The majority of *Mammillaria* species are endemic to Mexico (Ortega-Baes and Godínez-Alvarez 2006). These globular cacti are small and low-growing with distinctly tubercular stems (Martorell and Peters 2005). *M. hernandezii* and *M. dixanthocentron* are included in the International Union for the Conservation of Nature and Natural Resources (IUCN) Red List of Threatened Species (Anonymous 2016).

Lázaro-Castellanos et al. (2018) reported efficient use of *m*T in the micropropagation of *Mammillaria* species. In vitro germinated seedlings were used as the explant. To produce three types of explants, apical section, basal section, and two lateral sections of the stem were dissected from plantlets with a height of about1.0 cm. For shoot multiplication, MS medium supplemented with BA, KIN, *m*T, and TDZ at various concentrations (0.0, 0.4, 1.1, 2.2, 4.4, and 8.9 μM) was tested. Most of the shoot regeneration was obtained in the lateral stem section explants. In *M. hernandezii* MS medium supplemented with 2.2 μM *m*T gave 7.4 shoots per explant. The presence of *m*T (2.2 and 4.4 μM) in the medium gave the greatest average shoot number, height, and callus production for *M. hernandezii* compared with other treatments. In *M. dixanthocentron*, similar results were observed at 1.1 µM (16.7 shoots per explant) and 8.9 µM (15.8 shoots per explant) *m*T treatments. For lateral explants of *M. lanata*, all the tested variables were significantly higher in the 1.1 and 4.4 µM *m*T treatments. However, for apical explants TDZ (2.2, 4.4, and 8.9 µM) was found to be most effective.

10.4.17 *Persea americana* Mill.

P. americana (avocado) is a tropical Native American fruit, belonging to the family Lauraceae. It is cultivated for its food and medicinal purposes due to the high nutrient and therapeutic potential. It is used for the treatment of scabies, dander, and ergotism (Ranade and Thiagarajan 2015). The extracts from this plant have the potential to induce apoptosis in human breast cancer cells (Butt et al. 2006). It is a grafted crop which comprises a fruiting scion grafted to a rootstock cultivar. The rootstocks are selected for their beneficial field characteristics. They can be propagated by seed or by clonal propagation. Tissue culture can be used as an efficient alternative for clonal rootstock production of avocado. However, avocado has performed disappointedly in culture environments (Bairu and Kane 2011). Several problems like poor elongation, tip dieback, high rates of defoliation, stunting, and vitrification have been observed during the micropropagation of avocado (Barceló-Muñoz et al. 1999).

Bandaralage et al. (2015) reported the use of *m*T, instead of the mostly used cytokinin BA, for the micropropagation of nodal explants in avocado. Most of the problems like high callusing, necrosis, and negative effects at the rooting stage associated with BA-containing medium can be overcome using *m*T (Gentile et al. 2014). GA_3 is known for its influence on internode extension, leaf growth, and release of apical dominance (Gonbad et al. 2014). WPM supplemented with 0.4 µM and 4.1 µM *m*T or GA_3 alone or both *m*T (0.4, 0.8, 1.2, 4.1, and 8.2 µM) in combination with different concentrations of GA_3 (0.3, 0.6, 0.9, 0.3, and 0.3 µM, respectively) was tested for micropropagation. Budbreak, shoot tip dieback, defoliation, and necrosis were significantly different between experiments. There was 100% budbreak when 0.4 µM *m*T alone or in combination with GA_3 was employed. GA_3 promoted shoot growth compared to controls without PGRs, but at a high concentration (2.9 µM) resulted in very thin elongated leaves relative to *m*T-treated shoots. *m*T effectively eliminated shoot tip dieback and defoliation due to hormone stress. Compared to individual hormone or no hormone control treatments, combined hormone treatments resulted in broader and healthier leaves. Treatment with 0.4 µM *m*T along with 0.3 µM GA_3 produced significantly taller shoots. This indicates the clear synergistic effect of *m*T and GA_3 on shoot length in *P. americana*.

10.4.18 *Juglans nigra* L.

J. nigra is a highly valued multipurpose tree, as it is used for making veneer, cabinets, gunstocks, and furniture (Michler et al. 2007). Due to its high timber value, efforts are made to improve the genetic characteristics of the tree. After the development of genetically superior varieties, the commercial demand for elite clones has greatly increased.

Significant differences in shoot elongation between half-sibling *J. regia* genotypes when treated with same concentrations of PGRs have been reported (Scaltsoyiannes et al. 1998). The application of commonly used cytokinins such as BA, KIN, and *t*Z in concentrations different than optimal concentrations has led to deleterious effects such as stunted growth, shoot hyperhydricity, shoot tip necrosis, chlorosis, fasciation, and higher phenolic exudation on micropropagated plants (Heile-Sudholt et al. 1986; Revilla et al. 1989; Van Sambeek et al. 1997; Bosela and Michler 2008). Against this background, Stevens and Pijut (2018) conducted a study with the aim of evaluating whether manipulating PGR and the culture environment using novel techniques could improve black walnut in vitro shoot proliferation, independent of genotype, and generate plantlets from in vitro grown shoots. Nodal sections were placed on a semisolid basal DKW medium for shoot initiation. To investigate the influence of different cytokinins in shoot induction, 8.9 μM BA in combination with 0.0, 4.1, 6.2, 8.3, and 10.4 μM *m*T or *m*T alone in the same concentrations (0.0, 4.1, 6.2, 8.3, and 10.4 μM) or 4.1 μM *m*T in combination with 2.5, 5.0, or 12.5 μM *t*Z was tested. Shoots from juvenile explants cultured without PGR were shorter, had fewer nodes than the optimum treatment, and often failed to elongate. Shoots cultured with 8.9 μM BA in combination with 4.1 μM *m*T produced shoots longer (1.7 and 1.3 cm) than all other treatments and with more nodes (6.4 and 5.1) for genotype No. 55 and No. 189, respectively. A synergistic effect of BA and *m*T was observed in this study. Shoots exposed to 8.9 μM BA or 4.1 μM *m*T alone elongated normally, but the combination of the two resulted in longer shoots with more nodes per shoot. According to this study, shoots from mature explants often failed to elongate regardless of cytokinin type and concentration.

10.5 Conclusion

This chapter was intended to provide a compilation of existing information on *m*T with its effects on plant physiology, interaction with other PGRs, and recent micropropagation studies in various genera using *m*T. Studies on the application of *m*T are limited, and it has been restricted as an alternative cytokinin to BA in micropropagation. In PTC, the choice of cytokinin to be used is decided by its cumulative efficiency in induction and multiplication of normal shoots and successful rooting and acclimatization. In various studies, *m*T induced direct and indirect shoot organogenesis in different plants without any abnormalities. The response in terms of shoot induction and multiplication by using *m*T was found to be superior to other cytokinins in some systems. It is recommended as a solution for common problems associated with micropropagation such as difficulties in rooting, acclimatization, hyperhydricity, shoot tip necrosis, and genetic instability. *m*T and auxins act synergistically in different plants resulting in high shoot induction and growth of the shoots. We believe that *m*T will play a major role in plant morphogenesis in future.

References

Abrie AL, Van Staden J (2001) Micropropagation of the endangered *Aloe polyphylla*. Plant Growth Regul 33:19–23

Ahmad A, Anis M (2019) *Meta*-topolin improves *in vitro* morphogenesis, rhizogenesis and biochemical analysis in *Pterocarpus marsupium* Roxb.: a potential drug-yielding tree. J Plant Growth Regul 38:1007–1016

Amoo SO, Van Staden J (2013) Influence of plant growth regulators on shoot proliferation and secondary metabolite production in micropropagated *Huernia hystrix*. Plant Cell Tissue Organ Cult 112:249–256

Amoo SO, Finnie JF, Van Staden J (2009) *In vitro* propagation of *Huernia hystrix*: an endangered medicinal and ornamental succulent. Plant Cell Tissue Organ Cult 96:273–278

Amoo SO, Aremu AO, Van Staden J (2012a) *In vitro* plant regeneration, secondary metabolite production and antioxidant activity of micropropagated *Aloe arborescens* Mill. Plant Cell Tissue Organ Cult 111:345–358

Amoo SO, Finnie JF, Van Staden J (2012b) Acetylcholinesterase inhibition, antioxidant, antiinflammatory, antimicrobial and phytochemical properties of *Huernia hystrix*. Phytother Res 26:639–645

Andre CM, Hausman JF, Guerriero G (2016) *Cannabis sativa*: the plant of the thousand and one molecules. Front Plant Sci 7:19

Anonymous (2016) IUCN Red List of threatened species. Vesrsion 2016.1. International Union for the Conservation of Nature

Aremu AO, Bairu MW, Szucova L, Dolezal K, Finnie JF, Van Staden J (2012a) Shoot and root proliferation in 'Williams' banana: are the topolins better cytokinins? Plant Cell Tissue Organ Cult 111:209–218

Aremu AO, Bairu MW, Szucova L, Dolezal K, Finnie JF, Van Staden J (2012b) Assessment of the role of meta-topolins on *in vitro* produced phenolics and acclimatization competence of micropropagated 'Williams' banana. Acta Physiol Plant 34:2265–2273

Aremu AO, Plačková L, Bairu MW, Novak O, Plihalova L, Dolezal K, Finnie JF, Van Staden J (2014) How does exogenously applied cytokinin type affect growth and endogenous cytokinins in micropropagated *Merwilla plumbea*? Plant Cell Tissue Organ Cult 118:245–256

Asghari F, Hossieni B, Hassani A, Shirzad H (2012) Effect of explants source and different hormonal combinations on direct regeneration of basil plants (*Ocimum basilicum* L.). Aus J Agric Eng 3:12–17

Bairu MW, Kane ME (2011) Physiological and developmental problems encountered by *in vitro* cultured plants. Plant Growth Regul 63:101–103

Bairu MW, Stirk WA, Doležal K, Van Staden J (2007) Optimizing the micropropagation protocol for the endangered *Aloe polyphylla*: can *meta*-topolin and its derivatives serve as replacement for benzyladenine and zeatin? Plant Cell Tissue Organ Cult 90:15–23

Bairu MW, Kulkarni MG, Street RA, Mulaudzi R, Van Staden J (2009) Studies on seed germination, seedling growth, and *in vitro* shoot induction of *Aloe ferox* Mill., a commercially important species. Hortic Sci 44:751–756

Bairu MW, Aremu AO, Van Staden J (2011a) Somaclonal variation in plants: causes and detection methods. Plant Growth Regul 63:147–173

Bairu MW, Novák O, Doležal K, Van Staden J (2011b) Changes in endogenous cytokinin profiles in micropropagated *Harpagophytum procumbens* in relation to shoot-tip necrosis and cytokinin treatments. Plant Growth Regul 63:105–114

Bandaralage JCAH, Hayward A, O'Brien C, Mitter N (2015) Gibberellin and cytokinin in synergy for a rapid nodal multiplication system of avocado. In: Proceedings of the World Avocado Congress VIII, Lima, pp 13–18

Barceló-Muñoz A, Encina CL, Simón-Pérez E, Pliego-Alfero F (1999) Micropropagation of adult avocado. Plant Cell Tissue Organ Cult 58:11–17

Baskaran P, Van Staden J (2012) Somatic embryogenesis of *Merwilla plumbea* (Lindl.) Speta. Cell Tissue Organ Cult 109:517–524

Baskaran P, Van Staden J (2013) Rapid *in vitro* micropropagation of *Agapanthus praecox*. S Afr J Bot 86:46–50

Baskaran P, Ncube B, Van Staden J (2012) *In vitro* propagation and secondary product production by *Merwilla plumbea (*Lindl.) Speta. Plant Growth Regul 67:235–245

Bhattacharyya P, Kumaria S, Tandon P (2016) High frequency regeneration protocol for *Dendrobium nobile*: a model tissue culture approach for propagation of medicinally important orchid species. S Afr J Bot 104:232–243

Bhattacharyya P, Kumar V, Van Staden J (2017) Assessment of genetic stability amongst micropropagated *Ansellia africana*, a vulnerable medicinal orchid species of Africa using SCoT markers. S Afr J Bot 108:294–302

Bhattacharyya P, Paul P, Kumaria S, Tandon P (2018) Transverse thin cell layer (t-TCL)-mediated improvised micropropagation protocol for endangered medicinal orchid *Dendrobium aphyllum* Roxb: an integrated phytomolecular approach. Acta Physiol Plant 40:137–151

Bose B, Kumaria S, Choudhury H, Tandon P (2016) Assessment of genetic homogeneity and analysis of phytomedicinal potential in micropropagated plants of *Nardostachys jatamansi,* a critically endangered, medicinal plant of alpine Himalayas. Plant Cell Tissue Organ Cult 124:331–349

Bose B, Kumaria S, Choudhury H, Tandon P (2017) Insights into nuclear DNA content, hydrogen peroxide and antioxidative enzyme activities during transverse thin cell layer organogenesis and *ex vitro* acclimatization of *Malaxis wallichii*, a threatened medicinal orchid. Physiol Mol Biol Plants 23:955–968

Bosela MJ, Michler CH (2008) Media effects on black walnut (*Juglans nigra* L.) shoot culture growth *in vitro*: evaluation of multiple nutrient formulations and cytokinin types. In Vitro Cell Dev Biol Plant 44:316–329

Butt AJ, Roberts CG, Seawright AA, Oelrichs PB, MacLeod JK, Liaw TY, Kavallaris M, Somers-Edgar TJ, Lehrbach GM, Watts CK, Sutherland RL (2006) A novel plant toxin, persin, with *in vivo* activity in the mammary gland, induces Bim-dependent apoptosis in human breast cancer cells. Mol Cancer Ther 5:2300–2309

Chan PC, Xia Q, Fu PP (2007) *Ginkgo biloba* leave extract: biological, medicinal, and toxicological effects. J Environ Sci Health C 25:211–244

Chauhan RS, Nautiyal MC (2005) Commercial viability of cultivation of an endangered medicinal herb *Nardostachys jatamansi* at three different agroclimatic zones. Curr Sci 89:1481–1488

Chauhan RD, Taylor NJ (2018) *Meta*-topolin stimulates de novo shoot organogenesis and plant regeneration in cassava. Plant Cell Tissue Organ Cult 132:219–224

Cheruvathur MK, Thomas TD (2011) An efficient plant regeneration system through callus for *Pseudarthria viscida* (L.) Wright and Arn., a rare ethnomedicinal herb. Physiol Mol Biol Plants 17:395–401

Cheruvathur MK, Thomas TD (2014) Shoot organogenesis from root-derived callus of *Rhinacanthus nasutus* (L.) Kurz. and assessment of clonal fidelity of micropropagated plants using RAPD analysis. Biotechnol Appl Biochem 172:1172–1182

Chukwujekwu JC, Fennell CW, Van Staden J, Bornman CH (2002) Optimization of the tissue culture protocol for the endangered *Aloe polyphylla*. S Afr J Bot 68:424–429

Clapa D, Fira A, Simu M, Vasu LB, Buduroi D (2014) Improved *in vitro* propagation of *Paulownia elongata, P. fortunei* and its interspecific hybrid *P. elongata* x *P. fortunei*. Bull Univ Agric Sci Vet Med Cluj-Napoca Hortic 71:422–427

Coenen C, Lomax TL (1997) Auxin-cytokinin interactions in higher plants: old problems and new tools. Trends Plant Sci 2:351–356

Corroon J, Phillips JA (2018) A cross sectional study of cannabidiol users. Cannabis Cannabinoid Res 3:152–161

Devi SP, Kumaria S, Rao SR, Tandon P (2014) Single primer amplification reaction (SPAR) methods reveal subsequent increase in genetic variations in micropropagated plants of *Nepenthes khasiana* Hook. f. maintained for three consecutive regenerations. Gene 538:23–29

Dewitte W, Chiappetta A, Azmi A, Witters E, Strnad M, Rembur J, Noin M, Chriqui D, Van Onckelen H (1999) Dynamics of cytokinins in apical shoot meristems of a day-neutral tobacco during floral transition and flower formation. Plant Physiol 119:111–122

Dimitrova N, Nacheva L, Berova M (2016) Effect of *meta*-topolin on the shoot multiplication of pear rootstock OHF-333 (*Pyrus communis* L.). Acta Sci Pol Hortorum Cultus 15:43–53

Driver JA, Kuniyuki AH (1984) *In vitro* propagation of Paradox walnut *Juglans regia* X *Juglans hindsii* rootstock. Hortic Sci 19:507–509

Eldeen IMS, Van Staden J (2008) Anti-inflammatory and mycobacterial activity of leaf extracts of *Coleonema album*. S Afr J Bot 74:345–347

Esterhuizen LL, Meyer R, Dubery IA (2006) Antimicrobial compounds from *Coleonema album* (Rutaceae). Z Naturforsch C 61:489–498

Fajinmi OO, Amoo SO, Finnie J, Van Staden J (2014) Optimization of *in vitro* propagation of *Coleonema album*, a highly utilized medicinal and ornamental plant. S Afr J Bot 94:9–13

Feria MJ, Garcia JC, Zamudio MAM, Gomide JL, Colodette JL, López F (2013) Kraft pulping and bleaching of paulownia wood. Cell Chem Technol 47:595–601

Gaspar T, Kevers C, Debergh P, Maene L, Paques M, Boxus PH (1987) Vitrification: morphological, physiological, and ecological aspects. In: Bonga JM, Durzan DJ (eds) Cell and tissue culture in forestry. Springer, Dordrecht, pp 152–166

Gentile A, Gutierrez MJ, Martinez J, Frattarelli A, Nota P, Caboni E (2014) Effect of *meta*-Topolin on micropropagation and adventitious shoot regeneration in *Prunus* rootstocks. Plant Cell Tissue Organ Cult 118:373–381

Gentile A, Frattarelli A, Nota P, Condello E, Caboni E (2017) The aromatic cytokinin *meta*-topolin promotes *in vitro* propagation, shoot quality and micrografting in *Corylus colurna* L. Plant Cell Tissue Organ Cult 128:693–703

Gonbad RA, Sinniah UR, Abdul Aziz M, Mohamad R (2014) Influence of cytokinins in combination with GA_3 on shoot multiplication and elongation of tea clone Iran 100 (*Camellia sinensis* L. O. Kuntze). Sci World J 2014:943054, 9 p

Gopi C, Ponmurugan P (2006) Somatic embryogenesis and plant regeneration from leaf callus of *Ocimum basilicum* L. J Biotechnol 126:260–264

Grace OM (2011) Current perspectives on the economic botany of the genus *Aloe L.* (Xanthorrhoeaceae). S Afr J Bot 77:980–987

Grace OM, Simmonds MSJ, Smith GF, Van Wyk AE (2008) Therapeutic uses of *Aloe* L. (Asphodelaceae) in Southern Africa. J Ethnopharmacol 119:604–614

Haisel D, Hofman P, Vágner M, Lipavska H, Ticha I, Schafer CA, Capkova V (2001) *Ex vitro* phenotype stability is affected by *in vitro* cultivation. Biol Plant 44:321–324

Heile-Sudholt C, Huetteman CA, Preece JE, Van Sambeek JW, Gaffney GR (1986) *In vitro* embryonic axis and seedling shoot tip culture of *Juglans nigra* L. Plant Cell Tissue Organ Cult 6:189–197

Holub J, Hanus J, Hanke DE, Strnad M (1998) Biological activity of cytokinins derived from *ortho*- and *meta*-hydroxybenzyladenine. Plant Growth Regul 26:109–115

Horgan R, Hewett EW, Horgan JM, Purse J, Wareing PF (1975) A new cytokinin from *Populus* x *robusta*. Phytochemistry 14:1005–1008

Howell SH, Lall S, Che P (2003) Cytokinins and shoot development. Trends Plant Sci 8:453–459

Huyluoğlu Z, Ünal M, Palavan-Ünsal N (2008) Cytological evidences of the role of meta-topolin and benzyladenin in barley root tips. Adv Cell Mol Biol 1:31–37

Igwe EO, Charlton KE (2016) A systematic review on the health effects of plums (*Prunus domestica* and *Prunus salicina)*. Phytother Res 30:701–731

Jackson AL, Chen R, Loeb LA (1998) Induction of microsatellite instability by oxidative DNA damage. Proc Natl Acad Sci U S A 95:12468–12473

Kamínek M, Vanek T, Motyka V (1987) Cytokinin activities of N 6-benzyladenosine derivatives hydroxylated on the side-chain phenyl ring. J Plant Growth Regul 6:113–120

Kevers C, Franck T, Strasser RJ, Dommes J, Gaspar T (2004) Hyperhydricity of micropropagated shoots: a typically stress-induced change of physiological state. Plant Cell Tissue Organ Cult 77:181–191

Koduru S, Grierson DS, Afolayan AJ (2007) Ethnobotanical information of medicinal plants used for treatment of cancer in the Eastern Cape Province, South Africa. Curr Sci 92:906–908

Kőszeghi S, Bereczki C, Balog A, Benedek K (2014) Comparing the effects of benzyladenine and *meta*-topolin on sweet basil (*Ocimum basilicum*) micropropagation. Not Sci Biol 6:422–427

Kubalakova M, Strnad M (1992) The effects of aromatic cytokinins (populins) on micropropagation and regeneration of sugar beet *in vitro*. Biol Plant 34:578–579

Kupchan SM, Baxter RL (1975) Mezerein: antileukemic principle isolated from *Daphne mezereum* L. Science 187:652–653

Lata H, Chandra S, Techen N, Khan IA, ElSohly MA (2016) *In vitro* mass propagation of *Cannabis sativa* L.: a protocol refinement using novel aromatic cytokinin *meta*-topolin and the assessment of eco-physiological, biochemical and genetic fidelity of micropropagated plants. J Appl Res Med Aromat Plants 3:18–26

Lázaro-Castellanos JO, Mata-Rosas M, González D, Arias S, Reverchon F (2018) *In vitro* propagation of endangered *Mammillaria* genus (Cactaceae) species and genetic stability assessment using SSR markers. In Vitro Cell Dev Biol Plant 54:518–529

Madzikane-Mlungwana O, Moyo M, Aremu AO, Plíhalová L, Doležal K, Van Staden J, Finnie JF (2017) Differential responses to isoprenoid, N 6-substituted aromatic cytokinins and indole-3-butyric acid in direct plant regeneration of *Eriocephalus africanus*. Plant Growth Regul 82:103–110

Mahomed IM, Ojewole JAO (2003) Hypoglycemic effect of *Hypoxis hemerocallidea* corm (African potato) aqueous extract in rats. Meth Find Exp Clin Pharmacol 25:617–624

Martorell C, Peters EM (2005) The measurement of chronic disturbance and its effects on the threatened cactus *Mammillaria pectinifera*. Biol Conserv 124:199–207

Masondo NA, Aremu AO, Finnie JF, Van Staden J (2015) Growth and phytochemical levels in micropropagated *Eucomis autumnalis* subspecies *autumnalis* using different gelling agents, explant source, and plant growth regulators. In Vitro Cell Dev Biol Plant 51:102–110

McCartan SA, Van Staden H (1998) Micropropagation of the medicinal plant, *Scillana talensis* planch. Plant Growth Regul 25:177–180

McCown BH, Lloyd G (1981) Woody Plant medium (WPM) - a mineral nutrient formulation for microculture of woody plant species. HortScience 16:453–453

McGaw BA, Horgan R, Heald JK (1985) Cytokinin metabolism and the modulation of cytokinin activity in radish. Phytochemistry 24:9–13

Mechoulam R, McCallum NK, Burstein S (1976) Recent advances in the chemistry and biochemistry of cannabis. Chem Rev 76:75–112

Michler CH, Woeste KE, Pijut PM (2007) Black walnut. In: Kole C (ed) Forest trees. Springer, Berlin, pp 189–198

Miller CO, Skoog F, Von Saltza MH, Strong FM (1955) Kinetin, a cell division factor from deoxyribonucleic acid. J Am Chem Soc 77:1392–1392

Miller CO, Skoog F, Okumura FS, Von Saltza MH, Strong FM (1956) Isolation, structure and synthesis of kinetin, a substance promoting cell division. J Am Chem Soc 78:1375–1380

Mok MC (1994) Cytokinins and plant development. In: Cytokinins: chemistry, activity and function. CRC Press, Boca Raton, pp 155–166

Mok MC, Mok DW (1985) The metabolism of [^{14}C]-thidiazuron in callus tissues of *Phaseolus lunatus*. Physiol Plant 65:427–432

Moyo M, Amoo SO, Aremu AO, Gruz J, Subrtova M, Dolezal K, Van Staden J (2014) Plant regeneration and biochemical accumulation of hydroxybenzoic and hydroxycinnamic acid derivatives in *Hypoxis hemerocallidea* organ and callus cultures. Plant Sci 227:157–164

Murashige T, Skoog F (1962) A revised medium for rapid growth and bio assays with tobacco tissue cultures. Physiol Plant 15:473–497

Naaz A, Hussain SA, Anis M, Alatar AA (2019) Meta-topolin improved micropropagation in *Syzygium cumini* and acclimatization to *ex vitro* conditions. Biol Plant 63:174–182

Nacheva L, Gercheva P, Ivanova V, Ibrahim O (2017) *Meta*-topolin improves lateral bud proliferation in micropropagation of *Ginkgo biloba* L. In: VI International symposium on production and establishment of micropropagated plants, vol 1155, pp 355–360

Naidoo D, Aremu AO, Van Staden J, Finnie JF (2017) *In vitro* plant regeneration and alleviation of physiological disorders in *Scadoxus puniceus*. S Afr J Bot 109:316–322

Ncube B, Finnie JF, Van Staden J (2011) Seasonal variation in antimicrobial and phytochemical properties of frequently used medicinal bulbous plants from South Africa. S Afr J Bot 77:387–396

Ndhlala AR, Amoo SO, Ncube B, Moyo M, Nair JJ, Van Standen J (2013) Antibacterial, antifungal, and antiviral activities of African medicinal plants. In: Kuete V (ed) Medicinal plant research in Africa. Elsevier Science, Amsterdam, pp 621–659

Nirmalram JS, Tamilselvi D, Muthiah PT (2011) Hydrogen bonded supramolecular assembly in N^6-benzyladeninium nitrate and N^6-benzyladeninium 3-hydroxy picolinate: a synthetic cytokinin. J Chem Crystallogr 41:864–867

Noodeen LD, Leopold AC (eds) (1988) Senescence and aging in plants. Academic Press, San Diego, p 525

Nowakowska K, Pacholczak A, Tepper W (2019) The effect of selected growth regulators and culture media on regeneration of *Daphne mezereum* L. 'Alba'. Rend Lincei Sci Fis Nat 30:197–205

Ojewole JA (2006) Antinociceptive, anti-inflammatory and antidiabetic properties of *Hypoxis hemerocallidea* Fisch. & CA Mey. (Hypoxidaceae) corm ['African Potato'] aqueous extract in mice and rats. J Ethnopharmacol 103:126–134

Okie WR, Ramming DW (1999) Plum breeding worldwide. HortTechnology 9:162–176

Ornellas TS, Werner D, Holderbaum DF, Scherer RF, Guerra MP (2015) Effects of vitrofural, BAP and *meta*-Topolin in the *in vitro* culture of *Dendrocalamus asper*. In: VI International symposium on production and establishment of micropropagated plants, vol 1155, pp 285–292

Ortega-Baes P, Godínez-Alvarez H (2006) Global diversity and conservation priorities in the Cactaceae. Biodivers Conserv 15:817–827

Owira PM, Ojewole JA (2009) 'African potato' (*Hypoxis hemerocallidea* corm): a plant-medicine for modern and 21st century diseases of mankind? - a review. Phytother Res 23:147–152

Pâques M (1991) Vitrification and micropropagation: causes, remedies and prospects. In: International symposium on plant biotechnology and its contribution to plant development, multiplication and improvement, vol 289, pp 283–290

Pérez-Tornero O, Burgos L (2000) Different media requirements for micropropagation of apricot cultivars. Plant Cell Tissue Organ Cult 63:133–141

Piagnani C, Zocchi G, Mignani I (1996) Influence of Ca^{2+} and 6-benzyladenine on chestnut (*Castanea sativa* Mill.) *in vitro* shoot-tip necrosis. Plant Sci 118:89–95

Radha MH, Laxmipriya NP (2015) Evaluation of biological properties and clinical effectiveness of *Aloe vera*: a systematic review. J Tradit Complement Med 5:21–26

Rafighi A, Tabarsa T (2011) Manufacturing high performance wood composite panel from *Paulownia*. Key Eng Mater 471:1091–1094

Ranade SS, Thiagarajan P (2015) A review on *Persea americana* Mill. (avocado) - its fruits and oil. Int J Pharm Tech Res 8:72–77

Rekha K, Rao RR, Pandey R, Prasad KR, Babu KS, Vangala JR, Kalidevi SV, Rao JM (2013) Two new sesquiterpenoids from the rhizomes of *Nardostachys jatamansi*. J Asian Nat Prod Res 15:111–116

Revilla MA, Majada J, Rodriguez R (1989) Walnut (*Juglansregia* L.) micropropagation. Ann Sci For 46:149–151

Saha S, Dey T, Ghosh P (2010) Micropropagation of *Ocimum kilimandscharicum* Guerke (Labiatae). Acta Biol Crac Ser Bot 52:50–58

Sahoo Y, Pattnaik SK, Chand PK (1997) *In vitro* clonal propagation of an aromatic medicinal herb *Ocimum basilicum* L. (sweet basil) by axillary shoot proliferation. In Vitro Cell Dev Biol-Plant 33:293–296

Scaltsoyiannes A, Tsoulpha P, Panetsos KP, Moulalis D (1998) Effect of genotype on micropropagation of walnut trees (*Juglansregia*). Silvae Genet 46:326–331

Shahzad A, Akhtar R, Bukhari NA, Perveen K (2017) High incidence regeneration system in *Ceratonia siliqua* L. articulated with SEM and biochemical analysis during developmental stages. Trees 31:1149–1163

Sharaibi OJ, Afolayan AJ (2017) Phytochemical analysis and toxicity evaluation of acetone, aqueous and methanolic leaf extracts of *Agapanthus praecox* Willd. Int J Pharm Sci Res 8:5342–5348

Smulders MJM, De Klerk GJ (2011) Epigenetics in plant tissue culture. Plant Growth Regul 63:137–146

Stevens ME, Pijut PM (2018) Rapid *in vitro* shoot multiplication of the recalcitrant species *Juglans nigra* L. In Vitro Cell Dev Biol Plant 54:309–317

Strnad M, Peters W, Hanuš J, Beck E (1994) *Ortho*-topolin-9-glucoside, an aromatic cytokinin from *Populus* x *canadensis* cv *Robusta* leaves. Phytochemistry 37:1059–1062

Strnad M, Hanuš J, Vaněk T, Kaminek M, Ballantine JA, Fussell B, Hanke DE (1997) Meta-topolin, a highly active aromatic cytokinin from poplar leaves (*Populus canadensis* Moench., cv. *Robusta*). Phytochemistry 45:213–218

Trávníček Z, Marek J, Doležal K, Strnad M (1997) The structure of 6-(2-hydroxybenzylamino) purine acetic acid solvate. Z Kristallogr Cryst Mater 212:538–541

Van Sambeek JW, Lambus LJ, Khan SB, Preece JE (1997) *In vitro* establishment of tissues from adult black walnut. General Technical Report-North Central Forest Experiment Station, USDA Forest Service, NC191, pp 78–92

Varga CA, Veale DJH (1997) Isihlambezo: utilization patterns and potential health effects of pregnancy-related traditional herbal medicine. Soc Sci Med 44:911–924

Vijayakumar J, Ponmanickam P, Samuel P, Sudarmani DNP, Pandyarajan J (2017) Influence of *Meta*-Topolin on efficient plant regeneration via micropropagation and organogenesis of safflower *(Carthamus tinctorius* L.) cv. NARI-H-15. Am J Plant Sci 8:688–705

Werbrouck SPO (2010) Merits and drawbacks of new aromatic cytokinins in plant tissue culture. Acta Hortic 865:103–107

Werbrouck SPO, Van der Jeugt B, Dewitte W, Princen E, Van Onckelen HA, Debergh PC (1995) The metabolism of benzyladenine in *Spathiphyllum floribundum* schott 'petite' in relation to acclimatization problems. Plant Cell Rep 14:662–665

Werbrouck SPO, Strnad M, Van Onckelen HA, Debergh PC (1996) Meta-topolin, an alternative to benzyladenine in tissue culture? Physiol Plant 98:291–297

Werner T, Motyka V, Strnad M, Schmulling T (2001) Regulation of plant growth by cytokinin. Proc Nat Acad Sci U S A 98:10487–10492

Meta-topolin: Advantages and Disadvantages for In Vitro Propagation 11

Yulianna G. Zaytseva, Elena V. Ambros, and Tatyana I. Novikova

Contents

Abstract

Discoveries of new plant growth regulators are significantly enriching the range of tools used by researchers in plant tissue culture. *Meta*-topolin (*m*T, 6-(3-hydroxybenzylamino) purine), firstly isolated from poplar leaves, is a highly active natural aromatic cytokinin successfully used over the last two decades in many different biological fields. The main criteria of its efficiency in plant micropropagation are high multiplication rate, shoot health, good rooting, and acclimatization ability. The absence of abnormalities and somaclonal variation are further advantages. For this reason, *m*T is examined in various regeneration systems in comparison with isoprenoid, aromatic, as well as phenylurea derivatives with cytokinin-like activity. This chapter will review the advantages and disadvantages of *m*T application as an inducer of different morphogenetic

Y. G. Zaytseva (✉) · E. V. Ambros · T. I. Novikova
Central Siberian Botanical Garden, Siberian Branch of the Russian Academy of Sciences, Novosibirsk, Russian Federation

N. Ahmad, M. Strnad (eds.), *Meta-topolin: A Growth Regulator for Plant Biotechnology and Agriculture*, https://doi.org/10.1007/978-981-15-9046-7_11

pathways of plant regeneration in vitro, including axillary shoot proliferation, de novo organogenesis, and somatic embryogenesis.

Keywords

Meta-topolin · Axillary shoot proliferation · Adventitious shoot regeneration · Somatic embryogenesis · In vitro abnormalities

11.1 Introduction

Cytokinins (CKs) are one of the most important groups of plant growth regulators (PGRs), which play a crucial role in a plethora of developmental events in plants including cell division and differentiation, initiation and development of shoots, vascular system formation, chloroplast biogenesis, as well as coordination of many physiological processes (Krikorian 1995). Despite their functional diversity, there is still much to be studied, and plant tissue culture (PTC) represents an interesting experimental system for revealing their possibilities. Modeling morphogenetic processes in PTC in the absence of organismic control opens new horizons in exploring CK regulation mechanisms in plant systems. Nowadays, there are a number of natural and synthetic CKs, which can be divided into two groups of compounds: adenine derivatives with isoprenoid or aromatic side chains and phenylurea derivatives (Van Staden et al. 2008). Of adenine types the most commonly used in PTC is N^6-benzyladenine (BA) characterized by high efficiency and affordability. At the same time, application of BA for stimulating shoot production in a wide range of woody and herbaceous species sometimes results in hyperhydricity, shoot-tip necrosis, poor rooting, and low survival rate (Bairu et al. 2009; Amoo et al. 2011). Further, high levels of BA have been frequently considered to be responsible for genetic variability of regenerated plantlets, although clonal propagation and conservation of elite genotypes with superior characteristics as well as rare and endangered plants require the clonal fidelity of progeny (Oono 1985; Giménez et al. 2001; Krishna et al. 2016).

With the discovery of the natural aromatic cytokinin, *meta*-topolin (*m*T) isolated from poplar leaves (topol-czech for poplar tree), an effective tool for inducing a wide array of morphogenic responses, was made available (Strnad et al. 1997). In a comprehensive review, Aremu et al. (2012a) evaluated the use of topolins in PTC with focus on structure–activity relations and metabolism. The authors also provided a detailed analysis of the effects of topolins in comparison with other CKs on a wide range of species. This chapter reviews the advantages and disadvantages of *m*T application in various PTC systems to better understand its mode of action and function during morpho- and organogenesis.

11.2 Shoot Production

11.2.1 Axillary Shoot Proliferation

Currently, the most frequently used micropropagation technique for commercial mass production of plants is axillary shoot proliferation from isolated apical or axillary buds using relatively high CK concentrations. Search for the most effective CK is thus a major task in plant micropropagation (Werbrouck 2010; Amoo et al. 2011). However, apart from high reproduction rate, it is important to avoid or, at least, minimize in vitro abnormalities at morphological, physiological, and molecular genetic levels caused by in vitro conditions.

BA is the most commonly used adenine-type CK for proliferation of axillary meristems. However, the problems associated with using the PGR has resulted in searching for alternative branching triggers. The discovery of *m*T, as a cytokinin with a BA-like aromatic structure, has stimulated a great deal of research on *m*T use for micropropagation through activation of axillary meristem proliferation (Table 11.1). Some researchers have found *m*T to be twice as effective as BA for inducing shoot growth (Nowakowska et al. 2019). *m*T is the preferred PGR in terms of increasing shoot proliferation in a number of plant genotypes (Table 11.1). It offers major advantages over BA and even thidiazuron (TDZ) for inducing axillary shoot production (Waman et al. 2016). Moreover, in some instances, *m*T has been shown to be necessary for micropropagation of recalcitrant plant genotypes (Wojtania 2010). From the data of Table 11.1, it can be concluded that the effective concentrations of *m*T for axillary shoot production fall within a range of 1–20μM with 5μM as the most frequently used concentration. Exceeding these levels can lead to loss of shoot quality and development of shoot anomalies. The duration of cultivation on media with *m*T may vary, but in most cases, this does not diminish the quality of the microshoot (Table 11.1).

Recent studies have shown that *m*T possesses synergistic effects with auxins and some synthetic PGRs with cytokinin activity. The synergistic effects of *m*T in combination with α-naphthaleneacetic acid (NAA) appeared to be more effective than with BA in the induction of multiple shoots and proliferation from cotyledonary node explants of *Pterocarpus marsupium* (Ahmad and Anis 2019). A synergistic effect of *m*T with indole-3-butyric acid (IBA) was found by Fajinmi et al. (2014) during micropropagation of *Coleonema album* from shoot tips. Moreover, it was shown that *m*T enhanced shoot proliferation from shoot-tip explants of *C. album* compared to traditionally used cytokinins such as BA, kinetin (KIN), and TDZ. The benefits of using *m*T and its derivatives for conservation of encapsulated protocorm-like bodies induced from nodal segments of *Ansellia africana* seedlings have been shown (Bhattacharyya et al. 2018). This study also revealed successful proliferation of topolin-derived shoot and synergistic effects of IBA in regenerating artificial seed-derived plantlets in orchids. Moreover, Baskaran et al. (2018a) reported 91% regeneration from encapsulated shoot tips of *Urginea altissima* based on the synergistic action of *m*T and NAA.

Table 11.1 *m*T-induced axillary shoots proliferation

Plant species or cultivars	Explant type	*m*T-induced axillary shoot proliferation			References
		Optimal PGR concentrations	*m*T exposure	Beneficial effects	
Aloe arborescens	Shoot tips	7.5 μM *m*T	6 weeks	Shoots proliferation Rooting Antioxidant activity Photosynthetic pigment content Secondary metabolite production	Amoo et al. (2014)
Actinidia chinensis	Nodal segments	2.64 μM *or* 1.8 μM *m*T	12 weeks	Shoot proliferation Increase in shoot weight, leaf number, and area Acclimatization capacity	Saeiahagh et al. (2019)
Amelanchier alnifolia	Microshoots	20 μM *m*T	8 weeks	Shoot proliferation High levels of endogenous CK metabolites and free IAA	Moyo et al. (2018)
Cannabis sativa	Nodal segments	2.0 μM *m*T	4–6 weeks	Shoot proliferation Rooting Acclimatization capacity Secondary metabolite production	Lata et al. (2016)
Carthamus tinctorius	Shoot tips and nodes	3.0 mg/L *m*T + 0.5 mg/L CPPU	45 days	Shoot proliferation Shoot growth	Vijayakumar et al. (2017)
Citrus 'Ruby Blood'	Microcuttings Encapsulation of microcuttings	1.0 mg/L *m*T + 1 mg/L NAA 10 mg/L *m*T + 5 mg/L IBA	45 days	Sprouting Shoot proliferation Shoot elongation	Chiancone et al. (2016)
Coleonema album	Shoot tips	5 μM *m*T	12 weeks	Shoots proliferation	Fajinmi et al. (2014)

Corylus colurna	Microshoots	0.049 µM IBA, 0.29 µM GA_3 + 8.2 µM *m*T	60 days	Shoot proliferation Shoot elongation Absent tip necroses and hyperhydricity Reduced symptoms of apical necrosis of rooted plants Higher chlorophyll content Modulate antioxidant enzymes activity	Gentile et al. (2017)
Dendrobium nobile	Nodal segments	1 mg/L *m*T + 0.8 mg/L putrescine	8 weeks	Shoot induction	Bhattacharyya et al. (2016)
Eucalyptus grandis × *Eucalyptus urophylla* hybrids	Shoot tips and nodes	0.2 mg/L *m*T	5 weeks	Less vitrified shoots Rooting	Van der Westhuizen (2014)
Harpagophytum procumbens	Shoot tips	2.5 µM *m*T	6 weeks	Shoots proliferation Rooting Antioxidant activity Photosynthetic pigment content Secondary metabolite production	Amoo et al. (2014)
Hypoxis hemerocallidea	Shoots	10 µM *m*T	8 weeks	Shoot proliferation Shoot elongation	Moyo et al. (2014)
Lycopersicon esculentum	Shoot tips	1.0 mM *m*T	12 weeks	Shoot proliferation Leaves number Shoot elongation	Al-Kaaby (2016)
Malaxis wallichii	Transverse thin cell layer of pseudostem segments with one or two nodes	1.0 mg/L *m*T + 0.5 mg/L NAA	4 weeks	Regeneration frequency Plant height Plant growth Rooting Acclimatization capacity	Bose et al. (2017)

(continued)

Table 11.1 (continued)

Plant species or cultivars	Explant type	*m*T-induced axillary shoot proliferation			References
		Optimal PGR concentrations	*m*T exposure	Beneficial effects	
Malus x domestica 'Golden Delicious'	Meristems (0.3–0.4 mm)	1 mg/L *m*T	8 weeks	Increase of survival and regeneration after cryopreservation Decrease of hyperhydricity	Poisson et al. (2016)
Mesomelaena pseudostygia	Zygotic embryos	1.0 µM *m*T or 1.0 µM BA	6 weeks	–	Lai et al. (2014)
Musa spp. AAA cv. 'Williams'	Shoot tips	10 µM *m*T	70 days	Higher total chlorophyll/ carotenoid ratios Anti-senescence Acclimatization capacity	Aremu et al. (2012b)
Nardostachys jatamansi	Shoot tips	1.0 mg/L *m*T	4 weeks	Shoot proliferation Photosynthetic pigment contents Decrease of hyperhydricity	Bose et al. (2016)
Pelargonium x hortorum	Axillary bud Shoot tips	0.5–1 mg *m*T	24 weeks	Inhibition of senescence Shoot proliferation For recalcitrant genotypes	Wojtania (2010)
Pistacia vera	Nodal segments	2 mg/L *m*T	30 days	Shoot proliferation	Benmahioul et al. (2016)
	Shoot tips	5 µM *m*T + 2 µM KIN	6 weeks	Shoot quality Shoot elongation Inhibition of shoot-tip necrosis	Marìn et al. (2016)
Prunus domestica L. and *Prunus insititia* × *domestica*	Microshoots	2.1 µM *m*T	45 days	Shoot quality Rooting	Gentile et al. (2014)
Prunus insititia × *domestica* 'Ferdor' *P. domestica* 'Torinel'	Shoots	2.1 µM *m*T	63 days	Large leaves Shoot elongation Increase of internodes	Monticelli et al. (2017)

Pterocarpus marsupium	Nodal segments	5.0 µM *m*T + 1.0 µM NAA	4 weeks	Shoot elongation after TDZ treatment Shoot multiplication	Ahmad et al. (2018)
Pterocarpus marsupium	Cotyledonary nodes	7.5 µM *m*T + 1.0 µM NAA	12 weeks	Shoot multiplication Shoot proliferation Rooting Acclimatization capacity Biomass content Photosynthetic pigments	Ahmad and Anis (2019)
Sisyrinchium laxum	In vitro germinated seedlings	3.3 µM *m*T	30 days	Shoot proliferation Shoot quality	Ascough et al. (2011)
Syzygium cumini	Nodal segments	5.0 µM *m*T + 2.0 µM NAA	12 weeks	Shoot proliferation Activities of antioxidant enzymes Rooting Acclimatization capacity Decrease of hyperhydricity	Naaz et al. (2019)
Ulmus glabra	Multi-apex cultures	2.1 µM *m*T	12 weeks	Shoot proliferation Secondary metabolite production (higher concentrations of *O*-glucosides) Decrease of ethylene content (senescence) Increase efficiency of photosynthesis	Malá et al. (2013)
Urginea altissima	Synthetic seeds from shoot tips	10 µM *m*T + 2 µM NAA	4 weeks	Increase of shoot-sprouting frequency Shoot proliferation Shoot elongation	Baskaran et al. (2018a)
Withania coagulans	Nodal segments	2.50 mg/L *m*T + 50 mg/L AdS + 0.1 mg/L NAA	21 days	Shoot elongation	Joshi et al. (2016)

The synthetic PGRs based on phenylurea (TDZ and *N*-(2-chloro-4-pyridyl)-*N*-phenylurea, CPPU) are considered to be powerful tools for inducing proliferation of axillary meristem without using exogenous auxins or CKs of the adenine type. However, during micropropagation of *Carthamus tinctorius*, it has been shown that the combination of CPPU and *m*T promotes high plant regeneration from shoot-tip explants (Vijayakumar et al. 2017). Transferring TDZ-pretreated nodal explants of *Pterocarpus marsupium* to the medium augmented with *m*T (5.0μM) in combination with NAA (1.0μM) is also the best way to enhance the proliferation rate as well as average shoot length (Ahmad et al. 2018). It is well-known that TDZ-induced shoots demonstrate abnormalities such as shoot shortening and fasciation, hyperhydricity, abnormal leaf morphology, as well as inhibition of root development (Novikova and Zaytseva 2018; Dewir et al. 2018). Thus, the use of *m*T in terms of its ability to minimize tissue culture-derived disorders and promote shoot elongation may be a new approach for overcoming the physiological and developmental aberrations caused by TDZ.

Transverse thin cell layer (TCL) culture of preexisting meristems is a powerful technique for use in plant micropropagation. BA and TDZ are of very effective for inducing morphogenesis from explants of small size. However, *m*T in combination with NAA was found to be capable of inducing shoot proliferation in TCL culture of *Malaxis wallichii* and more effective than conventional cytokinins (BA, KIN, N^6-isopentenyladenine—iP) for improving regeneration frequency, plant height and growth of regenerated shoots, as well as in vitro rooting and acclimatization (Bose et al. 2017). Moreover, Poisson et al. (2016) showed that *m*T promotes micropropagation of *Malus x domestica* 'Golden Delicious' and its survival, regeneration, and hyperhydricity after cryopreservation. Thus, *m*T due to its ability to reduce tissue hyperhydration may be the best choice for improving shoot proliferation from meristems after cryopreservation.

However, the success of *m*T application has been shown to be strongly dependent on the plant genotype. For instance, in banana tissue culture, the use of *m*T promoted shoot proliferation in the genome group of CEMSA 3/4, Williams, and Grande Naine (Bairu et al. 2009; Escalona et al. 2003), but it was not suitable for the AB genome of 'Elakki Bale' where it resulted into poor multiplication (Bohra et al. 2016). The role of the genotypes and variety in response to *m*T treatment has also been shown to be important and highly species-dependent in the cultivars of *Prunus* spp. and *Pelargonium x hortorum* (Monticelli et al. 2017; Wojtania 2010). Moreover, there was no increase in proliferation frequency or shoot number per explant under *m*T treatment compared with other adenine or phenyl urea CKs in *Eucomis autumnalis* (Ascough et al. 2011; Masondo et al. 2014), *Daphne mezereum* (Nowakowska et al. 2019), and *Pinus radiata* (Montalbán et al. 2011). The effects were comparable with those obtained using BA in *Mesomelaena pseudostygia* (Lai et al. 2014) and *Pelargonium x hortorum* (Wojtania 2010). In conclusion, the concentrations and combinations of different PGRs must be tested with respect to plant genotype in order to achieve the maximum level of axillary shoot production, and *m*T is not an exception.

11.2.2 Adventitious Shoot Organogenesis

De novo shoot organogenesis via direct or indirect pathways from various types of explants including leaves, petioles, hypocotyl segments, roots, and other type of explants cultured on different basal media supplemented with *m*T alone or in combination with other PGRs has been well demonstrated for various species (Table 11.2). Although the formation of adventitious buds is not desirable for clonal propagation, de novo organogenesis is a key step in the application of genetic engineering techniques. For example, Chauhan and Taylor (2018) developed simple and rapid plant regeneration systems for de novo shoot organogenesis from cassava leaf, petiole, and stem internode explants using *m*T. Application of *m*T alone was capable of inducing shoot regeneration, whereas a two-stage system combining 1μM *m*T with 1μM 2,4-D in a first stage medium, followed by subculture at elevated levels of *m*T (6μM), was superior for inducing multiple shoot regeneration events in more than 35% of explants in the cultivar TME 7. Furthermore, *m*T was also found to be beneficial for stimulating shoot regeneration from somatic embryos and cotyledon explants. Developed protocols are applicable for potential genetic transformation and gene editing technologies in cassava (Chauhan and Taylor 2018).

Vijayakumar et al. (2017) also reported optimized *m*T-based regeneration protocols for new genotypes and cultivars of safflower (*Carthamus tinctorius*) cv. NARI-H-15 to improve agronomic properties through an *Agrobacterium*-mediated genetic transformation system. They established a protocol in which addition of 4.0 mg/L *m*T and 1.5 mg/L CPPU to Murashige and Skoog basal medium (MS) stimulated 100% organogenic response (74.7 shootlets) without formation of hyperhydricity from immature leaf calli, as well as cotyledon and stem-derived calli formation after 45 days of culture.

In vitro regeneration systems via organogenesis (direct and indirect) from semi-solid and liquid culture and synthetic seed technology were developed for a threatened perennial bulbous medicinal plant *Urginea altissima* (Baskaran et al. 2018a). The highest number of shoots through direct adventitious regeneration was achieved on liquid MS medium supplemented with 10μM *m*T in combination with 2μM BA followed by transfer of shoots to semi-solid MS media. Further, application of 10μM *m*T and 2μM NAA proved to be the best for sprouting *U. altissima* synthetic seeds and resulted in significantly higher shoot numbers (12.6) and their lengths (7.4 cm) than any other treatment (Baskaran et al. 2018a).

A number of studies have reported the greater efficacy of *m*T alone or in combinations with various CKs or auxins in comparison with BA or other classical CKs for inducing de novo shoot organogenesis (Table 11.2). However, some contradictory results after applying *m*T for shoot regeneration have also been found (Akhtar and Shahzad 2019; Košmrlj et al. 2015; Montalbán et al. 2011; Naidoo et al. 2017). These data confirm that even a slight structural difference between BA and *m*T can have a profound impact on de novo plant regeneration.

Table 11.2 *m*T-induced adventitious shoot regeneration

Plant species/cultivars	Explant type	Morphogenesis pathway	*m*T-induced adventitious shoot regeneration: Optimal PGR concentrations	Exposure	Beneficial effects	References
Carthamus tinctorius	Cotyledons, leaves, and stems	Indirect organogenesis	1. 1.5 mg/L NAA or 1.5 mg/L CPPU	60 days	Shoot regeneration	Vijayakumar et al. (2017)
			2. 4.0 mg/L *m*T + 1.5 mg/L CPPU	45 days	Shoot growth	
					Decrease of hyperhydricity	
					In vitro rooting acclimatization capacity	
Eucomis autumnalis	Leaves	Direct organogenesis	10 μM *m*T	12 weeks	Shoot elongation	Baskaran et al. (2018b)
					Root elongation	
	Leaves and bulbs	Direct organogenesis	4 μM *m*T and 4 μM *m*T + 5 μM NAA	10 weeks	Shoot elongation	Masondo et al. (2015)
					Root elongation	
					Increase of root number	
					Secondary metabolite production (flavonoid content)	
	Leaves	Direct organogenesis	5 μM NAA +	10 weeks	Shoot regeneration	Aremu et al. (2016)
			5 μM *m*T		Shoot number	
					Increase of endogenous CKs content	
Eucomis zambesiaca	Leaves	Direct organogenesis	5 μM NAA +	15 weeks	Shoot regeneration	Aremu et al. (2016)
			5 μM *m*T		Shoot elongation	
					Shoot number	
					Root quality	
					Increase of endogenous CKs content	

Manihot esculenta	Leaf lobes, leaf petioles, petioles, and stem internodes	Direct organogenesis		4 weeks	Shoot regeneration	Chauhan and Taylor (2018)
				3 weeks		
	Leaf petioles	Indirect organogenesis		4 weeks	Shoot regeneration	
			2. 6 µM *m*T	4 weeks	Shoot number	
					Acclimatization capacity	
	Somatic embryo-derived cotyledons	Direct organogenesis	1. 1 µM 2,4-D + 1 µM *m*T	7 days	Shoot regeneration	
			2. 6 µM *m*T	8 weeks		
Merwilla plumbea	Leaves	Direct organogenesis	1 µM *m*T	12 weeks	Shoot regeneration	Aremu et al. (2013)
					Increase of antioxidant activity	
Mammillaria hernandezii	Lateral stem sections	Indirect organogenesis and direct organogenesis	2.2 µM *m*T	4 weeks	Shoot regeneration	Lázaro-Castellanos et al. (2018)
					Shoot elongation	
					Clonal fidelity	
M. dixanthocentron and *M. lanata*	Lateral stem sections	Indirect organogenesis and direct organogenesis	1.1 µM *m*T	4 weeks		
Nardostachys jatamansi	Leaves and petioles	Indirect organogenesis	1. 1.5 mg/L NAA + 1.0 mg/L *m*T	3 weeks	Callus growth	Bose et al. (2016)
			2. 1.0 mg/L *m*T + 0.5 mg/L NAA	4 weeks	Shoot regeneration	
					Decrease of hyperhydricity	
					High photosynthetic pigment contents	
					Clonal fidelity	

(continued)

Table 11.2 (continued)

Plant species/cultivars	Explant type	Morphogenesis pathway	*m*T-induced adventitious shoot regeneration: Optimal PGR concentrations	Exposure	Beneficial effects	References
Prunus domestica and *Prunus insititia* × *domestica*	Leaves	Indirect organogenesis	1. 2.1 μM *m*T	40 days	Shoot regeneration	Gentile et al. (2014)
			2. 11.35 μM TDZ		Shoot number	
Scadoxus puniceus	Leaves	Direct organogenesis	10 μM *m*T	12 weeks	Shoot elongation	Naidoo et al. (2017)
					Decrease of hyperhydricity	
	Bulb halves derived from the seedlings	Direct organogenesis	5 μM GCA + 5 μM *m*T	12 weeks	Shoot regeneration	
					Shoot growth	
					Reduce exudation of phenolic substances into the medium	
Urginea altissima	Leaves	Direct organogenesis	10 μM *m*T + 2 μM NAA	12 weeks	Shoot regeneration	Baskaran et al. (2018a)
					Shoot number	
					Shoots elongation	
					Roots regeneration	
	Leaves	Indirect organogenesis	10 μM *m*T + 2 μM IAA	12 weeks		
	Longitudinal thin cell layers	Direct organogenesis	10 μM *m*T + 2 μM BA	8 weeks		

11.3 Somatic Embryogenesis

Regardless of the plant species, somatic embryogenesis (SE) occurs through three distinct stages which are independent of each other and are controlled by different factors: embryogenic callus induction, maturation of somatic embryos, and conversion of somatic embryos into plantlets.

The most widely applied and efficient inducer of SE at the initiation stage is an auxinic herbicide, 2,4-dichlorophenoxyacetic acid (2,4-D) (Nic-Can and Loyola-Vargas 2016). This substance possesses not only strong auxin-like activity but also being a herbicide acts as a stress inducer. 2,4-D induces DNA hypermethylation and, therefore, maintains the cells in a highly active mitotic stage and mediates the transition from somatic to embryonic state (Endress 1994). However, a combination of auxin and CK, generally in a 4:1 ratio, successfully induces SE in numerous plant species (Van Staden et al. 2008). Raemakers et al. (1995) revealed that SE was induced on CK supplemented media: most often with BA (57%), followed by KIN (37%), zeatin (3%), and TDZ (3%). Currently, TDZ is applied more actively for triggering direct and indirect SE in many plant species (Novikova and Zaytseva 2018). However, there are very few published reports where *m*T is used in the combination with auxins to stimulate SE. Saharan et al. (2011) reported that semi-solid MS medium supplemented with 0.5 mg/L *m*T and 2.5 mg/L 2,4-D initiated embryogenic callus from the root explant of *Balanites aegyptiaca*. Somatic embryos obtained using liquid medium supplied by *m*T were of significantly high plant regeneration frequency (78%) with well-developed shoots and roots (Saharan et al. 2011). The ability of *m*T to induce gametic embryogenesis through microspore reprogramming was observed in an isolated microspore culture in *Citrus clementina* cvs. ‘Monreal Rosso’ and ‘Nules’ (Chiancone et al. 2015). This technology is an efficient approach to obtaining homozygous individuals which can be used for various breeding applications (Germanà et al. 2013). Chiancone et al. (2015) revealed genotype-dependent response to *m*T treatment in an isolated microspore culture of *Citrus clementina*. Further, the authors connected the *m*T effect on the microspore embryogenesis induction with its anti-senescence activity. Although it is not clear how anti-senescence substances influence induction of microspore embryogenesis, this approach can be used to increase SE efficiency (Chiancone et al. 2015). Additionally, *m*T demonstrated some proembryogenic activity in inducing the germination stage in isolated microspore culture of recalcitrant barley genotypes (Esteves et al. 2014). Replacement of BA by *m*T in the regeneration medium was found to result in a 2.9-fold increase in embryo differentiation into green plantlets which indicated their acquisition of photosynthetic capacity. *m*T was shown to be translocated faster in plant tissues in comparison with BA, which prevented its localized accumulation (Kamínek et al. 1997) and reduced callusing and hyperhydricity (Esteves et al. 2014). The use of alternative PGR in both the induction (TDZ and dicamba) and regeneration (*m*T) SE stages resulted in substantial improvement in the efficiency of protocols for barley genotypes. To promote the germination of papaya somatic embryos at the expression stage, various CKs were used. Sprouting was higher in the somatic embryos obtained under 10 μM *m*T compared to the other concentrations evaluated. However, somatic embryos showed very low rooting ability regardless of the CK type (BA or *m*T) and concentration used (Solorzano-Cascante et al. 2018).

High frequency of plant regeneration in *Albizia lebbeck* was developed via cyclic SE resulting in the formation of the secondary somatic embryos originating from the primary ones (Saeed and Shahzad 2015). The authors enhanced the maturation of primary somatic embryos (up to 53.67%) and induced cyclic SE on MS medium containing 5.0μM *m*T, 2.5μM NAA, and 75.0μM glutamine. Consequently, the results indicate that *m*T can be successfully employed to promote various processes of SE including induction, maturation, and germination of somatic embryos.

11.4 Rooting and Acclimatization

The final result of micropropagation depends largely on successful in vitro rooting and ex vitro acclimatization. Rooting of microshoots is a complex and critical process controlled by both endogenous and environmental factors including culture medium, growth regulators, and their combination, genotypes, types of explant, and various culture conditions. Synergistic action of CKs and auxins was reported to control the formation of adventitious roots (Della Rovere et al. 2013). However, in vitro adventitious rooting capacity can be negatively affected by the residual effects of CKs applied during the multiplication phase (Van Staden et al. 2008). This effect relates to the type and the concentration of CKs, which can differently affect the endogenous plant hormone levels. For instance, the positive effect of *m*T versus BA is partly associated with structural advantages of *m*T and resulted not only in different metabolism of the CK but also in increase of rooting and acclimatization competence of micropropagated plantlets (Werbrouck et al. 1996; Aremu et al. 2012a). It was reported that the application of *m*T induced spontaneously rooting (Wojtania 2010) as well as decreased a negative carryover effect on ex vitro acclimatization (Aremu et al. 2012b). The best acclimatization ability of *m*T-treated plantlets could be due to the high shoot quality at multiplication stage. High rhizogenetic response and acclimatization competence (78%) of *C. colurna* shoots were obtained under 4.1μM or 8.2μM *m*T in comparison with BA. Besides, *m*T-induced shoots did not exhibit any signs of shoot-tip necrosis of rooted plants and possessed more functional photosynthetic apparatus during the multiplication stage. These results confirmed a positive role of *m*T on plant survival during acclimatization and on promotion of transition to autotrophic condition through enhancement of chloroplast differentiation, reduction of chlorophyll degradation, and modulation of antioxidant enzyme activities (De Oliveira et al. 2008; Gentile et al. 2017). The best rhizogenetic competency of *m*T in treated microshoots in comparison with other CKs was observed in *Prunus domestica* and *Prunus insititia* × *domestica* (Gentile et al. 2014), *Eucalyptus* species (Van der Westhuizen 2014), and *Syzygium cumini* (Naaz et al. 2019). Moreover, synergistic use of *m*T with some auxins (IBA, NAA) resulted in not only the best shoot proliferation but also a successful in vitro rooting and ex vitro acclimatization of some species, such as

Ansellia africana (Bhattacharyya et al. 2018), *Malaxis wallichii* (Bose et al. 2017), *Pterocarpus marsupium* (Ahmad and Anis 2019), *Urginea altissima* (Baskaran et al. 2018a), *Carthamus tinctorius* (Vijayakumar et al. 2017), and *Eucomis autumnalis* (Masondo et al. 2015) (Tables 11.1 and 11.2).

The indisputable advantage of *m*T is given by the ability to combine of in vitro multiplication and rooting stages. For example, use of 2.0μM *m*T at multiplication stage from *Cannabis sativa* nodal explants provided one-step regeneration system of rapid shoot proliferation, in vitro rooting without application of auxins, and 100% survival frequency of acclimatized plants (Lata et al. 2016). However, the *m*T effect on in vitro rooting and ex vitro acclimatization is highly genotype-specific. For example, it was reported that in some cases *m*T failed to improve rooting and acclimatization capacity. Thus, the number of roots of *Eucomis autumnalis* plantlets was significantly higher on BA and NAA combinations compared to *m*T variants (Baskaran et al. 2018b). Aremu et al. (2016) revealed that iP compared to *m*T had the least inhibitory effect on rooting of bulbous perennial, ornamental, and medicinal species such as *E. zambesiaca*. Chiancone et al. (2016) have not observed positive effect of *m*T on in vitro rooting of *Citrus* 'Ruby Blood' encapsulated microcuttings.

11.5 Overcoming of In Vitro Abnormalities by *m*T

It is known that in vitro cultivation process, as a rule, is accompanied by various types of stresses, such as mechanical damage, high humidity, lack of oxygen and non-compliance with the natural light cycle, increased osmotic pressure, and high concentrations of exogenous PGRs in the nutrient medium (De Klerk 2007). As a result, the unusual in vitro conditions can lead to the formation of specific microplant phenotypes which are characterized by abnormal morphology and anatomy (Pospisilova et al. 1999; Hazarika 2006). The physiological and anatomical anomalies occurring during this process include decreased photosynthetic efficiency, malfunctioning stomata, hyperhydricity (HH), poor cuticle, and epicuticular wax formation (Hazarika 2006). It was reported that *m*T is one of the most promising PGRs among CKs in alleviating shoot abnormalities during the micropropagation of some angiosperms (Aremu et al. 2012b; Amoo and Van Staden 2013; Malá et al. 2013; Amoo et al. 2015).

Hyperhydricity is characterized by large intercellular spaces in the mesophyll cell layer, reduced number of palisade cells, thinning cuticular layer or its lack, less epicuticular wax, fewer stomata on leaves, chloroplasts with small grana, and a lack of starch grains (Smith 2013; Bhatia and Sharma 2015). In general, the main symptoms of HH are chlorophyll *deficiency and* high water content (Kevers et al. 2004). Additionally, HH may also lead to a loss of apical dominance in the shoots due to necrosis of leaf tips and shoots. Frequently, h*yperhydrated* plants demonstrate very low survival rate at the acclimatization step. Treatments that decreased HH are connected with adjusting the relative humidity in the vessel, use of higher concentration of a gelling agent, growth retardants and osmotic agents, combination of lower CK content and ammonium nitrate in the media, use of nitrate or glutamine as the sole nitrogen source, use of effective CKs, etc. (Bhatia and Sharma 2015). The application of *m*T reduced shoot-tip necroses and HH of *Corylus colurna* microshoots in comparison with equimolar BA concentrations, which induced morphological abnormalities in around 40% of plantlets (Gentile et al. 2017). *m*T was shown to be effective in reducing HH in herbal and woody in vitro-derived plants, such as *Malus* x *domestica* cvs. 'Golden Delicious', 'Royal Gala' (Dobránszki et al. 2005; Poisson et al. 2016), *Sisyrinchium laxum* (Ascough et al. 2011), *Huernia hystrix* (Amoo and Van Staden 2013), *Eucalyptus* spp. (Van der Westhuizen 2014), etc. (Tables 11.1 and 11.2). In *Prunus* spp. genotypes *m*T decreased the rate of HH by half in comparison with BA (Gentile et al. 2014). Hyperhydrated shoots of *Nardostachys jatamansi* were not observed even at high concentrations (up to 2.0 mg/L) of *m*T in the culture medium (Bose et al. 2016). On the other hand, increasing *m*T level up to 1.5 mg/L decreased the shoot quality in *Pelargonium* cultures (Wojtania 2010). Bairu et al. (2007) found that the occurrence of HH was enhanced with increasing concentrations (0.5–15μM) of *m*T, zeatin, and BA in *Aloe polyphylla's* PTC. The dependence of HH on *m*T concentrations was also observed in *Pinus radiata*. The highest HH percentage (over 40%) was noticed in some of the BA or *m*T treatments at 10 and 40μM concentration (Montalbán et al. 2011). Zeng (2016) revealed that adenine-type CKs produced less HH than phenylurea derivatives. These results also indicated that application of zeatin can lead to lower HH than in the case of *m*T and BA which showed a similar ability to induce HH in *Arabidopsis thaliana*. Thus, acceptable concentrations minimizing HH are up to 10μM, and this effect is species-dependent.

HH is *associated with* oxidative stress, and this inhibits the growth and development of plantlets. The activities of antioxidant enzymes, such as superoxide dismutase (SOD), catalase (CAT), ascorbate peroxidase (APX), and glutathione reductase (GR), were found to be higher in hyperhydrated than in healthy leaves (Cutler et al. 1991; Papadakis et al. 2001). HH causes a significant decrease in leaf photosynthesis because the antioxidant enzyme system accumulates reactive oxygen species (ROS), such as superoxide anion, singlet oxygen, hydrogen peroxide, and hydroxyl radical, which suppress metabolism in PTC (Cassells and Curry 2001).

The efficiency of photosystem II, quantum efficiency and photochemical quenching were found to be lower in hyperhydrated leaves of apple than healthy ones. This aside, the imbalance between ROS production and antioxidant defense activity resulted in low morphogenetic capacity under in in vitro conditions (Chakrabarty et al. 2005). Several studies have confirmed the antioxidant effects of *m*T in in vitro plant organogenesis. The best effect of *m*T on physiological characteristics in comparison with equimolar BA concentrations was observed in *C. colurna* microshoots where high chlorophyll content, chlorophyll a/b ratio, low total carotenoid content, and CAT activity were revealed (Gentile et al. 2017). Increased total chlorophyll *a* and *a/b* ratio indicated high-intensity photosynthesis due to *m*T application. The highest chlorophyll *a/b* ratio in axillary shoots obtained via regeneration from in vitro apple tree leaves was at 2.0 and 6.0μM *m*T, respectively. Moreover, use of BA (2.2μM) and *m*T (2.2μM) resulted in the highest chlorophyll *a* and chlorophyll *b* content in apple leaves and maximum efficiency of the photochemical processes (*Fv*/*Fm* increased significantly) (Dobránszki and Mendler-Drienyovszki 2014). *Ulmus glabra* explants cultured on the *m*T medium also had a higher *Fv*/*Fm* than those on the BA medium (Malá et al. 2013). Levels of chlorophyll *a* and *b* in leaves of *m*T-treated *Prunus* shoots were also higher at the multiplication stage than BA-treated leaves (Gentile et al. 2014). Amoo et al. (2014) revealed that in *Aloe arborescens* plantlets, the levels of chlorophyll *a*, chlorophyll *b*, total chlorophyll, and total carotenoid decreased with increase in *m*T in nutrient medium from 2.5 to 10.0μM. In *Cannabis sativa*, there were no *statistically* significant differences in photosynthetic pigment content after *m*T application (Lata et al. 2016). In in vitro regenerated *Harpagophytum procumbens*, levels of chlorophyll *a*, chlorophyll *b*, total chlorophyll, and carotenoid in 5.0μM *m*T treatment were significantly higher than control (Amoo et al. 2014). Studies of *Malaxis wallichii* in vitro organogenesis have revealed that *m*T was three times more effective than any other CK tested (BA, KIN, iP) in the ability to protect in vitro-derived plantlets from oxidative stress (Bose et al. 2017). Gentile et al. (2017) showed that the chlorophyll *a* content and chlorophyll *a/b* ratio significantly increased in leaves of *m*T-derived (8.2μM) shoots of *Corylus colurna* over BA-derived shoots, while the chlorophyll *b* content was unaffected by the CK tested. High CAT activity was determined in BA-treated microshoots but not in the *m*T-treated ones; no significant differences were found in POD activity (Gentile et al. 2017). Amoo et al. (2015) observed that unsuitable type or high concentrations of CK could induce high CAT activity in acclimatized plants during the micropropagation of *Merwilla plumbea*. In vitro propagation protocols based on *m*T application can be useful for avoiding morpho-physiological disorders assisting successful ex vitro acclimatization.

11.6 Somaclonal Variability

True-to-type in vitro-derived plants is a priority in clonal micropropagation. The most likely factors causing genetic variability are stress caused by in vitro conditions, the type of explants and morphogenesis pathways (direct or indirect), number of subcultures, as well as the effect of exogenous growth regulators. PGRs significantly alter the metabolism of endogenous phytohormones and antioxidant system function, which can result in somaclonal variability (SV) (Cassells and Curry 2001). The effect of PGRs on SV is closely related to their type, concentration in nutrient medium, and time of exposure. Phenylurea derivatives such as TDZ were found to induce SV in *Cymbidium giganteum* (Roy et al. 2012), *Pilosocereus robinii* (Khattab et al. 2014), and *Teucrium scorodonia* (Makowczyńska et al. 2016). A high concentration (up to 15 mg/L) of BA resulted in an increase in chromosome number in micropropagated banana cultivar 'Williams' (Giménez et al. 2001) and genetic variability of rice callus cultures (Oono 1985). The impact of *m*T causes stress reactions to a lesser extent and, therefore, may reduce the risk of SV occurrence. For instance, the genetic fidelity of *m*T-propagated plants of *Cannabis sativa* was confirmed by Inter Simple Sequence Repeats (ISSR; Lata et al. 2016). High degree of genetic homogeneity among the microshoots of *Ansellia africana* was shown after treatment with *meta*-methoxy-topolin-9-tetrahydropyran-2yl (Me*m*TTHP; Bhattacharyya et al. 2018). Polymorphism was not confirmed during the *Random Amplification of Polymorphic DNA* (RAPD) analysis in *m*T-treated plantlets of *Pterocarpus marsupium* (Ahmad and Anis 2019). The lowest SV levels were detected among *m*T-derived *Actinidia chinensis* var. *chinensis* plantlets versus those obtained from zeatin or BA treatments (Saeiahagh et al. 2019). Thus, substitution of traditionally used CKs by *m*T may be a key to minimizing the appearance of SV. However, in order to reliable estimate *m*T effects on SV, detailed comprehensive investigation is necessary regarding all factors regulating the genetic changes.

11.7 Conclusions

The choice of PGRs, including CKs is reported to play a crucial role in the development of micropropagation systems that assure effective culture establishment, multiplication, rooting, and plantlet acclimatization. The application of BA, a highly potent and widely used CK in vitro, frequently gives rise to morpho-physiological disorders such as hyperhydricity, shoot-tip necrosis, poor rooting ability, oxidative stress, and genetic variability in a number of plant species. *m*T which is slightly different in structure from BA is found to be not only more efficient for shoot induction but does not have these in vitro drawbacks. Further, *m*T enhances chloroplast differentiation, reduces chlorophyll degradation, modulates antioxidant enzyme activities, and, for these reasons, improves rooting and acclimatization capacity. On the other hand, less *m*T efficiency in the clonal micropropagation of some species in comparison with other CKs is also presented. Therefore, even though *m*T is a potent CK for initiating various morphogenic pathways in PTC, all interacting factors need to be considered to achieve optimal results for commercial micropropagation of valuable genotypes.

Acknowledgments The study was carried out with financial support of the budgetary project of Central Siberian Botanical Garden, SB RAS "Assessment of morphogenetic potential of North Asian plant populations by experimental methods" (No. 0312-2016-0001), within the framework of the State Assignment.

References

Ahmad A, Anis M (2019) Meta-topolin improves in vitro morphogenesis, rhizogenesis and biochemical analysis in *Pterocarpus marsupium* Roxb.: a potential drug-yielding tree. J Plant Growth Regul 38:1007–1016. https://doi.org/10.1007/s00344-018-09910-9

Ahmad A, Ahmad N, Anis M (2018) Preconditioning of nodal explants in thidiazuron-supplemented liquid media improves shoot multiplication in *Pterocarpus marsupium* (Roxb.). In: Ahmad N, Faisal M (eds) Thidiazuron: from urea derivative to plant growth regulator. Springer, Singapore, pp 175–187. https://doi.org/10.1007/978-981-10-8004-3_8

Akhtar R, Shahzad AJ (2019) Morphology and ontogeny of directly differentiating shoot buds and somatic embryos in *Santalum album* L. J For Res 30:1179–1189. https://doi.org/10.1007/s11676-018-0679-5

Al-Kaaby NKZ (2016) *In Vitro* regeneration of tomato (*Lycopersicon esculentum* Mill.) role of meta-topolin, benzyladenine and casein hydrolysate. J Biol Agric Health 6:7–12
Amoo S, Van Staden J (2013) Influence of plant growth regulators on shoot proliferation and secondary metabolite production in micropropagated *Huernia hystrix*. Plant Cell Tissue Organ Cult 112:249–256
Amoo S, Finnie JF, Van Staden J (2011) The role of *meta*-topolins in alleviating micropropagation problems. Plant Growth Regul 63:197–206
Amoo SO, Aremu AO, Moyo M et al (2014) Physiological effects of a novel aromatic cytokinin analogue in micropropagated *Aloe arborescens* and *Harpagophytum procumbens*. Plant Cell Tissue Organ Cult 116:17–26
Amoo SO, Aremu AO, Moyo M et al (2015) Physiological and biochemical effects of a tetrahydropyranyl-substituted *meta*-topolin in micropropagated *Merwilla plumbea*. Plant Cell Tissue Organ Cult 121:579–590
Aremu AO, Bairu MW, Doležal K et al (2012a) Topolins: a panacea to plant tissue culture challenges? Plant Cell Tissue Organ Cult 108:1–16
Aremu AO, Bairu MW, Szüčová L et al (2012b) The role of *meta*-topolins on the photosynthetic pigment profiles and foliar structures of micropropagated 'Williams' bananas. J Plant Physiol 169:1530–1541
Aremu AO, Gruz J, Subrtov M et al (2013) Antioxidant and phenolic acid profiles of tissue cultured and acclimatized Merwilla plumbea plantlets in relation to the applied cytokinins. J Plant Physiol 170:1303–1308
Aremu AO, Plačková L, Pěnčík A et al (2016) Auxin-cytokinin interaction and variations in their metabolic products in the regulation of organogenesis in two *Eucomis* species. New Biotechnol 33:883–890
Ascough GD, Swart PA, Finnie JF et al (2011) Micropropagation of *Romulea minutiflora, Sisyrinchium laxum* and *Tritonia gladiolaris* - *Iridaceae* with ornamental potential. S Afr J Bot 77:216–221
Bairu MW, Stirk WA, Doležal K et al (2007) Optimizing the micropropagation protocol for the endangered *Aloe polyphylla*: can *meta*-topolin and its derivatives serve as replacement for benzyladenine and zeatin? Plant Cell Tissue Organ Cult 90:15
Bairu MW, Jain N, Stirk WA et al (2009) Solving the problem of shoot-tip necrosis in *Harpagophytum procumbens* by changing the cytokinin types, calcium and boron concentrations in the medium. S Afr J Bot 75:122–127
Baskaran P, Kumari A, Van Staden J (2018a) In vitro propagation via organogenesis and synthetic seeds of *Urginea altissima* (L.f.) Baker: a threatened medicinal plant. 3 Biotech 8:18
Baskaran P, Kumari A, Van Staden J (2018b) Analysis of the effect of plant growth regulators and organic elicitors on antibacterial activity of *Eucomis autumnalis* and *Drimia robusta ex vitro*-grown biomass. Plant Growth Regul 85:143–151
Benmahioul B, Kaïd-Harche M, Daguin F (2016) In vitro regeneration of *Pistacia vera* L. from nodal explants. J For Sci 62:198–203
Bhatia S, Sharma K (2015) Technical glitches in micropropagation. In: Bhatia S, Sharma K, Dahiya R, Bera T (eds) Modern applications of plant biotechnology in pharmaceutical sciences. Academic Press, Boca Raton, pp 393–404. https://doi.org/10.1016/B978-0-12-802221-4.00013-3
Bhattacharyya P, Kumaria S, Tandon P (2016) High frequency regeneration protocol for *Dendrobium nobile*: a model tissue culture approach for propagation of medicinally important orchid species. S Afr J Bot 104:232–243
Bhattacharyya P, Kumar V, Van Staden J (2018) *In vitro* encapsulation based short term storage and assessment of genetic homogeneity in regenerated *Ansellia africana* (Leopard orchid) using gene targeted molecular markers. Plant Cell Tissue Organ Cult 133:299–310
Bohra P, Waman AA, Sathyanarayana BN et al (2016) Influence of different growth regulators on *in vitro* multiplication of mixed diploid banana (*Musa* AB). Proc Natl Acad Sci India Sect B Biol Sci 86:179–185

Bose B, Kumaria S, Choudhury H et al (2016) Assessment of genetic homogeneity and analysis of phytomedicinal potential in micropropagated plants of *Nardostachys jatamansi*, a critically endangered, medicinal plant of alpine Himalayas. Plant Cell Tissue Organ Cult 124:331–349

Bose B, Kumaria S, Choudhury H et al (2017) Insights into nuclear DNA content, hydrogen peroxide and antioxidative enzyme activities during transverse thin cell layer organogenesis and ex vitro acclimatization of *Malaxis wallichii*, a threatened medicinal orchid. Physiol Mol Biol Plants 23(4):955–968

Cassells AC, Curry RF (2001) Oxidative stress and physiological, epigenetic and genetic variability in plant tissue culture: implications for micropropagators and genetic engineers. Plant Cell Tissue Organ Cult 64:145–157

Chakrabarty D, Park SY, Ali MB et al (2005) Hyperhydricity in apple: ultrastructural and physiological aspects. Tree Physiol 26:377–388

Chauhan RD, Taylor NJ (2018) Meta-topolin stimulates de novo shoot organogenesis and plant regeneration in cassava. Plant Cell Tissue Organ Cult 132:219–224

Chiancone B, Karasawa MMG, Gianguzzi V et al (2015) Early embryo achievement through isolated microspore culture in *Citrus clementina* Hort. ex Tan., cvs. 'Monreal Rosso' and 'Nules'. Front Plant Sci 6:413. https://doi.org/10.3389/fpls.2015.00413

Chiancone B, Martorana M, Casales F et al (2016) The effects of benzyladenine and meta-topolin on in vitro sprouting and regrowth after encapsulation of C35 citrange [*Citrus sinensis* (L.) Osb. × *Poncirus trifoliata* (L.) Raf] microcuttings. Acta Hortic 1113:99–106. https://doi.org/10.17660/ActaHortic.2016.1113.14

Cutler A, Saleem M, Wang H (1991) Cereal protoplast recalcitrance. In Vitro Cell Dev Biol 27:104–111

De Klerk GJ (2007) Stress in plants cultured in vitro. Propag Ornam Plants 7:129–137

De Oliveira LM, Paiva R, de Santana JRF et al (2008) Effect of cytokinins on in vitro development of autotrophism and acclimatization of *Annona glabra* L. In Vitro Cell Dev Biol Plant 44:128–135

Della Rovere F, Fattorini L, D'Angeli S et al (2013) Auxin and cytokinin control formation of the quiescent centre in the adventitious root apex of Arabidopsis. Ann Bot 112(7):1395–1407. https://doi.org/10.1093/aob/mct215

Dewir YH, Nurmansyah, Naidoo Y et al (2018) Thidiazuron-induced abnormalities in plant tissue cultures. Plant Cell Rep 37:1451–1470. https://doi.org/10.1007/s00299-018-2326-1

Dobránszki J, Mendler-Drienyovszki N (2014) Cytokinin-induced changes in the chlorophyll content and fluorescence of in vitro apple leaves. J Plant Physiol 17:1472–1478. https://doi.org/10.1016/j.jplph.2014.06.015

Dobránszki J, Jámbor-Benczúr E, Reményi ML et al (2005) Effects of aromatic cytokinins on structural characteristics of leaves and their post-effects on subsequent shoot regeneration from in vitro apple leaves of 'Royal Gala'. Int J Hortic Sci 11:41–46

Endress R (1994) Plant cell biotechnology. Springer, Berlin

Escalona M, Cejas I, Gonzalez-Olmedo J et al (2003) The effect of *meta*-topolin on plantain propagation using a temporary immersion bioreactor. Infomusa 12:28–30

Esteves P, Clermont I, Marchand S et al (2014) Improving the efficiency of isolated microspore culture in six-row spring barley: II-exploring novel growth regulators to maximize embryogenesis and reduce albinism. Plant Cell Rep 33:871–879. https://doi.org/10.1007/s00299-014-1563-1

Fajinmi OO, Amoo SO, Finnie JF et al (2014) Optimization of in vitro propagation of *Coleonema album*, a highly utilized medicinal and ornamental plant. S Afr J Bot 94:9–13. https://doi.org/10.1016/j.sajb.2014.05.006

Gentile A, Jáquez Gutiérrez M, Martinez J et al (2014) Effect of meta-Topolin on micropropagation and adventitious shoot regeneration in *Prunus* rootstocks. Plant Cell Tissue Organ Cult 118:373–381. https://doi.org/10.1007/s11240-014-0489-1

Gentile A, Frattarelli A, Nota P et al (2017) The aromatic cytokinin meta-topolin promotes in vitro propagation, shoot quality and micrografting in *Corylus colurna* L. Plant Cell Tissue Organ Cult 128:693–703. https://doi.org/10.1007/s11240-016-1150-y

Germanà MA, Aleza P, Carrera E et al (2013) Cytological and molecular characterization of three gametoclones of *Citrus clementina*. BMC Plant Biol 13:129. https://doi.org/10.1186/1471-2229-13-129

Giménez C, de García E, de Enrech NX et al (2001) Somaclonal variation in banana: cytogenetic and molecular characterization of the somaclonal variant cien BTA-03. In Vitro Cell Dev Biol Plant 37:217. https://doi.org/10.1007/s11627-001-0038-6

Hazarika BN (2006) Morpho-physiological disorders in *in vitro* culture of plants. Sci Hortic 108:105–120. https://doi.org/10.1016/j.scienta.2006.01.038

Joshi DH, Nekkala S, Soner D et al (2016) *In vitro* shoot multiplication of *Withania coagulans* (stocks). Plant Tiss Cult Biotechnol 26:201–209. https://doi.org/10.3329/ptcb.v26i2.30569

Kamínek M, Motyka V, Vaňková R (1997) Regulation of cytokinin content in plant cells. Physiol Plant 101:689–700. https://doi.org/10.1111/j.1399-3054.1997.tb01053.x

Kevers C, Franck T, Strasser RJ et al (2004) Hyperhydricity of micropropagated shoots: a typically stress-induced change of physiological state. Plant Cell Tissue Organ Cult 77:181–191. https://doi.org/10.1023/B:TICU.0000016825.18930.e4

Khattab S, El Sherif F, El-Garhy HA et al (2014) Genetic and phytochemical analysis of the *in vitro* regenerated *Pilosocereus robinii* by ISSR, SDS–PAGE and HPLC. Gene 533:313–321. https://doi.org/10.1016/j.gene.2013.09.026

Košmrlj K, Kladnik A, Bohanec B (2015) Adventitious regeneration in styrian oil pumpkin in relation to the endoreduplication pattern and induced tetraploidy on fusaric acid-supplemented media. Plant Growth Regul 75:587–594. https://doi.org/10.1007/s10725-014-9961-5

Krikorian AD (1995) Hormones in tissue culture and micropropagation. In: Davies PJ (ed) Plant hormones: physiology, biochemistry and molecular biology. Kluwer Academic, Dordrecht, pp 774–796

Krishna H, Alizadeh M, Singh D et al (2016) Somaclonal variations and their applications in horticultural crops improvement. 3 Biotech 6:54. https://doi.org/10.1007/s13205-016-0389-7

Lai S, Menon A, Turner S et al (2014) Development of an in vitro protocol for a difficult-to-propagate endemic Australian dryland sedge species *Mesomelaena pseudostygia* (Cyperaceae). In Vitro Cell Dev Biol Plant 50:99–109. https://doi.org/10.1007/s11627-013-9542-8

Lata H, Chandra S, Techen N et al (2016) *In vitro* mass propagation of *Cannabis sativa* L.: a protocol refinement using novel aromatic cytokinin *meta*-topolin and the assessment of eco-physiological, biochemical and genetic fidelity of micropropagated plants. J Appl Res Med Aromat Plants 3:18–26. https://doi.org/10.1016/j.jarmap.2015.12.001

Lázaro-Castellanos JO, Mata-Rosas M, González D et al (2018) *In vitro* propagation of endangered *Mammillaria genus* (*Cactaceae*) species and genetic stability assessment using SSR markers. In Vitro Cell Dev Biol Plant 54:518–529. https://doi.org/10.1007/s11627-018-9908-z

Makowczyńska J, Sliwinska E, Kalemba D et al (2016) *In vitro* propagation, DNA content and essential oil composition of *Teucrium scorodonia* L. ssp. scorodonia. Plant Cell Tissue Organ Cult 127:1–13. https://doi.org/10.1007/s11240-016-1024-3

Malá J, Máchová P, Cvrčková H et al (2013) The role of cytokinins during micropropagation of wych elm. Biol Plant 57:174–178. https://doi.org/10.1007/s10535-012-0252-6

Marìn JA, Garcıá E, Lorente P et al (2016) A novel approach for propagation of recalcitrant pistachio cultivars that sidesteps rooting by *ex vitro* grafting of tissue cultured shoot tips. Plant Cell Tissue Organ Cult 124:191–200. https://doi.org/10.1007/s11240-015-0871-7

Masondo NA, Aremu AO, Finnie JF (2014) Plant growth regulator induced phytochemical and antioxidant variations in micropropagated and acclimatized *Eucomis autumnalis* subspecies autumnalis (*Asparagaceae*). Acta Physiol Plant 36:2467–2479. https://doi.org/10.1007/s11738-014-1619-4

Masondo NA, Aremu AO, Finnie JF et al (2015) Growth and phytochemical levels in micropropagated *Eucomis autumnalis* subspecies autumnalis using different gelling agents,

explant source, and plant growth regulators. In Vitro Cell Dev Biol Plant 51:102–110. https://doi.org/10.1007/s11627-014-9646-9
Montalbán IA, De Diego N, Moncaleán P (2011) Testing novel cytokinins for improved in vitro adventitious shoots formation and subsequent *ex vitro* performance in *Pinus radiate*. Forestry 84:363–373. https://doi.org/10.1093/forestry/cpr022
Monticelli S, Gentile A, Frattarelli A et al (2017) Effects of the natural cytokinin *meta*-Topolin on in vitro shoot proliferation and acclimatization of *Prunus* spp. Acta Hortic 1155:349–354. https://doi.org/10.17660/ActaHortic.2017.1155.51
Moyo M, Koetle MJ, Van Staden J (2014) Photoperiod and plant growth regulator combinations influence growth and physiological responses in *Pelargonium sidoides* DC. In Vitro Cell Dev Biol Plant 50:487–492. https://doi.org/10.1007/s11627-014-9594-4
Moyo M, Adeyemi O, Aremu AO et al (2018) Deciphering the growth pattern and phytohormonal content in Saskatoon berry (*Amelanchier alnifolia*) in response to *in vitro* cytokinin application. New Biotechnol 42:85–94. https://doi.org/10.1016/j.nbt.2018.02.001
Naaz A, Hussain SA, Anis M et al (2019) Meta-topolin improved micropropagation in *Syzygium cumini* and acclimatization to *ex vitro* conditions. Biol Plant 63:174–182. https://doi.org/10.32615/bp.2019.020
Naidoo D, Aremu AO, Van Staden J et al (2017) *In vitro* plant regeneration and alleviation of physiological disorders in *Scadoxus puniceus*. S Afr J Bot 109:316–322. https://doi.org/10.1016/j.sajb.2017.01.010
Nic-Can G, Loyola-Vargas VM (2016) The role of the auxins during somatic embryogenesis. In: Loyola-Vargas V, Ochoa-Alejo N (eds) Somatic embryogenesis: fundamental aspects and applications. Springer, Cham, pp 171–182. https://doi.org/10.1007/978-3-319-33705-0_10
Novikova TI, Zaytseva YG (2018) TDZ-induced morphogenesis pathways in woody plant culture. In: Ahmad N, Faisal M (eds) Thidiazuron: from urea derivative to plant growth regulator. Springer, Singapore, pp 61–94
Nowakowska K, Pacholczak A, Tepper W (2019) The effect of selected growth regulators and culture media on regeneration of *Daphne mezereum* L. 'Alba'. Rend Lincei Sci Fis 30:197–205. https://doi.org/10.1007/s12210-019-00777-w
Oono K (1985) Putative homozygous mutations in regenerated plants of rice. Mol Gen Genet 198:377–384
Papadakis AK, Siminis CI, Roubelakis-Angelakis KA (2001) Reduced activity of antioxidant machinery is correlated with suppression of totipotency in plant protoplasts. Plant Physiol 126:434–444. https://doi.org/10.1104/pp.126.1.434
Poisson A-S, Berthelot P, Le Bras C et al (2016) A droplet-vitrification protocol enabled cryopreservation of doubled haploid explants of *Malus* x *domestica* Borkh. 'Golden Delicious'. Sci Hortic 209:187–191. https://doi.org/10.1016/j.scienta.2016.06.030
Pospisilova J, Ticha I, Kadlecek S et al (1999) Acclimatization of micropropagated plants in *ex vitro* conditions. Biol Plant 42:481–497. https://doi.org/10.1023/A:1002688208758
Raemakers CJJM, Jacobsen E, Visser RGF (1995) Secondary somatic embryogenesis and applications in plant breeding. Euphytica 81:93–107
Roy AR, Sajeev S, Pattanayak A et al (2012) TDZ induced micropropagation in *Cymbidium giganteum* Wall. Ex Lindl. and assessment of genetic variation in the regenerated plants. Plant Growth Regul 68:435. https://doi.org/10.1007/s10725-012-9732-0
Saeed T, Shahzad A (2015) High frequency plant regeneration in Indian siris via cyclic somatic embryogenesis with biochemical, histological and SEM investigations. Ind Crop Prod 76:623–637. https://doi.org/10.1016/j.indcrop.2015.07.060
Saeiahagh H, Mousavi M, Wiedow C et al (2019) Effect of cytokinins and sucrose concentration on the efficiency of micropropagation of 'Zes006' *Actinidia chinensis* var. *chinensis*, a redfleshed kiwifruit cultivar. Plant Cell Tissue Organ Cult https://doi.org/10.1007/s11240-019-01597-4
Saharan V, Yadav RC, Yadav NR et al (2011) Somatic embryogenesis and plant regeneration of *Balanites aegyptiaca* Del (L.): an industrial important arid tree. J Cell Tissue Res 11 (1):2529–2534

Smith RH (2013) *In vitro* propagation for commercial production of ornamentals. In: Smith RH (ed) Plant tissue culture, 3rd edn. Elsevier, Amsterdam, pp 127–145. https://doi.org/10.1016/B978-0-12-415920-4.00012-8

Solorzano-Cascante P, Sánchez-Chiang N, Jiménez VM (2018) Explant type, culture system, 6-benzyladenine, meta-topolin and encapsulation affect indirect somatic embryogenesis and regeneration in *Carica papaya* L. Front Plant Sci. https://doi.org/10.3389/fpls.2018.01769

Strnad M, Hanuš J, Vaněk T et al (1997) Meta-topolin, a highly active aromatic cytokinin from poplar leaves (*Populus* x *canadensis* Moench., cv. Robusta). Phytochemistry 45:213–218. https://doi.org/10.1016/S0031-9422(96)00816-3

Van der Westhuizen A (2014) The use of meta-topolin as an alternative cytokinin in the tissue culture of *Eucalyptus* species. Acta Hortic 1055:25–28. https://doi.org/10.17660/ActaHortic.2014.1055.4

Van Staden J, Zazimalova E, George EF (2008) Plant growth regulators II: cytokinins, their analogues and antagonists. In: George EF, Hall MA, De Klerk G-J (eds) Plant propagation by tissue culture, 3rd edn. Springer, Dordrecht, pp 1–501

Vijayakumar J, Ponmanickam P, Samuel P et al (2017) Influence of meta-topolin on efficient plant regeneration via micropropagation and organogenesis of safflower (*Carthamus tinctorius* L.) cv. NARI-H-15. Am J Plant Sci 8:688–705. https://doi.org/10.4236/ajps.2017

Waman AA, Bohra P, Sathyanarayana BN (2016) Effect of various cytokinins and auxins on in vitro regeneration of plantlets from isolated bud clumps of Silk Banana var. Nanjanagud Rasabale (*Musa* AAB). Ind J Plant Physiol 21:64. https://doi.org/10.1007/s40502-016-0204-9

Werbrouck SPO (2010) Merits and drawbacks of new aromatic cytokinins in plant tissue culture. Acta Hortic 865:103–108

Werbrouck SPO, Strnad M, Van Onckelen HA et al (1996) Meta-topolin, an alternative to benzyladenine in tissue culture? Physiol Plant 98:291–297. https://doi.org/10.1034/j.1399-3054.1996.980210.x

Wojtania A (2010) Effect of meta-topolin on *in vitro* propagation of *Pelargonium* × *hortorum* and *Pelargonium* × *hederaefolium* cultivars. Acta Soc Bot Pol 79:101–106. https://doi.org/10.5586/asbp.2010.013

Zeng Z (2016) Cytokinins induce hyperhydricity in leaves of *in vitro* grown *Arabidopsis thaliana*. Msc thesis plant breeding, Wageningen UR Plant Breeding, Netherlands

Role of *Meta*-topolin on in Vitro Shoot Regeneration: An Insight

12

Saikat Gantait and Monisha Mitra

Contents

Abstract

Meta-topolin [6-(3-hydroxybenzylamino)purine)] is an aromatic cytokinin bearing a benzyl ring substituted by a hydroxyl group at *meta*-position. It was first isolated from the leaves of poplar tree. *Meta*-topolin (*m*T) has immense potential for shoot regeneration. Its other beneficial attributes include delaying senescence, preventing shoot-tip necrosis, and evading the effects of hyperhydricity. The concise results obtained from different studies, conducted over the past one decade, distinctively show that the effects of *m*T basically include in vitro shoot induction, shoot proliferation, and increase in shoot length. When used in

S. Gantait (✉)
Crop Research Unit (Genetics and Plant Breeding), Bidhan Chandra Krishi Viswavidyalaya, Mohanpur, Nadia, West Bengal, India
e-mail: saikatgantait@bckv.edu.in

M. Mitra
Department of Agricultural Biotechnology, Faculty of Agriculture, Bidhan Chandra Krishi Viswavidyalaya, Mohanpur, Nadia, West Bengal, India

N. Ahmad, M. Strnad (eds.), *Meta-topolin: A Growth Regulator for Plant Biotechnology and Agriculture*, https://doi.org/10.1007/978-981-15-9046-7_12

combination with auxins, it exhibits an ability to induce regenerative callus. Based on all the beneficial attributes of *m*T, it can be regarded as a potent aromatic cytokinin that can be utilized in micropropagation. Considering its significant application in plant tissue culture, the present chapter intricately describes the nature, usage, and advantages of *m*T on shoot regeneration, in particular.

Keywords

Auxins · Callus · Cytokinin · Hyperhydricity · *Meta*-topolin · Micropropagation · Senescence

12.1 Introduction

Cytokinins are a major class of plant growth hormones that induce cell division in tissues (Koshimizu and Iwamura 1986). Its major properties include impeding the senescence process mainly by preventing protein degradation that further increases the activity of RNase (Hall 1973). The primary function of cytokinins is suppressing the apical dominance and thus culminating a way for the development of buds at lateral positions (Leopold and Kriedemann 1975). Considering the extensive uses of cytokinins, there has been an increase in demand on the usage of synthetic cytokinins having analogous effects to naturally derived cytokinins. A novel class of cytokinins, i.e., topolins, is recently being used in plant tissue culture. Chemically, they are regarded as aromatic cytokinins (ARCKs), and structural studies showed that topolins consist of a hydroxylated benzyl that is attached at N^6-position of adenine (Aremu et al. 2012a). *m*T was first isolated from the leaves of *Populus* × *canadensis*. The nomenclature of the hormone has been derived from the Czech word 'topol' that signifies the plant poplar from where it was first isolated (Strnad et al. 1997). Furthermore, the cytokinin activity depends on the N1 position which should remain free since in the case of hydroxyl at the *ortho*-position, hydrogen bonding occurs between this group and the N1 atom, which in turn makes the *meta*-substituents highly potent (Holub et al. 1998). Henceforth, the name *meta*-topolin [6-(3-hydroxybenzylamino)purine)] is clearly justified. Generally, *m*T serves as substrate for *O*-glucosyltransferase (ZOG1 enzyme found in *Zea mays* and *Phaseolus lunatus*). Additionally, from the reports on correlation between the activity of *m*T and their ability to serve as substrate for ZOG1 enzyme, it can be concluded that there is a similarity between the receptors and the binding sites of *m*T (Mok et al. 2005). The physical attributes of *m*T are, namely, 'solid form' and 'off white' to 'white' colour. During preparation of stock solution, *m*T is readily soluble in water, but other solvents like KOH and dimethyl sulfoxide (DMSO) are also equivalently effective. *m*T also has many multifaceted properties like evading the ill effects of shoot-tip necrosis and hyperhydricity and causing delay in senescence that eventually facilitates multiple shoot regeneration and proliferation (Malá et al. 2013).

12.2 Principles of Regeneration via In Vitro Shoots

The development and regeneration of new organs from different explants are based on the phenomena known as 'totipotency', which means that a single cell can develop into a functional organ. The entire process of shoot organogenesis spans entirely over three broad aspects, which are competence, determination, and morphogenesis (Sugiyama 1999). Competence is regarded as the initial step of shoot organogenesis where cell signalling induced by hormones leads to dedifferentiation of cells (Howell et al. 2003). The next step is determining the identity of the organ fixed by the proportion of plant growth hormone; in case of shoot organogenesis, it is possible through cytokinins (Gahan and George 2008). Morphogenesis is the ultimate step where shoot induction finally occurs (Sugiyama 1999).

Furthermore, the phenomena of shoot induction can be categorized into two pathways, namely, direct and indirect pathway. The direct pathway involves the formation of shoot bud, and there is no formation of callus (Yancheva et al. 2003), whereas indirect pathway involves an intermittent step where formation of callus occurs and from adventitious shoots induces from the respective callus (Gahan and George 2008).

12.3 In Vitro Shoot Regeneration in Plant System

The regeneration of shoots from desired explants, mediated via direct or indirect organogenesis approach, serves as a steadfast methodology for micropropagation (Gahan and George 2008). The regeneration of shoots in multiple numbers depends on the type of explant utilized and also on an array of factors like the nutrients utilized in basal media, the environmental factors, and the type and dosage of plant growth hormone used (Gantait et al. 2014).

12.3.1 Influence of Explant

The fundamental criteria before establishment of in vitro culture is the proper selection of explant that should be free from contamination and can be imparted by physical factors like dust or dirt and biological factors like bacteria, fungi, and other microorganisms (Gantait and Kundu 2017). There are innumerable types of explants utilized in micropropagation for inducing shoot directly. Shoot tips and axillary buds are predominantly used explants in any micropropagation experiment since it is easily available and it maintains the genetic integrity in the in vitro-derived plantlets from the mother plant (Rout et al. 2006). Histological studies showed that the shoot-tip region contains a zone comprising of meristematic cells that exhibit the phenomena called 'totipotency' and possess the ability of accumulating plant growth hormones at an anticipated level (Akin-Idowu et al. 2009). Another desirable explant utilized is nodal segment, which has the capability to generate multiple axillary buds, since bud break can be easily induced in a short period of time with the application of desired level of plant hormones (Gantait et al. 2009). Some other explants apart from shoot tip and nodal segment such as leaf, corm, bulb, epicotyl, hypocotyledon, cotyledon, root, zygotic embryos, and tillers exhibited promising results and displayed a positive correlation with shoot regeneration (Meyer et al. 2009; Nas et al. 2010; Niedz and Evens 2010; Valero-Aracama et al. 2010; De Diego et al. 2011, Moyo et al. 2011; Niedz and Evens 2011a; Baskaran et al. 2012; Aremu et al. 2014; Moyo et al. 2014; Masondo et al. 2014; Ncube et al. 2015; Aremu et al. 2016; Chiancone et al. 2017; Baskaran et al. 2018a, b; Behera et al. 2018; Chauhan and Taylor 2018; Ahmad and Anis 2019).

12.3.2 Influence of Basal Media

In any micropropagation setup, the basal media enriched with macro- and micronutrients is essential for proper regeneration of explant. The popularly used basal media is Murashige and Skoog (MS) (Murashige and Skoog 1962) medium for shoot regeneration (Table 12.1). There are also other types of media apart from MS media, where the results achieved were quite promising, such as Le Poivre (LP) basal medium (Cortizo et al. 2009); Woody Plant (WP) medium (De Diego et al. 2010; Lattier et al. 2014; Mirabbasi and Hosseinpour 2014; Wen et al. 2016; Rakrawee et al. 2018; Tongsad et al. 2018; Nowakowska et al. 2019), comprising low concentration of nitrate and ammonium ions (Boscla and Michler 2008); Nas and Read medium (NMR) (Nas et al. 2010); Murashige and Tucker (MT) medium (Niedz and Evens 2011b); Quoirin and Lepoivre (QL) (Quoirin and Lepoivre 1977) medium where calcium nitrate is the sole nitrogen source (Lattier et al. 2013); and Driver and Kuniyuki (DKW) medium (Driver and Kuniyuki 1984) that possess minimum proportions of ammonium ions with calcium nitrate serving as nitrogen sources (Gentile et al. 2017; Nacheva et al. 2015; Stevens and Pijut 2018).

12.3.3 Influence of Carbon Source and Concentration

The explants that require regeneration under in vitro conditions possess partial autotrophic condition; hence, carbon source is mandatory (Van Huylenbroeck and Debergh 1996). Carbon source, mainly sugars, is required for fine-tuning the osmotic regulation (Hazarika 2003). Majority of research reports indicated the use of optimum amount of carbon source (2–3% sucrose), which is prevalently used (Table 12.1). However, as low as 0.57% sucrose was used to assess its effect on multiple shoot induction from shoot tip of *Asparagus officinalis* that eventually resulted in elongated multiple shoots (Hudák et al. 2013). It is noteworthy to mention that there is a single instance wherein glucose was used as a carbon source instead of sucrose (Nacheva et al. 2015).

Table 12.1 Influence of *meta*-topolin (*m*T) on induction, proliferation, and elongation of in vitro multiple shoots

Plant species	Explant	Basal media	Carbon source	Physical factors (temp/LI/PP/RH)	*m*T (mg/L* or µM)	Supplement	Gelling agent	References
Harpagophytum procumbens	NS	MS	3% (w/v) sucrose	25 ± 1 °C; 45 µmol/m²/s; 16 h; NM	5	None	0.9% (w/v) Agar	Bairu et al. (2009a)
Aloe ferox Mill	ST	MS	3% sucrose	24 ± 1 °C; 45 µmol/m²/s; 16 h; NM	5	0.01% (w/v) MI	1% Agar	Bairu et al. (2009b)
Pinus pinea L.	NS	LP	3% sucrose	22 °C; 120 µmol/m²/s; 16 h; NM	50	None	0.4% (w/v) Gelrite; 0.02% (w/v) Difco-Bacto agar	Cortizo et al (2009)
Hypericum	L	MS	3% sucrose	23 ± 1 °C; dark; NA; NM	15	1.25 µM IAA	0.8% Agar	Meyer et al. (2009)
Saccharum officinarum	Apical dome	MS	NM	22 ± 1 °C; 100 µmol/m²/s; 16 h; NM	10	None	0.8% Agar	Vinayak et al. (2009)
Pinus sylvestris L.	AB	DCR, LP, WPM	3% sucrose	22 °C; 120 µmol/m²/s; 16 h; NM	50	None	1% Difco-Bacto agar	De Diego et al. (2010)
Lycaste aromatica	Pseudo bulb	MS	3% sucrose	25 ± 1 °C; 50 µmol/	8.87	2 mg/L Glycine; 0.01% MI	0.3% (w/v) Gellan gum	Mata-Rosas et al. (2010)

(continued)

Table 12.1 (continued)

Plant species	Explant	Basal media	Carbon source	Physical factors (temp/LI/PP/RH)	*m*T (mg/L* or μM)	Supplement	Gelling agent	References
				m^2/s; 16 h; NM				
Prunus microcarpa	Cotyledon, hypocotyl, root	NRM	3% sucrose	23 ± 1 °C; 80 μmol/m^2/s; 16 h; NM	12.5	None	0.55% Microbiological agar	Nas et al. (2010)
Citrus reticulata × *Poncirus trifoliata*	Epicotyl		NM	27 °C; 30–35 μmol/m^2/s; 16 h; NM		None	0.8% Agar	Niedz and Evens (2010)
Uniola paniculata	Tiller	MS	87.6 μM sucrose	24 ± 1 °C; 40 μmol/m^2/s; 16 h; 58%	10	0.56 μM MI, 1.2 μM thiamine	0.8% Agar	Valero-Aracama et al. (2010)
Pelargonium sp.	ST	MS	3% sucrose	24 ± 1 °C; 40 μmol/m^2/s; 16 h; NM	1	None	0.55% Microbiological agar	Wojtania (2010)
Telekia speciosa (Schreb.) Baumg.	In vitro seedling	MS	3% sucrose	22 ± 2 °C; 105 μmol/m^2/s; 16 h; NM	5	KNO_3 (500 mg/L); casein (500 mg/L);	0.7% Agar	Csabai et al. (2011)
Pinus pinaster Ait.	Zygotic embryos	MS	3% sucrose	22 °C; 100 μmol/m^2/s; 16 h; NM	2.2	0.2% AC	1% Difco-Bacto agar	De Diego et al. (2011)

Sclerocarya birrea	ST, hypocotyl, epicotyl	MS	3% sucrose	25 ± 2 °C; 40 μmol/ m^2/s; 16 h; 58%	8	0.01% MI, 3 g/L polyvinylpyrrolidone	NM	Moyo et al. (2011)
Citrus sinensis	Epicotyl	MT	NM	27 °C; 30 μmol/ m^2/s; 16 h; NM	0.25	0.75 μM BA	0.8% Agar	Niedz and Evens (2011b)
Aloe arborescens	ST	MS	3% sucrose	25 ± 2 °C; 40 μmol/ m^2/s; 16 h; NM	15	0.01% MI	0.8% Agar	Amoo et al. (2012)
Musa spp. AAA cv. 'Williams'	ST	MS	3% sucrose	25 ± 2 °C; 45 μmol/ m^2/s; 16 h; NM	30	None	0.3% Gelrite	Aremu et al. (2012b)
Musa spp. AAA cv. 'Williams'	ST	MS	3% sucrose	25 ± 2 °C; 45 μmol/ m^2/s; 16 h; NM	30	None	0.3% Gelrite; INCYDE	Aremu et al. (2012c)
Merwilla plumbea (Lindl.) Speta	L	MS	NM	25 ± 2 °C; 40 μmol/ m^2/s; 16 h; NM	0.42	Trimethoprim (TMP), yeast extract (YE), and yeast malt broth	NM	Baskaran et al. (2012)
Pistacia vera L.	NS	MS	3% sucrose	22 ± 1 °C; 40 μmol/ m^2/s; 16 h; NM	2	0.01% Inositol, 500 mg/L casein hydrolysate	NM	Benmahioul et al. (2012)
Pelargonium sidoides	ST	MS	3% sucrose	25 ± 2 °C; 40 μmol/	2	0.01% (w/v) MI, 180 mg/L ascorbic acid	0.8% Agar	Moyo et al. (2012)

(continued)

Table 12.1 (continued)

Plant species	Explant	Basal media	Carbon source	Physical factors (temp/LI/PP/RH)	*m*T (mg/L* or μM)	Supplement	Gelling agent	References
				m^2/s; 16 h; NM				
Magnolia 'Ann'	ST	MS	3% sucrose	23 ± 2 °C; 70 μmol/m^2/s; 16 h; NM	8	0.01% inositol, MES monohydrate	0.8% Agar	Parris et al. (2012)
Romulea sabulosa	ST	MS	3% sucrose	NM	5	None	0.8% Agar	Swart et al. (2012)
Pelargonium × *hortorum* cv. 'Grand Prix'	ST	MS	3% sucrose	20 °C; 40 μmol/m^2/s; 16 h; NM	1	0.01% MI	0.55% Agar	Wojtania and Węgrzynowicz-Lesiak (2012)
Huernia hystrix	ST	MS	3% sucrose	25 ± 2 °C; 40 μmol/m^2/s; 16 h; NM	20	0.01% MI	0.8% Agar	Amoo and van Staden (2013)
Barleria argillicola	ST, NS	MS	3% sucrose	25 ± 2 °C; 40 μmol/m^2/s; 16 h; NM	7	0.01% MI	0.8% Agar	Amoo and van Staden (2013)
Agapanthus praecox	ST	MS	3% sucrose	25 ± 2 °C; 40 μmol/m^2/s; 16 h; NM	4.1	22.2 μM BA	0.8% Agar	Baskaran and van Staden (2013)
Asparagus officinalis	ST	MS	0.57% sucrose	22 ± 2 °C; 105 μmol/	2	25 mg/L Fe-EDTA, 0.01% MI	0.8% Agar	Hudák et al. (2013)

				m^2/s; 16 h; NM				
Acer platanoides	ST	MS, WPM, QL	3% sucrose	22 ± 2 °C; 75 µmol/ m^2/s; 16 h; NM	2*	0.01% MI, MES (100 mg/ L),	0.75% Agar	Lattier et al. (2013)
Aloe arborescens	ST	MS	3% sucrose	25 ± 2 °C; 30 µmol/ m^2/s; 16 h; NM	10	0.01% MI	0.8% Agar	Amoo et al. (2014)
Merwilla plumbea	L	MS	3% sucrose	25 ± 2 °C; 40 µmol/ m^2/s; 16 h; NM	1	0.01% MI	0.8% Agar	Aremu et al (2014)
Coleonema pulchellum	ST	MS	3% sucrose	25 ± 2 °C; 40 µmol/ m^2/s; 16 h; NM	2.1	13.6 µM BA	0.8% Agar	Baskaran et al. (2014)
Scutellaria barbata	ST, NS	MS	3% sucrose	25 ± 2 °C; 40 µmol/ m^2/s; 16 h; NM	5	34.2 g D-maltose; 18 g fructose; 10 g D-mannose; 10 g MI, 18 g D-mannitol; 0.1 µM NAA	0.7% Agar	Brearley et al. (2014)
Paulownia elongate	AB	MS	3% sucrose	NM	1	None	0.3% Agar	Clapa et al. (2014)
Coleonema album	ST	MS	3% sucrose	25 ± 2 °C; 30 µmol/ m^2/s; 16 h; NM	5	0.01% MI, 2 µM IBA	0.8% Agar	Fajinmi et al. (2014)
Lonicera kamtschatica	ST	MS	3% sucrose	23 ± 3 °C; 35 µmol/	1	None	5% (w/v) wheat starch	Al et al. (2014)

(continued)

Table 12.1 (continued)

Plant species	Explant	Basal media	Carbon source	Physical factors (temp/LI/PP/RH)	*m*T (mg/L* or μM)	Supplement	Gelling agent	References
				m^2/s; 16 h; NM				
Prunus rootstock (Torinel)	AB	MS	3% sucrose	24 ± 2 °C; 37.5 μmol/m^2/s; 16 h; NM	2.1	None	0.55% Agar	Gentile et al. (2014)
Ocimum basilicum	In vitro seedling	½MS	3% sucrose	22 °C; 36 μmol/m^2/s; 16 h; NM	2.07	10 mg/L Meso-inositol	0.62% Agar	Kőszeghi et al. (2014)
Cornus 'NCCH1'	ST	WPM	3% sucrose	NM	2.5	0.01% MI, MES (100 mg/L),	0.75% Agar	Lattier et al. (2014)
Eucomis autumnalis	ST	MS	3% sucrose	25 ± 2 °C; 45 μmol/m^2/s; 16 h; NM	2	0.01% MI, 10 μM NAA	0.3% Gelrite	Masondo et al. (2014)
Ulmus glabra	AB	WPM	3% sucrose	24 ± 2 °C; 2000 lux; 16 h; NM	0.2*	100 mg/L Sequestrene, 10 mg/L glutamine	0.57% Agar	Mirabbasi and Hosseinpour (2014)
Hypoxis hemerocallidea	Corm	MS	3% sucrose	25 ± 2 °C; 45 μmol/m^2/s; 16 h; NM	10	0.01% MI	0.8% Agar	Moyo et al. (2014)
Dioscorea trifida	ST	MS	3% sucrose	27 ± 1 °C; 50 μmol/	25	0.2% AC, 0.23 μM GA	0.7% Agar	Engelmann-Sylvestre and

				m^2/s; 12 h; NM				Engelmann (2014)
Pelargonium hortorum	ST	MS	3% sucrose	21 °C; 40 µmol/ m^2/s; 12 h; NM	2.07	0.01% MI, 3.78 µM ABA	0.6% Agar	Wojtania and Skrzypek (2014)
Malus domestica	ST	MS	3% sucrose	23 ± 2 °C; 5500 lux; 16 h; NM	1	None	0.25% (w/v) Phytagel	Al et al. (2015)
Syringa vulgaris	ST	MS	2% sucrose	23 ± 1 °C; 35 µmol/ m^2/s; 12 h; NM	5	None	0.8% Agar	Ilczuk and Katarzyna (2015)
Cyrtanthus guthrieae	Bulbs	MS	3% sucrose	25 ± 2 °C; 60 µmol/ m^2/s; 12 h; NM	5	0.01% MI	0.8% (w/v) Bacto-agar	Ncube et al. (2015)
Pelargonium × *hortorum* 'Bergpalais	ST	MS	3% sucrose	23 °C; 40 µmol/ m^2/s; 16 h; NM	2.07	0.01% MI	0.6% Bacto-agar	Wojtania et al. (2015)
Eucomis autumnalis	L	MS	3% sucrose	25 ± 2 °C; 45 µmol/ m^2/s; 16 h; NM	5	5 µM NAA	0.8% Agar	Aremu et al. (2016)
Pistacia vera	ST	MS	3% sucrose	22 ± 1 °C; 40 µmol/ m^2/s; 16 h; NM	2*	0.01% MI, 500 mg/L casein hydrolysate	0.7% Agar	Benmahioul et al. (2016)

(continued)

Table 12.1 (continued)

Plant species	Explant	Basal media	Carbon source	Physical factors (temp/LI/PP/RH)	*m*T (mg/L* or μM)	Supplement	Gelling agent	References
Dendrobium nobile	NS	MS	3% sucrose	25 ± 2 °C; 35–50 μmol/m^2/s; 12 h; NM	1*	0.8 mg/L putrescine	0.8% Agar	Bhattacharyya et al. (2016)
Mixed diploid Banana (*Musa* AB)	ST	MS	3% sucrose	28 ± 2 °C; NM; 14 h; NM	8.89	None	0.8% Agar	Bohra et al. (2016)
Nardostachys jatamansi	ST	MS	3% sucrose	22 ± 1 °C; 50 μmol/m^2/s; 14 h; NM	1*	None	0.8% Agar	Bose et al. (2016)
Pyrus communis Rootstock OHF-333	ST	MS	3% sucrose	22 ± 2 °C; 40 μmol/m^2/s; 14 h; NM	9*	None	0.65% (w/v) Phyto-agar	Dimitrova et al. (2016)
Withania coagulans	NS	MS	3% sucrose	25 ± 2 °C; 35 μmol/m^2/s; 14 h; NM	2.5*	0.1 mg/L NAA + 50 mg/L AdS	0.8% Agar	Joshi et al. (2016)
Clerodendrum phlomidis	NS	MS	3% sucrose	25 ± 2 °C; 35 μmol/m^2/s; 16 h; NM	8.28*	271 μM AdS	0.8% Agar	Kher et al. (2016)
Cannabis sativa	NS	MS	3% sucrose	25 ± 2 °C; 52 μmol/	2	None	0.8% Agar	Lata et al. (2016)

				m^2/s; 16 h; NM				
Allium achoenoprasum	ST	MS	2% sucrose	25 ± 2 °C; 100 μmol/m^2/s; 16 h; NM	10	0.01% MI, 2 mg/L adenine	0.7% Agar	Tubić et al. (2016)
Silk Banana var. Nanjanagud Rasabale (*Musa* AAB)	Sucker	MS	3% sucrose	28 ± 2 °C; 40 μmol/m^2/s; 14 h; NM	35.48	None	0.6% Agar	Waman et al. (2016)
Paeonia × *lemoinei* 'High Noon'	AB	WPM	3% sucrose	25 ± 1 °C; 25 μmol/m^2/s; 14 h; NM	5	1.5 μM GA_3	0.7% Agar	Wen et al. (2016)
Pistacia vera	AB	MS	3% sucrose	NM; 40 μmol/m^2/s; NM; NM	2	0.01% MI, 500 mg/L casein hydrolysate	0.7% Agar	Benmahioul (2017)
Ansellia africana	NS	MS	3% sucrose	25 ± 2 °C; 40 μmol/m^2/s; 12 h; 80%	10	None	0.8% Agar	Bhattacharyya et al. (2017)
Citrus sinensis [L.] *Osbeck* × *Poncirus trifoliata* [L.] Raf.	Epicotyl segment	MS	3% sucrose	27 ± 1 °C; 35 μmol/m^2/s; 16 h; NM	1*	500 mg/L ascorbic acid, 500 mg/L malt extract	0.85% Agar	Chiancone et al. (2017)
Corylus colurna	ST	DKW	3% sucrose	24 ± 1 °C; 37.5 μmol/m^2/s; 16 h; NM	8.2	None	0.7% Agar	Gentile et al (2017)

(continued)

Table 12.1 (continued)

Plant species	Explant	Basal media	Carbon source	Physical factors (temp/LI/PP/RH)	*m*T (mg/L* or μM)	Supplement	Gelling agent	References
Cannabis sativa	ST	MS	3% sucrose	24 ± 1 °C; NM; 18 h; NM	0.5*	0.1% AC	0.7% Agar	Grulichova et al. (2017)
Eriocephalus africanus	ST	MS	3% sucrose	25 ± 2 °C; 40–50 μmol/m^2/s; 16 h; NM	5	0.01% MI	0.8% Agar	Madzikane-Mlungwana et al. (2017)
Prunus domestica	ST	MS	3% sucrose	24 ± 2 °C; 37.5 μmol/m^2/s; 16 h; NM	2.1	None	0.55% Agar	Monticelli et al. (2017)
Merwilla plumbea	ST	DKW	3% glucose	22 ± 2 °C; 40 μmol/m^2/s; 16 h; NM	5.5	None	0.65% Agar	Nacheva et al. (2015)
Scadoxus puniceus	L	MS	NM	25 ± 2 °C; 30 μmol/m^2/s; 16 h; NM	5.5	None	–	Naidoo et al. (2017)
Dendrocalamus asper	NS	MS	3% sucrose	24 ± 2 °C; 40–50 μmol/m^2/s; 16 h; NM	20	None	0.25% Phytagel	Ornellas et al. (2017)

Cotula bipinnata, Chironia linoides	ST	MS	3% sucrose	24 ± 1 °C; 30 μmol/m²/s; 16 h; NM	0.3*	None	0.7% Agar	Sacco et al. (2017)
Ceratonia siliqua	ST	MS	3% sucrose	24 ± 2 °C; 50 μmol/m²/s; 16 h; NM	10	None	0.8% Agar	Shahzad et al. (2017)
Carthamus tinctorius (NARI-H-15)	ST, NS	MS	3% sucrose	25 ± 2 °C; 15 μmol/m²/s; 16 h; NM	3	0.5 μM CPPU	0.8% Agar	Vijayakumar et al. (2017)
Ficus carica	AB	MS	3% sucrose	NM; 35 μmol/m²/s; 16 h; NM		None	0.65% Agar	Yahyaoui et al. (2017)
Santalum album			NM			None		Akhtar and Shahzad (2019)
Epilobium canum garretti	NS	MS	3% sucrose	25 ± 1 °C; 38 μmol/m²/s; 16 h; NM	8.8	2 mg/L thiamine	0.7% Agar	Alosaimi et al. (2018)
Urginea altissima	ST	MS	3% sucrose	25 ± 2 °C; 50 μmol/m²/s; 16 h; NM	10	2 μM IAA	0.8% Agar	Baskaran et al. (2018a)
Eucomis autumnalis	L	MS	3% sucrose	25 ± 2 °C; 40 μmol/m²/s; 16 h; NM	10	None	0.8% Agar	Baskaran et al. (2018b)

(continued)

Table 12.1 (continued)

Plant species	Explant	Basal media	Carbon source	Physical factors (temp/LI/PP/RH)	*m*T (mg/L* or μM)	Supplement	Gelling agent	References
Hedychium coronarium	Rhizome	MS	NM	25 ± 1 °C; 50 μmol/m^2/s; 16 h; NM	3	None	–	Behera et al. (2018)
Dendrobium aphyllum	NS	MS	3% sucrose	25 ± 2 °C; 50 μmol/m^2/s; 16 h; NM	15	10 μM $AgNO_3$, 10 μM TDZ		Bhattacharyya et al. (2018a)
Ansellia africana	NS	MS	3% sucrose	25 ± 2 °C; 40 μmol/m^2/s; 12 h; NM	7.5	None	0.8% Agar	Bhattacharyya et al. (2018b)
Cassava	L	MS	2% sucrose	NM	7.5	None	0.8% Agar	Chauhan and Taylor (2018)
Kadsura heteroclita	NS	MS	3% sucrose	25 °C; NM; 16 h; NM	0.5*	None	0.26% Gellan gum	Jedoroh et al. (2018)
Mammillaria dixanthocentron	ST	MS	3% sucrose	25 ± 1 °C; 55 μmol/m^2/s; 16 h; NM	2.2	0.1% AC	0.8% Agar	Lázaro-Castellanos et al. (2018)
Mammillaria hernandezii, *M. lanata*	ST	MS	3% sucrose	25 ± 1 °C; 55 μmol/m^2/s; 16 h; NM	4.4	None	0.8% Agar	Lázaro-Castellanos et al. (2018)
Rose cv. *Frisco*	NS	MS	4% sucrose	24 ± 1 °C; 60 μmol/	2.1*	None	0.7% agar	Mahmood et al. (2018)

				m^2/s; 16 h; NM				
Amelanchier alnifolia	In vitro shoots	MS	3% sucrose	24 ± 1 °C; 40 µmol/ m^2/s; 16 h; NM	20	0.01% MI	NM	Moyo et al. (2018)
Stylosanthes hamata cv. *Verano*	In vitro seedling	MS	3% sucrose	25 ± 1 °C; 27 µmol/ m^2/s; 16 h; NM	3*	None	0.26% Phytagel	Ngoenngam et al. (2018)
Malus × domestica Borkh	ST	MS	3% sucrose	21 °C; 30 µmol/ m^2/s; 16 h; NM	8	None	0.6% Agar	Podwyszyńska and Cieślińska (2018)
Gluta usitata (217 Mae Ka)	ST	WPM	3% sucrose	25 ± 2 °C; NM; 16 h; NM	2*	None	0.2% Phytagel	Rakrawee et al. (2018)
Gluta usitata (Napong3)	NS	WPM	3% sucrose	25 ± 2 °C; NM; 16 h; NM	0.5*	0.2% AC	0.26% Gellan gum	Rakrawee et al. (2018)
Juglans nigra	NS	DKW	3% sucrose	24 ± 2 °C; 80 µmol/ m^2/s; 16 h; NM		None	0.22% Phytagel	Stevens and Pijut (2018)
Tectona griandis	ST	WPM	3% sucrose	25 ± 2 °C; NM; 16 h; NM	0.25*	None	0.26% Gellan gum	Tongsad et al. (2018)
Pterocarpus marsupium	Cotyledonary node	MS	3% sucrose	24 ± 2 °C; 50 µmol/	7.5	14 µM NAA	0.8% Agar	Ahmad and Anis (2019)

(continued)

Table 12.1 (continued)

Plant species	Explant	Basal media	Carbon source	Physical factors (temp/LI/PP/RH)	*m*T (mg/L* or μM)	Supplement	Gelling agent	References
				m^2/s; 16 h; NM				
Stylosanthes hamata cv. Verano	In vitro seedling	MS	3% sucrose	25 ± 1 °C; 27 μmol/m^2/s; 16 h; NM	3*	None	0.26% Phytagel	Ngoenngam et al. (2019)
Daphne mezereum	NS	MS	3% sucrose	NM	1*	0.01% MI	Bacto-agar	Nowakowska et al. (2019)
Daphne mezereum	NS	WPM	3% sucrose	NM	2*	None	Bacto-agar	Nowakowska et al. (2019)

2,4-D 2,4-Dichlorophenoxy acetic acid, *AB* axillary buds, *AC* activated charcoal, *BA* N^6-benzyladenine, *DKW* Driver and Kuniyuki walnut medium (Driver and Kuniyuki 1984), *IAA* indole-3-acetic acid, *IBA* indole-3-butyric acid, *INCYDE* 2-Chloro-6-(3-methoxyphenyl)aminopurine, *L* leaf, *LI* light intensity, *LP* Le Poivre basal medium, *MI* myo-inositol, *MS* Murashige Skoog medium (Murashige and Skoog 1962), *MT* Murashige and Tucker medium (Murashige and Tuker 1969), *NA* not applicable, *NAA* α-naphthalene acetic acid, *NM* not mentioned, *NRM* Nas and Read medium (Nas and Read 2004), *NS* nodal segments, *PP* photoperiod, *QL* Quoirin and Lepoivre, *RH* relative humidity, *ST* shoot tip, *TDZ* thidiazuron, *Temp* temperature, *WPM* Woody plant medium
* indicates that the plant growth regulator doses are in mg/L

12.3.4 Influence of Physical Environment

Light and temperature are the indispensable parameters for successful establishment of any in vitro culture. Light is dependent upon two factors, namely, intensity and duration. The source of light utilized in any in vitro culture room is fluorescent lamp that can be 'cool' or 'warm' (Gantait and Kundu 2017). Based on the research reports of past one decade, it was observed that the optimum light intensity ranging between 40 and 300 μmol/m^2/s was maintained for successful in vitro regeneration (Table 12.1). Additionally, temperature also plays an obligatory role for normal growth of in vitro culture, and it is mainly dependent on the basic design and structural organization of laboratory. The temperature regime adopted during establishment of in vitro multiple shoot culture protocols is summarized specifically in Table 12.1. Relative humidity (RH) also serves as a regulatory factor in any commercial tissue culture laboratory. Lower levels of RH result in instilling a positive growth in explants, since it strikes a perfect balance with the transpiration rate, whereas higher levels of humidity cause degenerative disorders in explants (Ghashghaie et al. 1992). Generally, the optimum level of RH recurrent in all the relevant literatures scrutinized is around 60%.

12.3.5 Influence of *Meta*-topolin

There are ~100 reports on the usage of *meta*-topolin as a plant growth hormone for shoot regeneration that have been documented and illustrated precisely in this chapter. For shoot induction and multiplication, proper dosage of cytokinins is essential (Gahan and George 2008). *m*T proved to be an effective cytokinin for multiple shoot culture in a number of medicinal, ornamental, and aromatic plants (Fig. 12.1). In majority of the cases, *m*T alone was sufficient to induce multiple shoots and also aid in shoot proliferation in many species summarized in Table 12.1. A range of 5–10 μM *m*T was employed to induce multiple shoot proliferation in various plant species through direct and indirect regeneration systems (Bairu et al. 2009a; Vinayak et al. 2009; Nas et al. 2010; Swart et al. 2012; Clapa et al. 2014; Al et al. 2014; Mukherjee et al. 2020). It was observed that even at a concentration lower than 5 μM, *m*T resulted in successful shoot induction (Wojtania and Węgrzynowicz-Lesiak 2012; Hudák et al. 2013; Aremu et al. 2014; Al et al. 2015; Lata et al. 2016).

12.3.6 Influence of *m*T in Combination with Other PGRs or Additives

*m*T was also coupled with various classes of plant hormones and also growth additives that gave fruitful results (Table 12.1). *m*T when used in combination with auxins provided positive results towards shoot induction; however, the formation of regenerative callus was observed when it was exogenously used with auxins like indole-3-acetic acid (IAA) (Meyer et al. 2009). Alternatively, multiple shoot was observed in *Coleonema album* when combined with indole-3-butyric acid

Fig. 12.1 Influence of *meta*-topolin (10 μM in semi-solid MS medium with 3% sucrose incubated at 25 ± 1 °C temperature, 50 μmol/m^2/s PPFD light intensity, 16 h photoperiod with 60% RH) on in vitro multiple shoot proliferation of several plant species. Multiple shoot cultures of (**a**) *Anthurium andreanum*, (**b**) *Musa* sp. cv. Grande Nine, (**c**) *Coleus forskohlii* Briq., (**d**) *Gerbera jamesonii* Bolus, (**e**) *Rauwolfia serpentina* (L.) Benth. ex Kurz., (**f**) *Sphagneticola* sp., (**g**) *Fragaria* × *ananassa* Dutch, (**h**) *Vanilla planifolia* Andrews (figures are not in scale). (*Source*: Unpublished photographs of Saikat Gantait)

(IBA), where myo-inositol (MI) was used as an additive (Fajinmi et al. 2014). Surprisingly, in combination with N^6-benzyladenine (BA), shoot induction occurred along with the formation of callus in the basal end, when shoot tips of *Coleonema pulchellum* were cultured (Baskaran et al. 2014). In combination with other cytokinins, a high synergistic effect was observed with *m*T. A high frequency of multiple shoot formation was observed when *m*T was used in combination with BA (Baskaran and van Staden 2013). Even the use of abscisic acid (ABA) along with *m*T gave positive results in multiple shoot regeneration in *Pelargonium hortorum* (Wojtania and Skrzypek 2014). Combination of *m*T with gibberellic acid (GA_3) resulted in shoot multiplication with increased shoot length (Engelmann-Sylvestre and Engelmann 2014; Wen et al. 2016). Similarly, with the application of additives like MI and casein hydrolysate along with *m*T resulted in longer shoot length in *Pistacia vera* (Benmahioul 2017). Additives like trimethoprim, yeast extract, and yeast malt broth when used in the medium together with *m*T also resulted in an increase in nodal length of multiple shoots in *Merwilla plumbea* (Baskaran et al. 2012). Synthetic cytokinins like thidiazuron (TDZ) when used in combination with *m*T and an additive $AgNO_3$ exhibited better results during in vitro culture of nodal segments of *Dendrobium aphyllum* (Bhattacharyya et al. 2018a). Activated charcoal serves as a potent additive for in vitro culture since it has the capability of absorbing hazardous phenolic compounds that are generated during in vitro regeneration (Gantait et al. 2009). Activated charcoal gave promising results with high frequency of shoot multiplication when used in combination with *m*T (De Diego et al. 2011; Engelmann-Sylvestre and Engelmann 2014; Grulichova et al. 2017).

12.4 Conclusion and Future Prospect of *m*T Use

In this chapter, the extensive use of *mT* in various plant species along with their appropriate doses has been scrutinized and elaborately described. The usage of *m*T in various in vitro cultures has shown promising results when used alone or in combination with other plant growth hormones or additives. The other facets of *m*T apart from effective shoot induction and multiplication include increase in shoot length, fresh weight, and increase in photosynthetic capacity of in vitro-derived plantlets (Ahmad and Anis 2019). It also has an additional capability of inducing regenerative callus when coupled with some auxins, thus highlighting the efficiency of *m*T. Considering these positive attributes, we can consider *m*T as an upcoming plant growth regulator that can be utilized commercially. The need for a comprehensive and reproducible protocol on in vitro culture of plantlets using *m*T is an utmost necessity for commercialization of plant tissue culture system globally. However, the utility of this compound needs to be harnessed in plants having medicinal and aesthetic value.

Acknowledgement Authors acknowledge the e-library assistance from Bidhan Chandra Krishi Viswavidyalaya, Mohanpur, West Bengal, India. We are further thankful to the anonymous reviewers and the editor of this chapter for their critical comments and suggestions on the manuscript.

In Compliance with Ethical Standards *Conflict of Interest*: none

References

Ahmad A, Anis M (2019) Meta-topolin improves in vitro morphogenesis, rhizogenesis and biochemical analysis in *Pterocarpus marsupium* Roxb.: a potential drug-yielding tree. J Plant Growth Regul 38:1007–1016

Akhtar R, Shahzad A (2019) Morphology and ontogeny of directly differentiating shoot buds and somatic embryos in *Santalum album* L. J For Res 30:1179–1189

Akin-Idowu PE, Ibitoye DO, Ademoyegun OT (2009) Tissue culture as a plant production technique for horticultural crops. Afr J Biotech 8:3782–3788

Al F, Clapa D, Cristea V, Plopa C (2014) In vitro propagation of *Lonicera kamtschatica*. Agric Sci Pract 1-2:89–90

Al F, Simu M, Vlaicu B, Clapa D (2015) Aspects regarding the in vitro propagation of 'Royal Gala' apple cultivar. Bull Univ Agric Sci Vet Med Cluj Napoca Hortic 72:342–349

Alosaimi AA, Tripepi RR, Love SL (2018) Micropropagation of *Epilobium canum garretti* (Firechalice) by axillary shoot culture. HortSci 53:62–66

Amoo SO, van Staden J (2013) Influence of plant growth regulators on shoot proliferation and secondary metabolite production in micropropagated *Huernia hystrix*. Plant Cell Tissue Organ Cult 112:249–256

Amoo SO, Aremu AO, van Staden J (2012) In vitro plant regeneration, secondary metabolite production and antioxidant activity of micropropagated *Aloe arborescens* Mill. Plant Cell Tissue Organ Cult 111:345–358

Amoo SO, Aremu AO, Moyo M et al (2014) Physiological effects of a novel aromatic cytokinin analogue in micropropagated *Aloe arborescens* and *Harpagophytum procumbens*. Plant Cell Tissue Organ Cult 116:17–26

Aremu AO, Bairu MW, Doležal K et al (2012a) Topolins: a panacea to plant tissue culture challenges? Plant Cell Tissue Organ Cult 108:1–16

Aremu AO, Bairu MW, Szüčová L et al (2012b) Shoot and root proliferation in 'Williams' banana: are the topolins better cytokinins? Plant Cell Tissue Organ Cult 111:209–218

Aremu AO, Bairu MW, Novák O et al (2012c) Physiological responses and endogenous cytokinin profiles of tissue-cultured 'Williams' bananas in relation to roscovitine and an inhibitor of cytokinin oxidase/dehydrogenase (INCYDE) treatments. Planta 236:1775–1790

Aremu AO, Plačková L, Bairu MW et al (2014) How does exogenously applied cytokinin type affect growth and endogenous cytokinins in micropropagated *Merwilla plumbea*? Plant Cell Tissue Organ Cult 118:245–256

Aremu AO, Plačková L, Pěnčík A et al (2016) Auxin-cytokinin interaction and variations in their metabolic products in the regulation of organogenesis in two *Eucomis* species. New Biotechnol 33:883–890

Bairu MW, Jain N, Stirk WA et al (2009a) Solving the problem of shoot-tip necrosis in *Harpagophytum procumbens* by changing the cytokinin types, calcium and boron concentrations in the medium. S Afr J Bot 75:122–127

Bairu MW, Kulkarni MG, Street RA et al (2009b) Studies on seed germination, seedling growth, and in vitro shoot induction of *Aloe ferox Mill.*, a commercially important species. HortSci 44:751–756

Baskaran P, van Staden J (2013) Rapid in vitro micropropagation of *Agapanthus praecox*. S Afr J Bot 86:46–50

Baskaran P, Ncube B, van Staden J (2012) *In vitro* propagation and secondary product production by *Merwilla plumbea* (Lindl.) Speta. Plant Growth Regul 67:235–245

Baskaran P, Moyo M, van Staden J (2014) *In vitro* plant regeneration, phenolic compound production and pharmacological activities of *Coleonema pulchellum*. S Afr J Bot 90:74–79

Baskaran P, Kumari A, van Staden J (2018a) *In vitro* propagation via organogenesis and synthetic seeds of *Urginea altissima* (Lf) Baker: a threatened medicinal plant. 3 Biotech 8:18

Baskaran P, Kumari A, van Staden J (2018b) Analysis of the effect of plant growth regulators and organic elicitors on antibacterial activity of *Eucomis autumnalis* and *Drimia robusta* ex vitro-grown biomass. Plant Growth Regul 85:143–151

Behera S, Kamila PK, Rout KK et al (2018) An efficient plant regeneration protocol of an industrially important plant, *Hedychium coronarium* J. Koenig and establishment of genetic & biochemical fidelity of the regenerants. Ind Crop Prod 126:58–68

Benmahioul B (2017) Factors affecting in vitro micropropagation of Pistachio (*Pistacia vera* L.). Agric For J 1:56–61

Benmahioul B, Dorion N, Kaid-Harche M et al (2012) Micropropagation and ex vitro rooting of pistachio (*Pistacia vera* L.). Plant Cell Tissue Organ Cult 108:353–358

Benmahioul B, Kaïd-Harche M, Daguin F (2016) In vitro regeneration of *Pistacia vera* L. from nodal explants. J For Sci 62:198–203

Bhattacharyya P, Kumaria S, Tandon P (2016) High frequency regeneration protocol for *Dendrobium nobile*: a model tissue culture approach for propagation of medicinally important orchid species. S Afr J Bot 104:232–243

Bhattacharyya P, Kumar V, van Staden J (2017) Assessment of genetic stability amongst micropropagated *Ansellia africana*, a vulnerable medicinal orchid species of Africa using SCoT markers. S Afr J Bot 108:294–302

Bhattacharyya P, Paul P, Kumaria Set al. (2018a) Transverse thin cell layer (t-TCL)-mediated improvised micropropagation protocol for endangered medicinal orchid *Dendrobium aphyllum* Roxb: an integrated phytomolecular approach. Acta Physiol Plant 40:137

Bhattacharyya P, Kumar V, van Staden J (2018b) In vitro encapsulation based short term storage and assessment of genetic homogeneity in regenerated *Ansellia africana* (Leopard orchid) using gene targeted molecular markers. Plant Cell Tissue Organ Cult 133:299–310

Bohra P, Waman AA, Sathyanarayana BN et al (2016) Influence of different growth regulators on in vitro multiplication of mixed diploid Banana (*Musa* AB). Proc Natl Acad Sci India Sect B 86:179–185

Bose B, Kumaria S, Choudhury H et al (2016) Assessment of genetic homogeneity and analysis of phytomedicinal potential in micropropagated plants of *Nardostachys jatamansi*, a critically endangered, medicinal plant of alpine Himalayas. Plant Cell Tissue Organ Cult 124:331–349
Bosela MJ, Michler CH (2008) Media effects on black walnut (*Juglans nigra* L.) shoot culture growth in vitro: evaluation of multiple nutrient formulations and cytokinin types. In Vitro Cell Dev Biol Plant 44:316–329
Brearley TA, Vaidya BN, Joshee N (2014) Cytokinin, carbon source, and acclimatization requirements for in vitro propagation of *Scutellaria barbata* D. Don and *Scutellaria racemosa* Pers. Am J Plant Sci 5:3662–3672
Chauhan RD, Taylor NJ (2018) Meta-topolin stimulates de novo shoot organogenesis and plant regeneration in cassava. Plant Cell Tissue Organ Cult 132:219–224
Chiancone B, Martorana L, Gianguzzi V et al (2017) The effects of meta-Topolin and benzyladenine on in vitro organogenesis from epicotyl cuttings of Troyer citrange (*Citrus sinensis* [L.] Osbeck × *Poncirus trifoliata* [L.] Raf.). Acta Hortic 1155:185–192
Clapa D, Fira A, Simu M et al (2014) Improved in vitro propagation of Paulownia elongata, P. fortunei and its interspecific hybrid P. *elongata* x P. *fortunei*. Bull Univ Agric Sci Vet Med Cluj Napoca Hortic 71:422–427
Cortizo M, de Diego N, Moncaleán P et al (2009) Micropropagation of adult stone pine (*Pinus pinea* L.). Trees 23:835–842
Csabai J, Nagy Z, Mándy A (2011) In vitro shoot proliferation of *Telekia speciosa* (Schreb.) Baumg. Induced by different cytokinins. Acta Biol Hung 62:453–462
De Diego N, Montalbán IA, Moncaleán P (2010) In vitro regeneration of adult *Pinus sylvestris* L. trees. S Afr J Bot 76:158–162
De Diego N, Montalbán IA, Moncaleán P (2011) Improved micropropagation protocol for maritime pine using zygotic embryos. Scand J For Res 26:202–211
Dimitrova N, Nacheva L, Berova M (2016) Effect of *meta*-topolin on the shoot multiplication of pear rootstock OHF-333 (*Pyrus communis* L.). Acta Sci Pol-Hortorum 15:43–53
Driver JA, Kuniyuki AH (1984) In vitro propagation of paradox walnut rootstock. HortSci 19:507–509
Engelmann-Sylvestre I, Engelmann F (2014) Effect of various growth regulators on growth of yam (*Dioscorea trifida* L.) in vitro shoot tips. Afr J Bot 13:1645–1649
Fajinmi OO, Amoo SO, Finnie JF et al (2014) Optimization of in vitro propagation of *Coleonema album*, a highly utilized medicinal and ornamental plant. S Afr J Bot 94:9–13
Gahan PB, George EF (2008) Adventitious regeneration. In: George EF, Hall MA, De Klerk GJ (eds) Plant propagation by tissue culture, 3rd edn. Springer, Dordrecht, pp 355–401
Gantait S, Kundu S (2017) *In vitro* biotechnological approaches on *Vanilla planifolia* Andrews: advancements and opportunities. Acta Physiol Plant 39:196
Gantait S, Mandal N, Bhattacharyya S, Das PK, Nandy S (2009) Mass multiplication of *Vanilla planifolia* with pure genetic identity confirmed by ISSR. Int J Dev Biol 3:18–23
Gantait S, Sinniah UR, Das PK (2014) *Aloe vera*: a review update on advancement of in vitro culture. Acta Agric Scand B Soil Plant Sci 64:1–12
Gentile A, Gutiérrez MJ, Martinez J et al (2014) Effect of meta-topolin on micropropagation and adventitious shoot regeneration in Prunus rootstocks. Plant Cell Tissue Organ Cult 118:373–381
Gentile A, Frattarelli A, Nota P et al (2017) The aromatic cytokinin meta-topolin promotes in vitro propagation, shoot quality and micrografting in *Corylus colurna* L. Plant Cell Tissue Organ Cult 128:693–703
Ghashghaie J, Brenckmann F, Saugier B (1992) Water relations and growth of rose plants cultured in vitro under various relative humidities. Plant Cell Tissue Organ Cult 30:51–57
Grulichova M, Mendel P, Lalge AB et al (2017) Effect of different phytohormones on growth and development of micropropagated *Cannabis sativa* L. MendelNet 24:618–623
Hall RH (1973) Cytokinins as a probe of developmental processes. Annu Rev Plant Physiol 24:415–444
Hazarika BN (2003) Acclimatization of tissue cultured plants. Curr Sci 85:1705–1712

Holub J, Hanus J, Hanke DE, Strnad M (1998) Biological activity of cytokinins derived from *ortho-* and *meta* hydroxybenzyladenine. Plant Growth Regul 26:109–115
Howell SH, Lall S, Che P (2003) Cytokinins and shoot development. Trend Plant Sci 8:453–459
Hudák I, Magyar-Tábori K, Zsombik L (2013) Effect of aromatic cytokinins and explant position on shoot multiplication of *Asparagus officinalis* L. Int J Hortic Sci 19:67–70
Ilczuk A, Katarzyna JK (2015) The effect of plant growth regulators and sucrose on the micropropagation of common lilac (*Syringa vulgaris* L.). Ann Warsaw Univ Life Sci SGGW Hortic Landsc Architect 36:3–12
Jedoroh N, Laipasu P, Chareonsap PP et al (2018) Induction of shoot and root from nodes of *Kadsura heteroclite*. Int J Agric Technol 14:1287–1292
Joshi H, Nekkala S, Soner D et al (2016) In vitro shoot multiplication of *Withania coagulans* (stocks) Dunal. Plant Tissue Cult Biotechnol 26:187–195
Kher MM, Soner D, Srivastava N et al (2016) Micropropagation of *Clerodendrum phlomidis* LF. J Hortic Res 24:21–28
Koshimizu K, Iwamura H (1986) Cytokinins. In: Takahashi N (ed) Chemistry of plant hormones. CRC Press, Florida, pp 153–159
Kőszeghi S, Bereczki C, Balog A et al (2014) Comparing the effects of benzyladenine and meta-topolin on sweet basil (*Ocimum basilicum*) micropropagation. Not Sci Biol 6:422–427
Lata H, Chandra S, Techen N et al (2016) In vitro mass propagation of *Cannabis sativa* L.: a protocol refinement using novel aromatic cytokinin meta-topolin and the assessment of eco-physiological, biochemical and genetic fidelity of micropropagated plants. J Appl Res Med Aromat Plants 3:18–26
Lattier JD, Touchell DH, Ranney TG et al (2013) Micropropagation and polyploid induction of *Acer platanoides* 'Crimson Sentry'. J Environ Hortic 31:246–252
Lattier J, Touchell DH, Ranney TG (2014) Micropropagation of an interspecific hybrid dogwood (*Cornus* 'NCCH1'). Propag Ornam Plants 14:184–190
Lázaro-Castellanos JO, Mata-Rosas M, González D et al (2018) In vitro propagation of endangered *Mammillaria* genus (Cactaceae) species and genetic stability assessment using SSR markers. In Vitro Cell Dev Biol Plant 54:518–529
Leopold AC, Kriedemann CE (1975) Plant growth and development. McGraw & Hill, New York
Madzikane-Mlungwana O, Moyo M, Aremu AO et al (2017) Differential responses to isoprenoid, N^6-substituted aromatic cytokinins and indole-3-butyric acid in direct plant regeneration of *Eriocephalus africanus*. Plant Growth Regul 82:103–110
Mahmood S, Reza MR, Hossain MG et al (2018) Response of cytokinins on in vitro shoot multiplication of Rose cv. Frisco. Res Rev J Agric Sci Technol 5:8–12
Malá J, Máchová P, Cvrčková H et al (2013) The role of cytokinins during micropropagation of wych elm. Biol Plant 57:174–178
Masondo NA, Aremu AO, Finnie JF et al (2014) Plant growth regulator induced phytochemical and antioxidant variations in micropropagated and acclimatized *Eucomis autumnalis* subspecies *autumnalis* (Asparagaceae). Acta Physiol Plant 36:2467–2479
Mata-Rosas M, Baltazar-García RJ, Moon P et al (2010) In vitro regeneration of *Lycaste aromatica* (Graham ex Hook) Lindl. (Orchidaceae) from pseudobulb sections. Plant Biotech Rep 4:157–163
Meyer EM, Touchell DH, Ranney TG (2009) In vitro shoot regeneration and polyploid induction from leaves of Hypericum species. HortScience 44:1957–1961
Mirabbasi SM, Hosseinpour B (2014) Prevention of shoot tip necrosis, hyperhydricity and callus production associated with in vitro shoot culture of *Ulmus glabra*. J Novel Appl Sci 3:683–689
Mok MC, Martin RC, Dobrev PI et al (2005) Topolins and hydroxylated thidiazuron derivatives are substrates of cytokinins O-glucosyltransferase with position specificity related to receptor recognition. Plant Physiol 137:1057–1066
Monticelli S, Gentile A, Frattarelli A et al (2017) Effects of the natural cytokinin meta-Topolin on in vitro shoot proliferation and acclimatization of *Prunus* spp. Acta Hortic 1155:375–380

Moyo M, Finnie JF, van Staden J (2011) Recalcitrant effects associated with the development of basal callus-like tissue on caulogenesis and rhizogenesis in *Sclerocarya birrea*. Plant Growth Regul 63:187–195

Moyo M, Finnie JF, van Staden J (2012) Topolins in *Pelargonium sidoides* micropropagation: do the new brooms really sweep cleaner? Plant Cell Tissue Organ Cult 110:319–327

Moyo M, Amoo SO, Aremu AO et al (2014) Plant regeneration and biochemical accumulation of hydroxybenzoic and hydroxycinnamic acid derivatives in *Hypoxis hemerocallidea* organ and callus cultures. Plant Sci 227:157–164

Moyo M, Aremu AO, Plačková L et al (2018) Deciphering the growth pattern and phytohormonal content in Saskatoon berry (*Amelanchier alnifolia*) in response to in vitro cytokinin application. New Biotechnol 42:85–94

Mukherjee E, Sarkar S, Bhattacharyya S, Gantait S (2020) Ameliorated reserpine production via in vitro direct and indirect regeneration system in *Rauvolfia serpentina* (L.) Benth. ex Kurz. 3 Biotech 10:294

Murashige T, Skoog F (1962) A revised medium for rapid growth and bioassay with tobacco tissue cultures. Physiol Plant 15:473–495

Murashige T, Tuker DPH (1969) Growth factor requirement of citrus tissue culture. In: International citrus symposium, Riverside, vol 1. Riverside: University of California, pp 1155–1169

Nacheva L, Gercheva P, Ivanova V et al (2015) *Meta*-topolin improves lateral bud proliferation in micropropagation of *Ginkgo biloba* L. Acta Hortic 1155:355–360

Naidoo D, Aremu AO, van Staden J et al (2017) In vitro plant regeneration and alleviation of physiological disorders in *Scadoxus puniceus*. S Afr J Bot 109:316–322

Nas MN, Read PE (2004) A hypothesis for the development of a defined tissue culture medium of higher plants and micropropagation of hazelnuts. Sci Hortic 101:189–200

Nas MN, Bolek Y, Sevgin N (2010) The effects of explant and cytokinin type on regeneration of *Prunus microcarpa*. Sci Hortic 126:88–94

Ncube B, Finnie JF, van Staden J (2015) In vitro regeneration of *Cyrtanthus* species: ornamental plants with medicinal benefits. In Vitro Cell Dev Biol Plant 51:42–51

Ngoenngam L, Pongtongkam P, Arananant J et al (2018) In vitro effect of plant growth regulators (PGRs) for callus induction and plant regeneration from suspension of Hamata (*Stylosanthes hamata* cv. Verano). Int J Agric Technol 14:1515–1524

Ngoenngam L, Pongtongkam P, Arananant J et al (2019) In vitro effect of gamma irradiation and plant growth regulators (PGRs) for induction and development of *Stylosanthes hamata* cv. Verano. Int J Agric Technol 15:63–74

Niedz RP, Evens TJ (2010) The effects of benzyladenine and meta-topolin on in vitro shoot regeneration of a *Citrus citrandarin* rootstock. Res J Agric Biol Sci 6:45–53

Niedz RP, Evens TJ (2011a) Mixture screening and mixture-amount designs to determine plant growth regulator effects on shoot regeneration from grapefruit (*Citrus paradisi* macf.) epicotyls. In Vitro Cell Dev Biol Plant 47:682–694

Niedz RP, Evens TJ (2011b) The effects of benzyladenine and meta-topolin on in vitro shoot regeneration of sweet orange. ARPN J Agric Biol Sci 6:64–73

Nowakowska K, Pacholczak A, Tepper W (2019) The effect of selected growth regulators and culture media on regeneration of *Daphne mezereum* L. 'Alba'. Rend Lincei Sci Fis Nat 30:197–205

Ornellas TS, Werner D, Holderbaum DF et al (2017) Effects of vitrofural, BAP and meta-Topolin in the in vitro culture of *Dendrocalamus asper*. Acta Hortic 1155:285–292

Parris JK, Touchell DH, Ranney TG et al (2012) Basal salt composition, cytokinins, and phenolic binding agents influence in vitro growth and ex vitro establishment of Magnolia 'Ann'. HortScience 47:1625–1629

Podwyszyńska M, Cieślińska M (2018) Rooting shoots of apple varieties and their tetraploids obtained by the in vitro technique. Acta Sci Pol Hortorum Cultus 17:49–62

Quoirin M, Lepoivre P (1977) Improved medium for in vitro culture of *Prunus* sp. Acta Hortic 78:437–442

Rakrawee R, Kittibanpacha K, Chareonsap PP et al (2018) Efficiency of antioxidant and absorbent on browning and the optimal factors of plant regeneration from young seed of *Gluta usitata* (217 Mae Ka) by tissue culture. Int J Agric Technol 14:911–922

Rout GR, Mohapatra A, Jain S (2006) Tissue culture of ornamental pot plant: a critical review on present scenario and future prospects. Biotechnol Adv 24:531–560

Sacco E, Mascarello C, Pamato M et al (2017) In vitro propagation of *Cotula bipinnata* Thunb and *Chironia linoides* L., South African ornamental species suitable for cultivation in Mediterranean area. Acta Hortic 1155:625–630

Shahzad A, Akhtar R, Bukhari NA et al (2017) High incidence regeneration system in *Ceratonia siliqua* L. articulated with SEM and biochemical analysis during developmental stages. Trees 31:1149–1163

Stevens ME, Pijut PM (2018) Rapid in vitro shoot multiplication of the recalcitrant species *Juglans nigra* L. In Vitro Cell Dev Biol Plant 54:309–317

Strnad M, Hanuš J, Vanek T et al (1997) Meta-topolin, a highly active aromatic cytokinin from poplar leaves (*Populus* × *canadensis* Moench., cv. Robusta). Phytochemistry 45:213–218

Sugiyama M (1999) Organogenesis in vitro. Curr Opin Plant Biol 2:61–64

Swart PA, Kulkarni MG, Bairu MW et al (2012) Micropropagation of *Romulea sabulosa* Schltr. ex Beg. - a potential ornamental plant. Sci Hortic 135:151–156

Tongsad P, Laipasu P, Chareonsap PP et al (2018) The effect of plant growth regulator and in vitro conservation of teak (*Tectona grandis* L.) by tissue culture. Int J Agric Technol 14:2171–2180

Tubić L, Savić J, Mitić N et al (2016) Cytokinins differentially affect regeneration, plant growth and antioxidative enzymes activity in chive (*Allium schoenoprasum* L.). Plant Cell Tissue Organ Cult 124:1–14

Valero-Aracama C, Kane ME, Wilson SB et al (2010) Substitution of benzyladenine with meta-topolin during shoot multiplication increases acclimatization of difficult-and easy-to-acclimatize sea oats (*Uniola paniculata* L.) genotypes. Plant Growth Regul 60:43

Van Huylenbroeck J, Debergh PC (1996) Impact of sugar concentration in vitro on photosynthesis and carbon metabolism during ex vitro acclimatization of *Spathiphyllum* plantlets. Physiol Plant 96:298–304

Vijayakumar J, Ponmanickam P, Samuel P et al (2017) Influence of *meta*-Topolin on efficient plant regeneration via micropropagation and organogenesis of safflower (*Carthamus tinctorius* L.) cv. NARI-H-15. Am J Plant Sci 8:688

Vinayak V, Dhawan AK, Gupta VK (2009) Efficacy of non-purine and purine cytokinins on shoot regeneration in vitro in sugarcane. Indian J Biotechnol 8:227–231

Waman AA, Bohra P, Sathyanarayana BN (2016) Effect of various cytokinins and auxins on in vitro regeneration of plantlets from isolated bud clumps of Silk Banana var. Nanjanagud Rasabale (*Musa* AAB). Ind J Plant Physiol 21:64–69

Wen S, Cheng F, Zhong Y (2016) Micropropagation of tree peony (*Paeonia* × *lemoinei* 'high noon') and the assessment of genetic stability by SSR analysis. Propag Ornam Plants 16:19–27

Wojtania A (2010) Effect of meta-topolin in vitro propagation of *Pelargonium* × *hortorum* and *Pelargonium* × *hederaefolium* cultivars. Acta Soc Bot Pol 79:101–106

Wojtania A, Skrzypek E (2014) Effects of cytokinins on antioxidant enzymes in in vitro grown shoots of *Pelargonium hortorum* LH Bayley. Acta Agrobot 4:33–42

Wojtania A, Węgrzynowicz-Lesiak E (2012) Ethylene and cytokinin interaction in the morphogenesis of *Pelargonium* × *hortorum* LH Bailey in vitro. Acta Physiol Plant 34:2407–2412

Wojtania A, Węgrzynowicz-Lesiak E, Dziurka M et al (2015) Sucrose and cytokinin interactions in relation to ethylene and abscisic acid production in the regulation of morphogenesis in *Pelargonium* × *hortorum* LH Bailey in vitro. Acta Biol Cracov Ser Bot 57:62–69

Yahyaoui E, Casamento D, D'Onghia et al (2017) Fig virus-free production and survival rate improvement using meristem tip culture technique associated with the encapsulation technology. Acta Hortic 1224:81–90

Yancheva SD, Golubowicz S, Fisher E et al (2003) Auxin type and timing of application determine the activation of the developmental program during in vitro organogenesis in apple. Plant Sci 165:299–309

Meta-topolin Promotes Improved Micropropagation, Photosynthetic Performances, Biomass and Proline Levels of an India Ipecac (*Tylophora indica* Burm f.)

13

Nigar Fatima, Naseem Ahmad, and Mohammad Anis

Contents

Abstract

The significance of a novel aromatic cytokinins, *meta*-topolin (*m*T), on in vitro morphogenesis, rhizogenesis and acclimatization competency of tissue culture raised plants of *Tylophora indica* have been studied. *m*T an advanced cytokinin

N. Fatima (✉) · M. Anis
Plant Biotechnology Laboratory, Department of Botany, Aligarh Muslim University, Aligarh, India

N. Ahmad
Department of Botany, Aligarh Muslim University, Aligarh, India

N. Ahmad, M. Strnad (eds.), *Meta-topolin: A Growth Regulator for Plant Biotechnology and Agriculture*, https://doi.org/10.1007/978-981-15-9046-7_13

was found potential for the development of the micropropagation system as well as growth and establishment of plants ex vitro. Increased multiplication rate was found using *m*T (5.0 μM) supplemented in Murashige and Skoog (MS 1962) medium, where stimulation of maximum shoot development (12.3 ± 0.22) with a shoot length (5.50 ± 0.00 cm)/explant after 4 weeks of culture duration was attained. The collective consequence of cytokinin-auxin (5.0 μM *m*T + 0.5 μM NAA) considerably showed the production of 18.2 ± 0.12 mean shoots/explant with shoot length of 5.40 ± 0.00 cm with 90% regeneration rate. The high rooting rates of microshoots (4 cm) were evident on 1/2 strength of MS medium amplified with petite intensity (0.1 μM) of indole-3-butyric acid (IBA) which yielded maximum mean roots (8.64 ± 0.04) and mean root length (4.12 ± 0.03 cm) in 92% cultures. Maximum Chl. *a* (0.45 ± 0.01 mg/g FW), Chl. *b* (0.40 ± 0.01 mg/g FW), and carotenoid (0.22 ± 0.02 mg/g FW) contents were recorded at optimal in *m*T (5.0 μM) concentration. However, the elevated intensity of proline level (0.26 ± 0.02 μg/g FW) was recorded on elevated application of *m*T (10.0 μM) in the basal medium. The survival rate 96.0 ± 0.91% was ascertained by the acclimatized plants in a planting substrate soilrite. *Tylophora* plantlets regenerated via the exposure of *m*T showed significantly improved magnitude of morpho-physiological quality of shoots intensification and root quality, staging of better biomass, pigments, and proline contents followed by inevitably enhanced adaptation potentiality. This standardized rejuvenation/re-establishment protocol may be supportive during the regeneration of novel genotypes of *T. indica* which could advance agronomic characters during in vitro selection practices along with genetic transformation system/genetic engineering.

Keywords

Plant growth regulators · *Meta*-topolin · Morphogenesis · Micropropagation · Acclimatization competency

Abbreviations

BA	N^6-benzyladenine
IAA	Indole-3-acetic acid
IBA	Indole-3-butyric acid
MS	Murashige and Skoog medium
*m*T	*Meta*-topolin
NAA	α-Naphthaleneacetic acid
NS	Nodal segment

13.1 Introduction

Tylophora indica (Burm. F.) Merrill (Asclepiadaceae) is an endangered forest species also known as *Tylophora asthmatica* due to its use in the treatment of asthma. The plant is native to southern and eastern Indian hills, plains, and forests. The plant was officially incorporated into Bengal Pharmacopoeia in 1884 called "Antamool," and the roots were used as a replacement for ipecac. The plant has extensive use in the treatment of a number of serious human diseases including asthma, bronchitis, rheumatism, and other respiratory problems, allergies, inflammations, dysentery, pertussis, diarrhea, etc. (CSIR 2003). In particular, the main active substance includes alkaloids such as tylophorine ($C_{24}H_{27}O_4N$), tylophorinine ($C_{23}H_{25}O_4N$), tylophrinidine ($C_{22}H_{22}O_4N$), and septidine, which have been isolated from both the leaves and roots of plants (Ali and Butani 1989). Tylophorine, one of the major alkaloids in the plant, exhibits immunosuppressive, anti-inflammatory, antitumor, anti-candidal, and anti-amoebic properties, while tylophrinidine has anti-leukemic properties (Donaldson et al. 1968; Bhutani et al. 1985; Mulchandani et al. 1971). Tribal and traditional healers treat *Tylophora* as the main source of drug supply for various diseases, especially in remote and forest areas. *T. indica* extracts stand for an anti-asthmatic herbal drug/formulation marketed by pharmaceutical groups (Acron Chemicals, Mumbai India, Sabinsa Corporation, Piscataway, NJ, USA).

Seeds of *T. indica* are primarily used for commercial plant propagation, even though they are too small and contain short viability and germination. Furthermore, propagation by vegetative cuttings is also in practice; however this is a relatively difficult process (Thomas and Philip 2005). On the other hand, the therapeutic achievement of the plant owing to the presence of valuable alkaloids anticipated for the treatment of various human diseases has directed to exploitation on extensive range. Naturally its ruthless and unrestricted harvest has therefore led to its categorization as an "endangered" plant species (Faisal et al. 2007). Hence, it is imperative to develop an innovative system for en masse propagation of this species, preserving and conserving wild-type genotypes in nature, for agricultural purposes and also for breeding new, high-content varieties for commercial/pharmaceutical fabrication.

In micropropagation, a range of features which are classified either environmental or hormonal/plant growth regulators (PGRs) manipulate in vitro propagation of plants. Besides, suitable relevance of exogenous cytokinins participates in a vital function in coordinating shoot propagation from axillary sprout (Werner et al. 2001). PGRs act as critical chemical component which regulates diverse physiological and developmental procedures throughout micropropagation (George et al. 2008). The biotechnological progresses within the area of phytohormones have extensively supported the exploration of novel composite (Strnad 1997; Tarkowski et al. 2009). During the past decades, a rise in the employment of topolins and their derivatives have been very exciting for the plant hormonal research worlwide. *Meta*-topolin, a class of topolin, is an aromatic cytokinin, primary isolated from poplar leaves which varies from isoprenoid cytokinins in its biological as well as biochemical activities (Strnad 1997). The *m*T has been confirmed to advance shoot proliferation, improve root efficiency, and relieve different physiological disarray during micropropagation programs (Aremu et al. 2012a, b). Research on the employment of *meta*-topolin (*m*T) in tissue culture has been carried out on several plant species, including *Aloe* polyphylla (Bairu et al. 2007a), *Musa acuminata* (Bairu et al. 2007b), *Syzygium cumini* (Naaz et al. 2019), *Pterocarpus marsupium* (Anees and Anis 2019), *Tecoma stans* (Hussain et al. 2019), *Allamanda cathartica* (Khanam et al. 2020), etc.

Cytokinins (CKs), a chief regulator of in vitro morphogenesis and growth in plants, play important functions in controlling many progression in plants (cell cycle, growth, etc.) and also affect function of the photosynthetic apparatus during shoots development. CKs are associated in the preservation and stimulation of chlorophyll biosynthesis, protein synthesis, sustain pigment concentration, support chloroplast replication, grana formation, etc. (Cag et al. 2003; Naaz et al. 2019; Hussain et al. 2019). However, *m*T functions in physiological parameters in plants have still not achieved much consideration.

Various tissue culture techniques have been contributed significantly for the successful multiplication, germplasm maintenance, and improvement programs of plant comprehensively. However, the detailed relevance of *meta*-topolin on regeneration frequency, rhizogenesis, carryover effect with acclimatization potency, and outcome on physiological parameters (photosynthetic pigments, proline) in tissue culture-raised plantlets is to be ascertained.

13.2 Materials and Methods

13.2.1 Establishment of Sterile Culture Source

Active growing juvenile shoots of *T. indica* were collected from the Botanical Garden, Department of Botany, Aligarh Muslim University, Aligarh. Nodal section containing axillary buds were dissected and washed under normal running tap water for 30 min. Five percent (v/v) "labolene," a laboratory detergent (Qualigens, Mumbai, India), were used for 5 min, cleanse systematically in sterile distilled water. Nodal explants were surface sterilized using 0.1% (w/v) mercuric chloride for 4–5 min and rinsed with sterilized distilled water below aseptic stipulation (Laminar Air Flow Cabinet) to eliminate the sterilant traces. Nodal explants (1–1.5 cm) were excised and inoculated vertically on the culture medium.

13.2.2 Medium Composition and Growth Conditions

Murashige and Skoog (1962) inorganic salts; a range of macroelements, microelements, organic supplements (vitamins and amino acid), and 3% (w/v) sucrose (carbon source); and 0.8% (w/v) agar-agar (Qualigens Fine Chemicals, Mumbai, India) were used to establish the in vitro culture. The MS medium was supplemented with varying concentrations of aromatic cytokinin *m*T (1.0, 2.5, 5.0, 7.5, and 10.0 M). The media pH was adjusted to 5.8 using 0.1 N NaOH/HCL before adding up of 0.8% agar-agar and autoclaved at 103 kPa and 121° for 15 min. Further, aseptic culture containing explants were placed in a growth chamber having controlled environmental provision of illumination intensity (50μmol m^{-2} s^{-1}), white fluorescent lamp of 2 × 40 W (Philips India Ltd. Kolkata), and 16 h photoperiod (25 ± 2 °C).

13.2.3 Rooting and Hardening Phase

The consequence of the *m*T on the rooting proficiency in *T. indica* shoots was estimated. The *m*T-derived microshoots were exposed to rooting medium MS full or half strength without growth regulators or complemented with IBA at different concentrations (0.1, 0.5, 1.0, and 2.0 μM). The number of rooted shoots was confirmed after 3 weeks duration. The experiment was repeated twice. The rooted plantlets were moistened with 1/4 MS basal salt solution and were planted in a pot containing soilrite, covered with polythene bags and kept inside growth chamber for acclimatization process. After 4 weeks of acclimatization phase, the microshoots were relocated to claypots holding garden soil/organic manure (1:1) and placed in greenhouse for hardening phase.

13.2.4 Estimation of Biomass

Fresh weights of in vitro-regenerated shoots maintained at different *meta*-topolin concentrations were recorded. Further, the shoots were dehydrated in shade for 24 h pursued by drying in an oven control at 45 °C for 48 h. The dry matter production was estimated by weighing the dried shoots. The mean standards were expressed in milligram (mg/FW).

13.2.5 Determination of Photosynthetic Pigments

Photosynthetic pigments (Chl *a* and *b* and carotenoid) were extracted from leaf tissues of *T. indica* cultures exposed to different *m*T levels and estimated by MacKinney (1941) and Maclachlan and Zalick (1963) techniques, respectively. The fresh heap of leaves (200 mg) was homogenized in mortar with 80% acetone solutions and filter with Whatman No. 1 filter paper. The obtained extract was diluted to make 10 ml final volume. The optical densities for the chlorophyll solutions at 645 and 663 nm and carotenoids at 480 and 510 nm wavelengths were measured using UV-Vis spectrophotometer (UV-1700 Pharma Spec, Shimadzu, Kyoto, Japan). These contents were indicated in mg/g fresh weight (FW).

13.2.6 Proline Determination

The proline content in the shoots of *Tylophora* regenerants was verified. About 300 g fresh leaf material was homogenized in 10 ml of 30% aqueous sulfosalicylic acid. Further the homogenate was centrifuged at 9000 rpm for 15 min, and about 2 ml supernatant aliquot was mixed with an equal volume of acetic acid and ninhydrin (1.25 g) in 30 ml acetic acid 20 ml of 6N H_3PO_4 and incubated for 1 h at 100 °C. The reaction was terminated in an ice bath and extracted with 4 ml of toluene and vortexed for 20 s. The chromatophore-containing toluene was further aspirated from aqueous phase, and the absorbance was analyzed at 520 nm using UV-Vis spectrophotometer (UV-1700 Pharma Spec, Shimadzu, Kyoto, Japan).

13.2.7 Statistical Investigation

Records were exposed to one-way analysis of variance (ANOVA) via SPSS program (SPSS Inc., edition 16.0, Chicago USA). Duncan's multiple range test (DMRT) at the statistical importance $P \leq 0.05$ and the evidence of significance of differences among mean values and the outcomes are articulated as the means ± SE of three experimentation. The morphological changes among the cultures were observed and documented at regular period.

13.3 Results and Discussion

13.3.1 The Ramification of *Meta*-topolin on Shoot Intensification

The selection and employment of cytokinins (CKs) in the plant tissue culture system are decisive through its combined competency in stimulating satisfactory rate of shoot multiplication, root initiation, and acclimatization proficiency of micropropagated plants. Besides, the critical factors like plant species, culture environment, etc. but also the type and the application of CKs could formulate or destroy the achievements of micropropagation procedures. The percentage response, shoot numbers, and shoot length were taken into consideration during the experiment. The micropropagation protocol was ascertained where MS medium was prepared with different concentrations (0.5–10.0 μM) of *meta*-topolin in *T. indica* plant through nodal explants. Shoot bud stimulation and development rate showed that *m*T (5.0 μM) gave highest quantity of shoots (12.3 ± 0.22) with a standard shoot length (5.50 ± 0.00 cm)/explant at 4 weeks incubation period (Table 13.1, Fig. 13.1a). Though, an enhancement in *m*T application above optimal level (5.0 μM) hinders the shoot initiation frequency and proved decline in multiplication rate as well as showed the development of callus masses at the base. Plant growth regulators signifying routes are not secluded; rather they are interrelated with a multifaceted regulatory arrangement concerning diverse defense indication pathways and developmental procedures. Besides, Criado et al. (2007) have well

Table 13.1 Effect of *meta*-topolin on shoot bud induction from nodal explants after 4 weeks of culture duration

Plant growth regulators *m*T (μM)	Regeneration (%)	Mean number of shoots	Mean shoot length (cm)
0.0	00	0.00 ± 0.00^{g}	0.00 ± 0.00^{f}
0.5	70	4.16 ± 0.06^{f}	2.34 ± 0.08^{e}
1.0	70	7.46 ± 0.288^{d}	3.48 ± 0.19^{d}
2.5	85	9.12 ± 0.04^{c}	4.10 ± 0.03^{c}
5.0	95	12.3 ± 0.22^{a}	5.50 ± 0.00^{a}
7.5	70	9.96 ± 0.02^{b}	3.94 ± 0.04^{c}
10.0	70	6.34 ± 0.14^{e}	4.64 ± 0.16^{b}

Values represent means ± standard error of 20 replicates per treatment in three repeated experiments. Means followed by the superscript alphabets are not significantly different ($P = 0.05$) using Duncan's multiple range test

discussed the managerial functions of CKs on diverse physiological and development progression in plants growth. Presently, several evidence showing the frequent employment of *meta*-topolin above the regularly employed CKs with a surge of promising results in tissue culture system (Aremu et al. 2012a, b). However, the effects of the arrangement precision in cytokinin action have been implicit. Substantial outcome on the action of cytokinins has been monitored by changing the arrangement of the hydroxyl set on the side chain (Kaminek et al. 1979). The occurrence of hydroxyl group in topolins grants them structural benefits, which help them to undertake O-glycosylation to outline storage forms (Bairu et al. 2008).

13.3.2 Positive Interactions of Auxin-Cytokinin

Auxin-cytokinin combination has a crucial role in several aspects of plant development and improvement. Their interactions are acknowledged to achieve the indispensable system of cell propagation and differentiation and manage morphological practices (meristem appearance and progress). The most favorable *m*T (5.0 μM) amalgamation with diverse auxins (NAA, IAA, and IBA) at a range of applications confirmed enhancement in shoot multiplication and propagation. After 8 weeks of culture stipulation, it was established that the utmost number of shoots (18.2 ± 0.12) with shoot length (5.40 ± 0.00 cm) was confirmed on MS medium holding the synergistic amalgamation of *m*T (5.0 μM) with low concentration/level of NAA (0.5 μM) (Table 13.2, Fig. 13.1b). The deviation in morphogenetic effects is supposed to be due to apical supremacy in plants which is directed by the cytokinin:auxin ratios and originated the action of basipetal transport of auxin from the apex which results in the reduction of axillary bud growth. The arrangement of *m*T with auxins have been accepted to prove the considerable effects designed for enhanced shoot multiplication in various plants like *Aloe polyphylla* (Bairu et al.

Fig. 13.1 (**a**) Stimulation of multiple shoots from nodal segments in *T. indica* on MS + *m*T (5.0 μM), after 4 weeks of culture maintenance. (**b**) Multiplication and proliferation of shoots from nodal segments in *T. indica* on MS + *m*T (5.0.μM) + NAA (0.5 μM), after 8 weeks of culture maintenance. (**c**) Rooting in in vitro-regenerated microshoots on 1/2 MS + IBA (1.0 μM), after 4 weeks duration. (**d**) In vitro-rooted plantlets of *T. indica* ready to transplant. (**e**) 4-week-old acclimatized plantlets of *T. indica* in soilrite

Table 13.2 Effect of optimal concentration of *m*T (5.0 μM) along with different auxins on shoot rejuvenation potentials in *T. indica* from nodal explants, after 8 weeks of culture condition

Plant growth regulators (μM)					
IAA	IBA	NAA	Regeneration (%)	Mean number of shoots	Mean shoot lengths
0.1			70	5.20 ± 0.05	2.08 ± 0.04^{g}
0.5			85	8.08 ± 0.04	4.62 ± 0.07^{c}
1.0			82	6.36 ± 0.18	4.00 ± 0.00^{d}
2.0			75	3.12 ± 0.04	3.64 ± 0.16^{e}
	0.1		75	5.24 ± 0.11	3.30 ± 0.12^{f}
	0.5		95	10.3 ± 0.15	5.30 ± 0.12^{a}
	1.0		85	7.40 ± 0.12	4.04 ± 0.02^{d}
	2.0		80	6.64 ± 0.16	3.64 ± 0.16^{e}
		0.1	85	6. 40 ± 0.10	4.96 ± 0.04^{b}
		0.5	90	18.2 ± 0.12	5.40 ± 0.00^{a}
		1.0	85	12.0 ± 0.05	4.00 ± 0.00^{d}
		2.0	80	9.06 ± 0.04	3.60 ± 0.00^{e}

Values represent means ± standard error of 20 replicates per treatment in three repeated experiments. Means followed by the superscript alphabets are not significantly different ($P = 0.05$) using Duncan's multiple range test

2007a, b), *Uniola paniculata* (Valero-Aracama et al. 2010), and *Pterocarpus marsupium* (Anees and Anis 2019).

However, the quality and quantity of shoots and shoot length were considerably low on medium supplemented with IBA and IAA when contrasted with effects of NAA on morphogenesis of plant. Throughout the phase of plants multiplication and growth in plant tissue culture regime, the low ratio of auxins accompanied by the high concentrations of cytokinins is employed for stimulation of successful shoot proliferation and subsequent development (Su and Zhang 2011). Auxins are involved in directing numerous characteristic of expansion and improvement in plants in vitro. The manifold functions of this plant hormone engage the improvement of shoot fabrication, cell divisions, and demarcation of conductive tissues, cell elongation, and encouraging apical dominance, root growth, flower initiation, etc. (Vogler and Kuhlemeier 2003; Santner et al. 2009).

13.3.3 Subculture Passages

Owing to the vigorous expansion of cultures onto the MS medium complemented with *m*T (5.0 μM), the shoot cultures were subculture on the freshly prepared MS medium at 3 weeks interval. Translocation and concentrated accumulation of *meta*-topolin in plant tissues are rapid that avert its confined accretion (Kaminek et al. 1987). The formation of *O*-glucoside metabolites is feasible by the action of hydroxyl cluster present in side chain of *m*T (Werbrouck et al. 1996). The *O*-glucosides are regarded as the storage forms of cytokinins which are stable under specific conditions but quickly transformed to active bases of cytokinins once required (Werbrouck et al. 1996). Moreover, the overturn isolation of *O*-glucoside consecutively assigns regular accessibility of cytokinins at physiological functions over a prolonged time period which returns in elevated in vitro shoot formation (Strnad 1997).

13.3.4 Root Progression

Formation of roots has major realistic implications in forestry, agronomy, and horticulture with commercial interest. The development of maximum shoots quantity which rooted easily, acclimatized, and established in the field conditions remains imperative for successful regeneration protocol. Moreover, it is well acknowledged that in vitro root arrangement in plants relies on the superiority of shoots. De Klerk (2002) has documented that application of exogenous cytokinin is important, inter alia, for the initiation of root pattern. *m*T has revealed an immense guarantee as a potent CK which might be supportive on the way to prevent rooting hindrance and acclimatization malfunctions within plant tissue culture system.

Meta-topolin functions as a plant growth hormone which carries out in vitro rooting capacity in the multiplication medium. It was observed that the emergence of roots appeared in the regeneration medium along with the shoot multiplication within the 2 weeks period. MS medium containing 5.0 μM *m*T showed rooting in 95% of the regenerated shoots. However, this response in root induction was found continued as soon as the cultures were further shifted to fresh MS medium containing *m*T. In vitro rooting achievements in the multiplication medium were previously observed in several plant species, namely, *Spathiphyllum floribundum* (Werbrouck et al. 1996), *Musa* spp. (Bairu et al. 2008), etc.

Further, a set of experiment was designed to evaluate the after effects on the rooting competency of the *m*T-initiated microshoots of *T. indica* onto the control MS medium (lacking of growth regulators) and in the existence of 1/2 strength of MS with IBA levels (0.1–2.0 μM). It was recorded that 1/2 MS stimulated the root initiation at a low concentration of IBA (0.1 μM) after 2 weeks of incubation. Auxins have been utilized as a growth regulators in micropropagation system designed for rooting process (Gaspar et al. 1996), particularly IBA reported in plant proliferation owing of its competency for encouraging adventitious roots (Pop et al. 2011). Nevertheless, the supremacy of IBA at lower concentrations was extremely practical in the current experiment. Half strength MS and a low concentration of IBA (1.0 μM) were found to be optimal for maximum regeneration of roots (8.64 $\pm$ 0.04) with root length 4.12 $\pm$ 0.03 cm when recorded after 4 weeks (Table 13.3, Fig. 13.1c). The efficacy of external relevance of IBA during root initiation and development has been acknowledged in many plants, namely, *Leptadenia reticulata* (Patel et al. 2014), *Embelia ribes* (Annapurna and Rathore 2010), and *Allamanda cathartica* (Khanam et al. 2020).

Table 13.3 Effect of different strength of MS medium in combination with different concentrations of auxin (IBA) on in vitro root initiations in *T. indica* microshoots after 4 weeks of culture

MS strength + plant growth regulator IBA	Regeneration (%)	Mean number of roots	Mean root length (cm)
MS	00	0.00 ± 0.00^{f}	$.00 \pm 0.00^{g}$
MS + IBA (0.1 μM)	85	6.16 ± 0.11^{b}	$3.52. \pm 0.13^{f}$
MS + IBA (0.5 μM)	80	5.42 ± 0.02^{c}	2.84 ± 0.02^{d}
MS + IBA (1.0 μM)	72	4.80 ± 0.00^{d}	2.50 ± 0.00^{e}
MS + IBA (2.0 μM)	70	4.00 ± 0.00^{e}	2.16 ± 0.04^{f}
1/2 MS + IBA (0.1 μM)	92	8.64 ± 0.16^{a}	5.18 ± 0.29^{a}
1/2 MS + IBA (0.5 μM)	85	6.12 ± 0.04^{b}	4.12 ± 0.03^{b}
1/2 MS + IBA (1.0 μM)	85	5.00 ± 0.00^{d}	3.40 ± 0.10^{c}
1/2 MS + IBA (2.0 μM)	80	4.12 ± 0.73^{e}	$3.00 \pm 0..00^{c}$

Values represent means ± standard error of 20 replicates per treatment in three repeated experiments. Means followed by the superscript alphabets are not significantly different ($P = 0.05$) using Duncan's multiple range test

The optimal level of cytokinins and other plant hormones (auxins) is critical for flourishing organ developments like root emergence, arrangement from multiplicated explants, etc. Subsequent outcomes regarding most favorable endogenous plant hormone levels and their reliance on diverse exogenous cytokinins are employed in the nurturing media which could facilitate progress in in vitro rooting effectiveness of the plant species.

13.3.5 Carryover Effects of *m*T on Acclimatization

Commercially, a satisfactory rooting and rapid acclimatization of in vitro regenerants can't be under prediction (Pospísilová et al. 2007). However, a low survival rate recorded in in vitro-rooted shoots during the acclimatization process might be owing to its frail character of the roots fashioned, susceptible to harm during roots transplant which could direct to plant loss (Deberg and Maene 1981). The tissue-raised plantlets showed about 96.0 ± 0.91% survival rate when transferred in soilrite planting substrate (Table 13.4, Fig. 13.1d, e). Soilrite's ultimate porosity and better water-holding capability make it a superior planting material. The hardened plantlets were shifted to the earthen pots holding garden soil and preserved into the polyhouse and watered regularly at frequent interval. No morphological variations were observed in acclimatization plants even after 3 months of adaptation in greenhouses. Adaptation of TC-raised plants to natural environment is an imperative and chief bottleneck towards flourishing micropropagation system and needs to be advanced cautiously.

Bairu et al. (2008) accounted that the category as well as quantity of cytokinins has a potent outcome on in vitro acclimatization capacity in plants. In the present investigation, *m*T is constantly maintained as the most promising cytokinin by

Table 13.4 Estimation of different planting substrates for hardening of in vitro raised plantlets of *T. indica* after 4 weeks of transfer planting substrates

Planting substrate	No. of plants transferred	No. of surviving plants	% survival rate
Vermicompost	50	38	70.2 ± 0.12^{c}
Soilrite™	50	46	96.0 ± 0.91^{a}
Garden soil: Vermicompost (3:1)	50	40	90.2 ± 0.26^{b}

Values represent means ± standard error of 20 replicates per treatment in three repeated experiments. Means followed by the superscript alphabets are not significantly different ($P = 0.05$) using Duncan's multiple range test

motivating improved acclimatization in micropropagation of *T. indica*. Werbrouck et al. (1996) detailed *meta*-topolin efficiency in improving peace lily multiplication of shoots and in vitro rooting at equimolar concentrations. The positive action of *m*T as a substitute of BA for the micropropagation of *Uniola paniculata* which progresses in rooting and acclimatization survival (Valero-Aracama et al. 2010). The amplified ex vitro adjustment may possibly be because of the outcome of an additional speedy yield of the *m*T chief imitative N^6-(3-*o*-β-D-glucopgyranosyl) benzyladenine-9-riboside (Strnad et al. 1997).

13.3.6 Effect on Biomass Content

A considerable enhancement in the total biomass of shoots exposed to different *m*T levels was observed in regenerated plants of *T. indica*. Maximum fresh biomass content (2.80 ± 0.02 g FW) was recorded on MS medium containing optimal *m*T (5.0 μM) (Fig. 13.2); however, this ratio was found to decline as the concentration of *m*T increases in the medium. Similar pattern in biomass content was reported in *Opuntia stricta* (de Souza et al. 2019).

13.3.7 Performance of Photosynthetic Pigments

Chlorophyll and carotenoids are the critical pigments which are involved in plants photosynthesis. *m*T has been established as an effective cytokinin in stimulation of chlorophyll levels in regenerated plants. The chlorophyll *a* and *b* contents were found to be increased in the *m*T-treated plants when compared with control. Chl *a* (0.45 mg/g FW) and Chl *b* (0.40 ± 0.01 mg/g FW) were monitored on *m*T (5.0 μM) (Fig. 13.3). The enhancement in photosynthetic pigments in plants was observed up to an optimal level of *m*T (5.0 μM); however, a prolonged culture period with increased *m*T concentrations reduced pigment contents. Though, a gradual breakdown in pigments was also recorded on higher concentrations. Carotenoid contents (0.22 ± 0.02 mg/g FW) enhanced considerably up to an optimal level of *m*T (5.0 μM) in the medium, and subsequently a decline was observed. Carotenoid

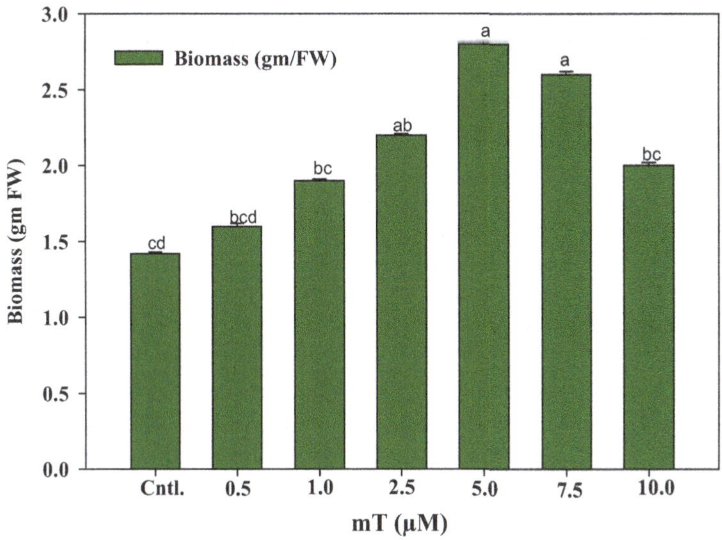

Fig. 13.2 Effect of *m*T on biomass contents in *T. indica* cultures. Lines represent means ± SE. Lines denoted by the same letter within response variables are not significantly different ($P = 0.05$) using Duncan's multiple range test

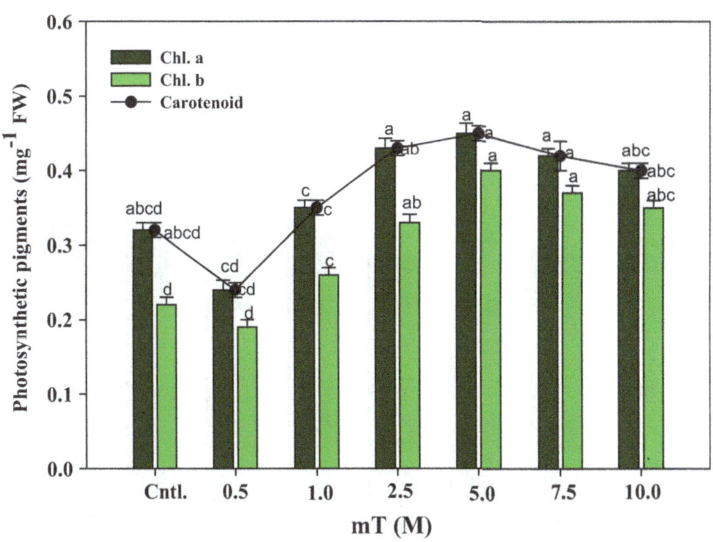

Fig. 13.3 Effect of *m*T on chlorophylls *a* and *b* and carotenoid contents in *T. indica* regenerants. Bars denoted by the same letter within response variables are not significantly different ($P = 0.05$) using Duncan's multiple range test

plays a revital role as a light harvesting molecules and perform an essential photoprotective functions by providing defense of the reaction centre from the damaging causes of excess light and oxygen (Lichtenthaler 1987; Young 2006). It has been recommended that cytokinin administers its effect on chloroplast existence, resolved by RNA synthesis in the nucleus (Legocka and Szweykowska 1981). Fakuda and Toyama (1982) reported that CKs increases the chlorophyll contents due to the reducing action of chlorophyllase.

13.3.8 Proline Contents

In the present research, the proline accretion was reviewed in the mature regenerants at various applications of *m*T (0.5–10.0 μM) in the medium. The maximum proline content (0.26 ± 0.02 μg/g FW) in plants was found on the higher concentrations of *m*T (10.0 μM) (Fig. 13.4). The enhancement in proline levels was associated to increased level of *m*T. The function of proline accumulation and its promising contribution in the adaptive mechanism under stress condition has been appraised (Ali et al. 1998; Fatima et al. 2011). Proline is a multifunctional amino acid that protects plant tissues against various stresses induced in vitro by acting as an osmoregulatory, N-storage compound, protectant of hydrophobic enzymes, stabilizes proline synthesis etc. (Greenway and Munns 1980). Earlier reports demonstrate

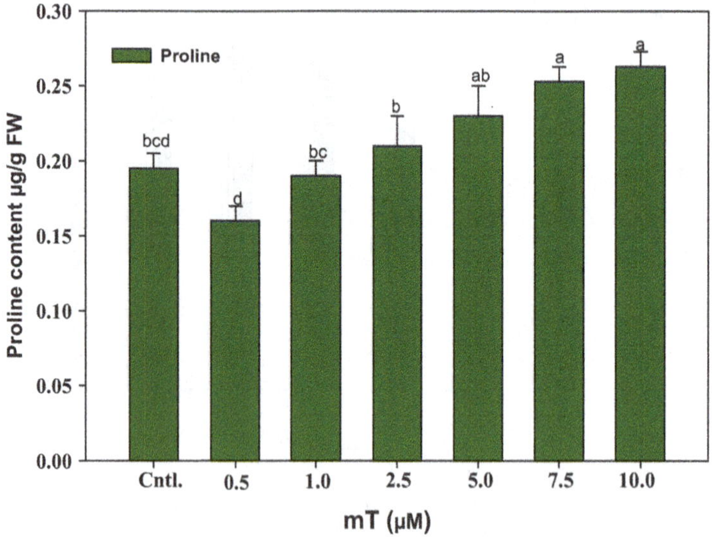

Fig. 13.4 Effect of *m*T on proline content in *T. indica* regenerants. Bars represent means. Bars denoted by the same letter within response variables are not significantly different ($P = 0.05$) using Duncan's multiple range test

the positive role of proline in the regenerated plants (Aremu et al. 2012a, b; Naaz et al. 2019; Maria de Souza et al. 2019).

13.4 Conclusion and Future Prospects Perpetuate

The present exploration reports the establishment of prosperous morphogenetic system in *T. indica* by the exposure of a BA analogue, i.e., *meta*-topolin. Verifying the determined optimal exogenous plant hormone levels and their interactions may progress the in vitro micropropagation efficiency and shows a significant carryover effect as well as excellency of acclimatized plants ex vitro. The micropropagation structure, germplasm conservation approach, and adjustment potential can be supportive in perpetuation of this species. Moreover, bioactive molecule is a potential indicator for diverse repository of essential compounds based on industrial produce without inducing any variability; also it necessitates to discover other effective phyto-constituents from the plant with valuable pharmacological properties which can serve as a source of novel high-quality formulations. A well-developed photosynthetic apparatus boosts the constant existence of in vitro plantlets during acclimatization phase. The role of *meta*-topolin contributed to the improved acclimatization competency in regenerated plants.

The employment of low levels of *m*T under in vitro environment emphasizes the immense potential requirements of plant compounds in plant tissue culture system. Thereby, it would be a valuable practice to significantly expose the primary mechanisms of actions of this compound. The region containing natural population of *T. indica* must be identified and extend as conservatories by condensing anthropogenic disturbance. Micropropagation and eco-restoration indeed sustain the in situ conservation (maintenance and management) behavior and ease population augmentation in species wherever natural proliferation is delayed because of destructive harvesting or reproductive obstructions. Confronted by such unprecedented genetic erosion and vanishing of species and ecosystems, conservation of natural resources assumes paramount urgency.

Acknowledgments The authors extended their gratitude to University Grants Commission, New Delhi, for supporting this Postdoctoral Fellowship (UGC-PDFWM 2017-22) to Dr. Nigar Fatima. UGC-BSR Faculty Fellowship Programme (2017) by University Grants Commission, New Delhi, to M. Anis is gratifying.

Statement of Contending of Interest The authors state that there is no argument of interest.

References

Ali M, Butani KK (1989) Alkaloids from *Tylophora indica*. Phytochemistry 28:3513–3517

Ali G, Shrivastava PS, Iqbal M (1998) Morphological response and proline content in *Bacopa monniera* cultures grown under copper stress. Plant Sci 138:190–197

Annapurna D, Rathore TS (2010) Direct adventitious shoot induction and plant regeneration of Embelia ribes F. Plant Cell Tissue Organ Cult 101:269–277

Anees A, Anis M (2019) Metatopolin improves *in vitro* morphogenesis, rhizogenesis and biochemical analysis in *Pterocarpus marsupium* Roxb.: a potential; drug-yielding tree. J Plant Growth Regul 38:1007–1016

Aremu AO, Bairu MW, Doležal K, Finnie JF, van Staden J (2012a) Topolins: a panacea to plant tissue culture challenges? Plant Cell Tissue Organ Cult 108:1–16

Aremu AO, Bairu MW, Szüčová L, Doležal K, Finnie JF, Van Staden J (2012b) Assessment of the role of meta-topolins on *in vitro* produced phenolics and acclimatization competence of micropropagated 'Williams banana'. Acta Physiol Plant 34:2265–2273

Bairu MW, Stirk WA, Doležal K, van Staden J (2007a) Optimizing the micropropagation protocol for the endangered *Aloe polyphylla*: can meta-topolin and its derivatives serve as replacement for benzyladenine and zeatin? Plant Cell Tissue Organ Cult 90:15–23

Bairu MW, Stirk WA, Doležal K, van Staden J (2007b) The role of topolins in micropropagation and somaclonal variation of banana cultivars 'Williams' and 'Grand Naine' (Musa spp. AAA). Plant Cell Tissue Organ Cult 95:373–379

Bairu MW, Stirk WA, Dolezal K, van Staden J (2008) The role of topolins in micropropagation and somaclonal variation of banana cultivars 'Williams' and 'Grand Naine' Musa spp. AAA. Plant Cell Tissue Organ Cult 95:373–379

Bhutani KK, Sharma GL, Sarin AN, Kaur R, Kumar V, Atal CK (1985) *In vitro* amoebicidal and bactericidal activities in medicinal plants. Indian J Pharm Sci 47:65–67

Cag S, Palavan-Ünsal N, Büyüktuncer D (2003) Comparison of the effects of meta-topolin and other cytokinins on chlorophyll and protein contents and peroxidase activity in cucumber cotyledons. Israel J Plant Sci 51:261–265

Criado M, Roberts I, Echeverria M, Barneix A (2007) Plant growth regulators and induction of leaf senescence in nitrogen-deprived wheat plants. J Plant Growth Regul 26:301–307

CSIR (2003) The wealth of India: a dictionary of Indian raw materials and industrial products, vol 10. Council of Scientific and Industrial Research, New Delhi, pp 398–399

De Klerk (2002) Rootings of microcuttings. Theory and practices. In Vitro Cell Dev Biol Plant 38:415–422

Deberg PC, Maene LJ (1981) A scheme for commercial propagation of ornamental plants by tissue culture. Sci Hortic 14:335–345

Donaldson GR, Atkinsn MR, Murray AW (1968) Inhibition of protein synthesis in Ehrlich ascites-tumor cells by the phenanthrene alkaloids tylophorine, tylocrebrine and cryptopleurine. Biophys Res Commun 31:104–109

Faisal M, Ahmad N, Mohd A (2007) An efficient micropropagation system of *Tylophoraindica*: an endangered, medicinally important plant. Plant Biotechnol Rep 1:155–161

Fakuda K, Toyama S (1982) Electron microscope studies on the morphogenesis of plastids XI. Ultrastructural changes of the chloroplasts in tomato leaves treated with ethylene and kinetin. Cytologia 47:725–736

Fatima N, Ahmad N, Anis M (2011) Enhanced *in vitro* regeneration and change in photosynthetic pigments, biomass and proline content in *Withania somnifera* L. (Dunal) induced by copper and zinc ions. Plant Physiol Biochem 49:1465–1471

Gaspar T, Kevers C, Penel C, Greppin H, Reid D, Thorpe T (1996) Plant hormones and plant growth regulators in plant tissue culture. In Vitro Cell Dev Biol Plant 32:272–289

George EF, Hall MA and De Klerk GJ (2008). Plant propagation by tissue culture. The background, 3rd Edn. Springer, Dordrecht. 1:149–190.

Greenway H, Munns R (1980) Mechanisms of tolerance in nanohalophytes. Annu Rev Plant Physiol 31:149–190

Hussain SA, Ahmad N, Anis M, Alatar AA (2019) Influence of metatopolin on *in vitro* organogenesis in *Tecoma stans* L., assessment of genetic fidelity and phytochemical profiling of wild and regenerated plants. Plant Cell Tissue Ogan Cult 138:339–351

Kaminek M, Paces V, Corse J, Challice JS (1979) Effect of stereo-specific hydroxylation of N6 (2-Isopentenyl) adenosine on cytokinin activity. Planta 145:239–243
Kaminek M, Vanek T, Motyka V (1987) Cytokinin activities of N6-benzyladenosine derivatives hydroxylated on the side-chain phenyl ring. J Plant Growth Regul 6:113–120
Khanam MN, Javed SB, Anis M, Alatar AA (2020) meta-Topolin induced in vitro rgeneration and metabolic profiling in Allamanda cathartica L. Ind Crop Prod 145:111944
Legocka J, Szweykowska A (1981) The role of cytokinins in the development and metabolism of barley leaves III, the effect of RNA metabolism in various cell compartments during senescence. Z Pflanzenphysiol 102:363–374
Lichtenthaler H (1987) Chlorophyll and carotenoids: pigments of photosynthetic biomembranes. Methods Enzymol 148:350–382
MacKinney G (1941) Absorption of light by chlorophyll solution. J Biol Chem 140:315–322
Maclachlan S, Zalick S (1963) Plastid structure, chlorophyll concentration and free amino acid composition of chlorophyll mutant barely. Can J Bot 7:1053–1062
Mulchandani NB, Iyer SS, Badheka LP (1971) Structure of tylophorinidine: a new potential antitumor alkaloid from *Tylophora indica*. Chem Ind 19:505–506
Murashige T, Skoog F (1962) A revised medium for rapid and bioassays with tobacco tissue cultures. Physiol Plant 15:473–497
Naaz A, Hussain SA, Anis M, Alatar AA (2019) *Meta*-topolin improved micropropagation in *Syzium cumini* and acclimatization to *ex vitro* conditions. Biol Plant 63:174–182
Patel AK, Agarwal T, Phulwari M, Vinod K, Shekhawat NS (2014) An efficient in vitro plant regeneration system from leaf of mature plant of Leptadenia reticulate (Jeewanti): A life giving endangered woody climber. Ind Crop Prod 52:499–505
Pop T, Pamfi D, Bellini C (2011) Auxin control in the formation of adventitious roots. Not Bot Hort Agrobot 39:307–316
Pospísilová J, Synková H, Haisel D, Semorádová S (2007) Acclimation of plantlets to *exvitro* conditions: effects of air humidity, irradiance, CO2 concentration and abscisic acid (a review). Acta Hortic 748:29–38
Rubio-Wilhelmi MM, Sanchez-Rodriguez E, Rosales MA, Begona B, Rios JJ (2011) Effect of cytokinins on oxidative stress in tobacco plants under nitrogen deficiency. Environ Exp Bot 72:167–173
Santner A, Calderon-Villalobos LIA, Estelle M (2009) Plant hormones are versatile chemical regulators of plant growth. Nat Chem 5:301–307
de Souza M, Barbosal MR, Rafael Zarate-Salazar J, Lozano-Isla F, Camara TR (2019) Use of meta-topolin, an unconventional cytokinin in the *in vitro* multiplication of *Opuntia stricta* Haw. Biotechnol Veg 19:85–96
Strnad M (1997) The aromatic cytokinins. Physiol Plant 101:674–688
Strnad M, Hanus J, Vanek T, Kaminek M, Ballantine JA, Fussell B, Hanke DE (1997) Meta-topolin, a highly active aromatic cytokinin from poplar leaves (*Populus canadensis* Moench, cv. Robusta). Phytochemistry 45:213–218
Su YH, Zhang XS (2011) Auxin-cytokinin interaction regulates meristem development. Mol Plant 4:616–625
Tarkowski P, Ge L, Yong JWH, Tan SN (2009) Trends in analytical chemistry, vol 28, pp 323–335
Thomas TD, Philip B (2005) Thidiazuron induced high frequency shoot organogenesis from leaf derived callus of a medicinal climber. *Tylophora indica* (Burm. f.) Merrill. In Vitro Cell Dev Biol Plant 41:124–128
Valero-Aracama C, Kane M, Wilson S, Philman N (2010) Substitution of benzyladenine with meta-topolin during shoot multiplication increases acclimatization of difficult- and easy-to acclimatize sea oats (*Uniola paniculata* L.) genotypes. Plant Growth Regul 60:43–49
Vogler H, Kuhlemeier C (2003) Simple hormone but complex signaling. Curr Opin Plant Biol 6:51–56

Wang Z, Pote J, Huang B (2003) Responses of cytokinins, antioxidant enzymes and lipid peroxidation in shoots of creeping bentgrass to high root-zone temperatures. J Am Soc Hortic Sci 128:648–655

Werbrouck SPO, Strnad M, Van Onckelen HA, Debergh PC (1996) Metatopolin an alternative to benzyladenine in tissue culture. Physiol Plant 98:291–297

Werner T, Motyka V, Strnad M, Schmulling T (2001) Regulation of plant growth by cytokinin. Proc Natl Acad Sci U S A 98:10487–10492

Young AJ (2006) The protective role of carotenoids in higher plants. Physiol Plant 83:702–708

Use of *Meta*-topolin in Somatic Embryogenesis

14

Carolina Sánchez-Romero

Contents

Abstract

Somatic embryogenesis is the developmental process through which a somatic cell or group of somatic cells give rise to an embryo, capable of developing into a whole plant. Somatic embryogenesis is a powerful in vitro technique, with multiple applications, including plant breeding by conventional or biotechnological means. Technologically, it is carried out through a sequence of in vitro culture steps. Although different physical and chemical factors influence the process, plant growth regulators play an essential role.

Cytokinins are one of the most important hormones regulating somatic embryogenesis, and, although hormone requirements vary greatly in different experimental systems, cytokinins are frequently included in different phases of this developmental process. Since their discovery as naturally occurring

C. Sánchez-Romero (✉)
Departamento de Botánica y Fisiología Vegetal, Universidad de Málaga, Málaga, Spain
e-mail: c.sanchez@uma.es

N. Ahmad, M. Strnad (eds.), *Meta-topolin: A Growth Regulator for Plant Biotechnology and Agriculture*, https://doi.org/10.1007/978-981-15-9046-7_14

cytokinins, *meta*-topolin and some of its derivatives have been increasingly used in plant tissue culture. Their use was initially limited to shoot multiplication, rooting, and seed germination, and only recently, this plant growth regulator has been applied to somatic embryogenesis protocols.

In the present chapter, the use of *meta*-topolin and its derivatives in somatic embryogenesis is revised. I examine their effects in the different phases of the somatic embryogenesis process and compare the results with those of other cytokinins.

Keywords

Biotechnology · Cytokinin · *Meta*-topolin · Plant growth regulators · Regeneration · Somatic embryogenesis

14.1 Introduction

Cytokinins are a group of plant hormones with an essential role in promoting cell division, a central process in plant growth and development. They are also involved in other physiological and developmental processes, including leaf senescence, apical dominance, formation and activity of apical meristems, promotion of sink activity, vascular development, and breaking of bud dormancy (Taiz et al. 2015).

Structurally, cytokinins can be divided into two main groups: adenine-type and phenylurea cytokinins. Adenine-type cytokinins can in turn be classified as aromatic or isoprenoid cytokinins, according to the nature of their N^6-side chain (Beyl 2011).

All the natural cytokinins are adenine derivatives, while phenylurea cytokinins are synthetic compounds exhibiting cytokinin activity, which have not been identified in plants (Bogaert et al. 2006; Taiz and Zeiger 2010).

Meta-topolin (*m*T) (N^6-(*meta*-hydroxybenzyl)adenine) is a naturally occurring aromatic cytokinin, which has been identified in different plant species (Strnad 1997; Beyl 2011). Chemically, *m*T is a hydroxylated N^6-benzyladenine (BA) analog, differing from BA only by the presence of an extra OH group in the *meta*-position on the aromatic ring of BA (Werbrouck 2010). A number of *m*T derivatives have also been identified in different plants species: 6-(3-methoxybenzylamino)purine (*meta*-methoxytopolin, Me*m*T), 6-(3-methoxybenzylamino)purine-9-riboside (Me*m*TR), 6-(3-fluorobenzylamino)purine (3FBA), and 6-(3-fluorobenzylamino) purine-9-riboside (3FBAR) (Bogaert et al. 2006).

As reported by Aremu et al. (2012), since the discovery of *m*T and its derivatives as natural aromatic cytokinins, their use in plant tissue culture has increased rapidly. Positive results have been obtained in different in vitro culture processes, such as shoot multiplication, rooting, and seed germination, improving shoot quality, shoot length, number of leaves, shoot dry weight, shoot/root ratio, and acclimatization competence. Interestingly, multiple investigations reported on a corrective role in some physiological disorders including abnormality reduction, hyperhydricity and shoot tip necrosis alleviation and delayed senescence (Aremu et al. 2012). In line with this, Solórzano-Cascante et al. (2018) reported that *m*T has less negative carryover effects, i.e., residual effects of the compound beyond the time period in which it was present.

Thus, *m*T has proved to be valuable in micropropagation systems and can be considered a suitable alternative to cytokinins traditionally used for this purpose. However, few investigations have described the use of this plant growth regulator (PGR) in somatic embryogenesis (SE). In the review of Aremu et al. (2012) on the application of topolins in plant tissue culture, no information is included on the utilization of *m*T in SE. However, this plant hormone has subsequently been repeatedly utilized in studies aiming to optimize SE protocols.

In the present chapter, the use of *m*T in SE is examined. The investigations in which this PGR or some of its derivatives have been used in some SE phases are revised, and their effects are compared with those of other cytokinins.

14.2 Somatic Embryogenesis

Somatic embryogenesis is the process through which differentiated somatic cells, single or in small groups, change their developmental program, giving rise to embryogenic cells. These embryogenic cells then follow a differentiation pathway to generate an embryo, which can further develop into a whole plant (Zimmerman 1993; Guan et al. 2016).

14.2.1 Phases of Somatic Embryogenesis

Although the initial morphological patterns of somatic embryo formation can be quite different from those of zygotic embryos and difficult to categorize (Elhiti and Stasolla 2016), subsequent steps followed in somatic embryo development are similar to those of zygotic embryos. Thus, in dicots, further development occurs through the typical consecutive stages, namely, globular, heart, torpedo, and cotyledonary (Winkelmann 2016).

In practice, SE is executed through a sequence of in vitro culture steps. Usually it includes induction of embryogenic cultures, proliferation, development and maturation of somatic embryos, and germination.

Induction of embryogenic cultures refers to all events that reprogram a differentiated somatic cell into an embryogenic cell (Winkelmann 2016). This process, which has recently been divided into different phases, i.e., dedifferentiation, acquisition of totipotency and commitment into embryogenic cells (Elhiti et al. 2013), implies a complete reorganization of the cellular state, including physiology, metabolism, and gene expression (Fehér et al. 2002).

Induction of embryogenic cultures can be accomplished through two pathways: direct SE, when somatic embryos arise directly from the initial explant, and indirect SE, when embryogenic differentiation is preceded by a phase of unorganized growth (Bhojwani and Razdan 1996).

Once established, embryogenic cells can be maintained in the embryogenic state under adequate culture conditions (Bhojwani and Razdan 1996). Proliferation of embryogenic cultures by repetitive subcultures allows multiplication of the original plant material (Lelu-Walter et al. 2013), an important quality of SE.

Embryogenic culture proliferation can occur by formation of proembryogenic masses, i.e., localized groups of meristematic cells (Bhojwani and Razdan 1996), or by secondary embryogenesis, a recurrent process in which new somatic embryos develop from previously existing embryos.

Maintenance of embryogenic cultures can be performed in both solid and liquid medium. Culture in liquid medium enables culture synchronization and multiplication at higher rates than on solid medium (von Arnold et al. 2002). It also permits scaling up in bioreactors and automation of the process (Egertsdotter et al. 2019).

Development and maturation of somatic embryos are usually induced in a separate culture phase by appropriate modifications of culture conditions. During this step, the developmental program switches from a proliferative pathway to a highly organized phase, in which somatic embryos arise and advance through successive developmental stages. Both events occurring during this phase, histodifferentiation and storage product accumulation, largely determine the quality of the obtained embryos and, consequently, the final performance of the SE process.

Finally, somatic embryos are induced to germinate, normally under culture conditions similar to those used for conversion of zygotic embryos. Development of root and shoot gives rise to a somatic plantlet that may exhibit the same characteristics as seedlings (Lelu-Walter et al. 2013).

To execute this developmental pathway efficiently, a number of critical physical and chemical treatments should be applied timely (von Arnold et al. 2005). Although hormones are not the only factors controlling SE, as in other morphogenic events in vitro, addition of PGRs to culture medium is the preferred way to manipulate SE (Jiménez 2005). According to Loyola-Vargas and Ochoa-Alejo (2016), these compounds regulate multiple genes temporally and spatially, which cause the changes in the genetic program of somatic cells and regulate the transition between each embryonic developmental stage.

Although there is no single mechanism for executing each SE step and large variability is found, some generalizations can be made. Thus, it is well established that auxin plays an important role in SE, both in induction of embryogenic cultures and in the subsequent elaboration of the proper morphogenetic events during embryo development (Karami et al. 2009). However, there are also species in which cytokinins, alone or in combination with auxins, induce SE (Altamura et al. 2016). According to Jiménez (2005), the importance of both auxin and cytokinin in the determination of embryogenic response can be explained by their determinant participation in cell division and cell cycle regulation. Nevertheless, in some species, other PGRs such as abscisic acid (ABA), gibberellins, or ethylene have been found to induce SE (Jiménez 2005) or SE has been initiated without hormone supplement.

Proliferation of embryogenic cultures is usually performed in culture media similar to those used for SE initiation (von Arnold 2008), although hormones are generally added at lower concentrations. Auxin is the main factor associated with proliferation and plays an important role in inhibiting development of proembryogenic masses into somatic embryos (von Arnold 2008). Nevertheless, cytokinins and auxin-cytokinin combinations have also been utilized in multiple cases (Jiménez 2005).

According to the role of auxin inhibiting somatic embryo development, development and maturation of these structures have commonly been addressed by reducing or removing this hormone from culture medium. In fact, this phase is normally induced in culture media lacking PGRs. Nevertheless, addition of a cytokinin, alone or in combination with an auxin, has been found to be beneficial for embryo development in some species (Jiménez 2005). In conifers, ABA is used to promote somatic embryo maturation, but its role on this phase is not clear in other plant groups (von Arnold et al. 2002).

Somatic embryo germination is usually induced in culture media without hormonal supplement, although auxins and cytokinins can promote this process (von Arnold 2008). In some cases, gibberellins, alone or in combination with cytokinins, have also been added to convert somatic embryos into plants (Jiménez 2005).

14.2.2 Applications of Somatic Embryogenesis

SE has multiple applications in different fields, such as basic research, genetic improvement, and commercial plant production.

As previously indicated, although the initial steps of somatic and zygotic embryogenesis present important differences, developing somatic embryos pass through stages similar to those described in zygotic embryos (Elhiti and Stasolla 2016). These similarities allow the utilization of SE as a model system to investigate the morphological, physiological, and molecular events occurring during plant embryogenesis (Elhiti and Stasolla 2016).

SE has important advantages in embryological studies as in vitro culture allows the targeted manipulation of culture conditions, which is difficult, if not impossible to perform in vivo. The selective addition or removal of specific chemicals to particular developmental stages is often used as a strategy to investigate the nature of the inductive conditions for the proper development of embryos (Elhiti and Stasolla 2016). Thus, SE may help us understand differentiation, as well as the biochemical and genetic mechanisms involved in the transition from one developmental stage to another.

Once embryogenic cultures have been established, they can produce a theoretically unlimited number of exact copies. This property makes SE an in vitro technique appropriate for large scale production of clonal plants. High proliferation rate, singulation (in most cases embryos can be separately handled), and the bipolar nature of embryos (simultaneous development of root and shoot meristem, which allows conversion into plants in a single step) are some of the advantages of SE over other clonal propagation methods (Janick 1993; Guan et al. 2016). This aside, the possibility of scaling up and automating in bioreactors contributes to reducing labor and costs and increases the reliability of the production process (Egertsdotter et al. 2019).

As pointed out by Murashige (1977), encapsulation of somatic embryos inside an artificial layer creating a capsule makes possible the production of synthetic seeds. The encapsulation technology provides to somatic embryos protection from mechanical damage and a supply of nutrients during germination. Synthetic seeds can be easily handled for storage, transport, and sowing, similar to their zygotic counterparts (Rai et al. 2009); and, therefore, they can be considered delivery systems for somatic embryos used as a means of clonal propagation (Janick 1993).

Cryopreservation, i.e., conservation in liquid nitrogen (−196 °C), is the only safe and cost-effective option for long-term conservation of plant material (Engelmann 2004).

Embryogenic cultures are composed of small, actively dividing cells, with few small vacuoles and a high nucleus-cytoplasmic ratio. These characteristics make them more able to withstand cryopreservation than large, highly vacuolated and differentiated cells (Engelmann 2004). In fact, embryogenic cultures are in general considered amenable tissues to cryopreservation (Bradaï et al. 2017) and have been repeatedly cryopreserved by one-step freezing or slow cooling techniques.

The joint use of SE and cryopreservation enables conservation of genetic resources, elite germoplasm, clonally propagated species, and biotechnological products, while in-field testing of the regenerated plants and other analyses are being carried out (Dunstan et al. 1995; Altamura et al. 2016).

However, the greatest potential use of SE is as regeneration method associated with biotechnological techniques for crop improvement (Janick 1993). In many biotechnological tools, plant improvement is achieved through manipulations at the cellular level, and, consequently, their applicability is only possible if a whole plant can be regenerated from a single cell (Bhojwani and Razdan 1996). Hence, the availability of an efficient regeneration protocol is a requisite for biotechnology exploitation, and, therefore, SE is at the base of some biotechnological applications, such as (1) production of transgenic plants; (2) generation of variant plants through somaclonal variation or by using mutagenic agents; (3) production of haploid and double-haploid plants, by induction of embryogenesis from microspores and subsequent chromosome doubling (Janick 1993); (4) production of somatic hybrids by protoplast fusion from intra- or intergeneric sources (Loyola-Vargas and Ochoa-Alejo 2016); and (5) cell selection against biotic or abiotic stressing agents (Janick 1993).

14.3 *Meta*-topolin and Somatic Embryogenesis

14.3.1 Embryogenic Cultures Initiation

Meta-topolin was first used for induction of SE by Lai et al. (2014). Using zygotic embryos and coleoptile segments from in vitro germinated embryos of *Mesomelaena pseudostygia*, these authors investigated the influence of different auxins and cytokinins on this process.

In a first experiment, Lai et al. (2014) cultured zygotic embryos on half-strength MS medium (Murashige and Skoog 1962) solidified with 6 g L^{-1} agar and supplemented with different auxins at various concentrations and several combinations of 2,4-dichlorophenoxyacetic acid (2,4-D), the auxin inducing higher callusing response, with different cytokinins. Specifically, they tested PGR-free medium, 1, 2, 5, or 10μM 2,4-D or α-naphthaleneacetic acid (NAA), 1, 10, or 40μM picloram; 2μM 2,4-D plus 0.5 or 1μM BA; 2μM 2,4-D plus 1, 2, or 5μM *m*T; 2μM 2,4-D plus 1, 5, or 10μM thidiazuron (TDZ); and 5μM 2,4-D plus 0.5 or 1μM BA.

Treatments in which basal medium was supplemented with 2μM 2,4-D and 2 or 5μM *m*T resulted in 10% callus formation, while no callus development was observed with 1μM *m*T. The results from other hormone treatments ranged from 0 to 20% callusing. The callus obtained was rarely friable, exhibiting in most cases a compact appearance and limited growth.

Due to limited callus material availability, the authors combined all calli obtained from the 2,4-D treatments and cultured them in PGR-free basal medium or supplemented with 1μM kinetin (KIN) or TDZ, to induce somatic embryo development and conversion into plants. A small proportion of calli transferred to medium lacking hormones gave rise to embryo-like structures. These were observed 2 weeks after calli transference to the medium and small plantlets developed 2 weeks later. No somatic embryos were observed in calli subcultured in cytokinin-containing media, although shoots and roots developed during the culture period. However, calli pooling makes it difficult to draw conclusions on the effect of *m*T on SE induction in *Mesomelaena pseudostygia*.

Callus was also initiated from coleoptiles 5–10 mm long excised from in vitro germinated embryos. Coleoptile segments were cultured on half-strength MS medium with different treatments of auxins and cytokinins, alone or in combination. Thus, PGR-free medium was compared with 1, 2, 5, or 10μM 2,4-D; 1, 5, or 20μM BA; 1, 2, or 5μM *m*T; 1μM NAA; 0.1, 0.5, 1, 2, or 10μM TDZ; 2μM 2,4-D plus 0.5 or 1μM BA; 5μM 2,4-D plus 0.5 or 1μM BA; 2μM 2,4-D plus 1, 2, or 5μM *m*T; and 2μM 2,4-D plus 1μM TDZ. Callus development was only observed in 2,4-D-supplemented media. The calli obtained were occasionally organogenic, but embryogenic characteristics were never evident.

Baskaran et al. (2015a) investigated the influence of *meta*-topolin riboside (*m*TR, 6-(3-hydroxybenzylamino)-9-ß-D-ribofuranosylpurine) on initiation of SE in an optimization experiment.

Expanding young leaves excised from in vivo grown plants of *Mondia whitei* were cultured on MS medium with 8 g L^{-1} agar and different concentrations of sucrose and PGR treatments. In the first experiment, the authors tested the effect of sucrose concentration (30, 35, 40, and 50 g L^{-1}) and the auxins 2,4-D and picloram (10, 15, 20 and 25μM). Second, various cytokinins (BA, *m*TR, KIN, or TDZ) at 1μM were combined with 20μM 2,4-D or picloram in MS medium containing 35 or 40 g L^{-1} sucrose. After 8 weeks, the calli obtained in the different treatments were transferred to MS medium with 20 g L^{-1} sucrose and 8 g L^{-1} agar to promote somatic embryo maturation and conversion into plants. Friable embryogenic callus developed over 2 weeks, and, 6 weeks later, differentiation of somatic embryos at different developmental stages could be observed in all treatments, except for the control, which lacked PGRs. Nevertheless, somatic embryo development and plantlet formation improved after transference to MS medium with 20 g L^{-1} sucrose and 8 g L^{-1} agar.

Although SE was induced at a high rate and acceptable numbers of somatic embryos at the different developmental stages were obtained with only auxins, the second experiment revealed that combination of 2,4-D or picloram with cytokinins was more effective in the production of somatic embryos at advanced stages and plantlets. Culture medium containing 40 g L^{-1} sucrose plus 20μM 2,4-D and 1μM TDZ gave rise to the highest production of somatic embryos at early stages and plant regeneration; however, higher frequency of embryogenesis (98.0 ± 0.20%) and enhanced development of cotyledonary-stage embryos were achieved when 20μM picloram was added jointly with 1μM *m*TR.

Following a similar experimental design, in which different sucrose concentrations and hormone combinations were tested, Baskaran et al. (2015b) initiated embryogenic cultures in *Aloe pruinosa*. Leaf explants excised from 20-day-old in vitro germinated seedlings were inoculated in solid (8 g L^{-1} agar) MS medium with 30–50 g L^{-1} sucrose, 20μM 2,4-D or picloram, and 20μM picloram plus 5μM BA, *m*TR, or zeatin (*t*Z). Although 6 weeks after culture initiation, embryogenic calli were observed in all PGR-containing treatments, the primary role in the production of friable embryogenic callus was only attributed to picloram, BA, and *t*Z.

14.3.2 Proliferation of Embryogenic Cultures

Baskaran et al. (2015b) also investigated the influence of different auxins and cytokinins on proliferation of embryogenic cultures and shoot regeneration of *Aloe pruinosa*. For this purpose, friable embryogenic callus obtained in the induction treatments previously indicated were transferred to embryogenic callus proliferation medium, consisting of solidified (8 g L^{-1} agar) MS medium with 30–50 g L^{-1} sucrose and a reduced auxin concentration (5μM), alone or combined with 5–15μM BA, *m*TR, *t*Z or TDZ. Twenty five μM phloroglucinol, a flavonoid used as PGR, was also tested in this phase.

Four weeks after culture initiation, the results revealed that BA and *tZ* were the most effective regulators for embryogenic callus proliferation and maturation, not being evident a relevant effect of *m*TR. Browning and necrosis of callus were a serious problem observed during this phase, conditioning embryo germination and conversion into plants. Interestingly, *m*TR and phloroglucinol appeared to be effective controlling phenolic activities, as both components delayed callus necrosis.

Saeed and Shahzad (2015) optimized a protocol for proliferation of *Albizia lebbeck* via secondary embryogenesis, utilizing *m*T as a cytokinin supplement. Optimization of secondary embryogenesis was addressed through a two-step culture sequence: firstly inducing maturation of the somatic embryoids in the embryogenic callus and later promoting the formation of secondary embryos through adventitious budding from primary embryos at advanced developmental stages.

As initial material, Saeed and Shahzad (2015) used embryogenic callus with somatic embryoids, initiated from shoot tips of an adult tree in WPM medium (Lloyd and Mc Cown 1981) solidified with 8 g L^{-1} agar and supplemented with 12.5μM KIN. In order to induce primary somatic embryo maturation, 100 mg of embryogenic tissues were cultured in solid (8 g L^{-1} agar) MS medium supplemented with different *m*T concentrations (2.5, 5.0 and 7.5μM), 5μM *m*T combined with various NAA concentrations (1.0, 2.5 and 5.0μM) or 5μM *m*T plus 2.5μM NAA, and 50, 75, or 100μM glutamine. Higher maturation rates, both in terms of maturation percentage and number of mature embryos per 100 mg embryogenic tissue, were achieved when 5μM *m*T was combined with 2.5 NAA and 75μM glutamine.

In a second experiment, approximately 50 mg of embryogenic tissues, mainly containing embryos at the cotyledonary stage, were cultured in solid MS with 5.0μM *m*T alone or in combination with NAA (1.0, 2.5 and 5.0μM) and glutamine (50, 75 and 100μM). Optimum results were newly achieved in MS medium with 5.0μM *m*T, 2.5μM NAA and 75μM glutamine. Under these conditions, 85.70 ± 0.67% secondary embryogenesis induction was accomplished, and 100 ± 1.15 secondary embryos were produced per 50 mg of primary embryogenic tissue.

Hence, for long-term maintenance, approximately 50 mg of embryogenic cultures of *Albizia lebbeck* were subcultured at 4-week intervals in MS medium plus 5.0μM *m*TR, 2.5μM NAA, and 75μM glutamine. Under these conditions, the embryogenic competence was maintained for at least 3 years (Saeed and Shahzad 2015).

Histological analysis revealed that development of somatic embryos occurred 2 weeks after transfer, from the peripheral region of preexisting embryos, at the base of their adaxial surface. Higher secondary embryogenesis was observed in embryos at the cotyledonary stage. Although a broad connection between the primary and the newly formed embryos was evident at the beginning of the secondary embryo development, these progressively separated becoming individualized as their development advanced. Secondary embryo development was similar to that observed in primary somatic embryos. Saeed and Shahzad (2015) identified six different stages: globular, elongated, early heart-shaped, heart-shaped, torpedo-shaped, and cotyledonary.

14.3.3 Somatic Embryo Development and Maturation

As previously indicated (Sect. 14.3.2), Saeed and Shahzad (2015) optimized maturation of primary somatic embryos as a first step to induce a cyclic SE system by secondary embryogenesis in *Albizia lebbeck*, obtaining the best results in MS medium with 5μM *m*TR, 2.5μM NAA, and 100μM glutamine. Interestingly, better germination (involving expansion of the two cotyledons and root formation) and conversion (involving differentiation of shoots and leaves and a developed root system) rates were achieved from secondary than from primary somatic embryos. Under optimum germination conditions (MS medium at half strength with 1.0μM gibberellic acid (GA_3)), 26.6 ± 0.88% of primary embryos and 46.7 ± 0.88% of secondary somatic embryos germinated, and 23.3 ± 0.88% and 41.7 ± 0.88% converted into plants, respectively. Saeed and Shahzad (2015) surmised that the positive role of *m*TR on the maturation of *Albizia lebbeck* somatic embryos may contribute to the development of a more efficient protocol for SE in this species.

Baskaran et al. (2015a) tested different PGR combinations for inducing somatic embryo development and conversion in *Mondia whitei*. Embryogenic callus initiated from leaf explants in solid (8 g L^{-1} agar) MS medium with 40 g L^{-1} sucrose, 20μM 2,4-D, and 1μM TDZ were transferred to solid MS medium with 20 g L^{-1} sucrose and 0.5μM of different cytokinins (BA, *m*TR, TDZ, and KIN), alone or in combination with 0.25μM indole-3-acetic acid (IAA) or NAA. Twelve weeks after culture initiation, somatic embryo development and plantlet formation were observed in all treatments. The best results were obtained in MS medium supplemented with 0.5μM *m*TR and 0.25μM IAA, with higher production of somatic embryos at heart, torpedo, and cotyledonary stages. More plantlets were also obtained in this culture medium, 20.40 ± 2.65 versus 9.00 ± 2.34–15.20 ± 2.59, in the rest of treatments. Regenerated plants exhibited good quality (2–3 cm shoot and 5–6 cm radicle) and were successfully acclimatized to ex vitro conditions, with a survival rate of 90%. According to the results, the combination of auxin and cytokinin, IAA and *m*TR in this case, proved to be essential to improve SE in this species.

Although embryogenic suspensions are highly productive, the development and maturation of somatic embryos have proven to be more problematic in liquid media (Timmis 1998; Gupta and Timmis 2005; Salaj et al. 2007). However, Baskaran et al. (2015b, 2017) tested the effect of *m*TR using cell suspension cultures for these purposes. Baskaran et al. (2015b) analyzed the effect of different PGRs, including *m*TR, on embryogenic suspension culture of *Aloe pruinosa*. Following the protocol of Baskaran and Van Staden (2012), suspension cultures were initiated by inoculating approximately 500 mg fresh weight of 3-week-old friable embryogenic callus selected from different proliferation media. Embryogenic tissues were cultured in 100 mL Erlenmeyer flasks containing 20 mL of liquid MS medium lacking hormones or supplemented with 0.5μM 2,4-D or picloram and 1–2μM BA or 1μM *t*Z, *m*TR, or TDZ. All culture media contained 30 g L^{-1} sucrose, and most PGR-containing treatments were supplemented with 10μM phloroglucinol. Four weeks later, significant differences were found in terms of somatic embryo development. The best results (38.7 ± 1.42 somatic embryos per settled cell volume at globular, club, and torpedo stages and 26.2 ± 0.87 at the cotyledonary stage) were obtained in liquid MS medium supplemented with 0.5μM picloram, 1μM TDZ, and 10μM phloroglucinol. Embryogenic tissue used in this case had been induced in MS medium with 40 g L^{-1} sucrose, 20μM picloram, and 5μM *tZ* and subsequently cultured in MS medium with 40 g L^{-1} sucrose, 5μM picloram, 5μM *tZ*, and 25μM phloroglucinol. Embryogenic callus derived from the same induction and proliferation culture conditions had significantly fewer embryos when suspension culture medium was supplemented with *m*TR: 11.4 ± 1.05 somatic embryos at early developmental stages and 4.7 ± 0.28 somatic embryos at late stages were obtained in liquid MS plus 0.5μM 2,4-D, 1μM *m*TR and 10μM phloroglucinol, and 16.9 ± 0.94 and 7.2 ± 0.43 embryos, respectively, in liquid MS plus 0.5μM picloram, 1μM *m*TR, and 10μM phloroglucinol. No germination was observed in suspension cultures or germination medium. Baskaran et al. (2015b) attributed this issue to oxidation of polyphenols, which play an inhibitory role in somatic embryo growth and germination.

As the frequency of SE and the number of somatic embryos obtained in *Mondia whitei* with their previous protocol (Baskaran et al. 2015a) was low, Baskaran et al. (2017) developed a new SE system, in which somatic embryo development and maturation were performed using cell suspension culture. In an investigation similar to that performed in *Aloe pruinosa* (Baskaran et al. 2015b), Baskaran et al. (2017) tested the effect of different auxin and cytokinin combinations on somatic embryo development in liquid medium. For this purpose, Baskaran et al. (2017) utilized 3-week-old friable embryogenic callus induced from expanding young leaves but with two different origins: initiation in MS medium with 35 g L^{-1} sucrose and 15μM 2,4-D and maintenance in MS medium supplemented with 5μM 2,4-D and 0.5μM TDZ and initiation in MS medium with 35 g L^{-1} sucrose and 15μM picloram and maintenance in MS medium supplemented with 5μM picloram and 0.5μM BA.

Somatic embryo development was performed by inoculating approximately 500 mg fresh weight of embryogenic callus in 20 mL liquid medium contained in 100 mL Erlenmeyer flasks. Liquid MS medium with 30 g L^{-1} sucrose was supplemented with different PGR combinations depending on culture origin. Thus, whereas embryogenic cultures derived from induction and proliferation media containing 2,4-D were transferred to liquid medium supplemented with 0.5μM 2,4-D and 1μM BA, TDZ, *m*TR, 6-(γ,γ-dimethylallylamino)purine (iP), or KIN, embryogenic cultures obtained from induction and proliferation media containing picloram were cultured in liquid medium supplemented with 0.5μM picloram and 1–2μM BA, TDZ, *m*TR, iP, or KIN as well as 0.5μM NAA and 1μM BA, TDZ, or *m*TR. One week after suspension culture initiation, 10 mL of culture medium was replaced by the same volume of freshly prepared medium, and 1 week later, embryogenic tissue was filtered through 200μm meshes and transferred to 250 mL Erlenmeyer flasks containing 30 mL of fresh medium. Incubation was carried out on an orbital shaker at 180 rpm. The effect of the different treatments was assessed by recording the number of somatic embryos at different developmental stages per settled cell volume, 3 weeks after culture initiation in embryo development medium. Production of somatic embryos at advanced developmental stages was significantly improved in *m*TR-containing media. Thus, 0.5μM picloram plus 1μM *m*TR produced the highest number of heart-shaped embryos, and 0.5 NAA plus 1μM *m*TR gave rise to the highest number of embryos at late developmental stages (torpedo and cotyledonary). However, higher globular-shaped embryo production was achieved in MS medium supplemented with 0.5μM 2,4-D and 1μM TDZ.

The germination capacity of embryos regenerated under optimum developmental conditions (0.5μM 2,4-D plus 1μM TDZ, 0.5μM picloram plus 1μM *m*TR, and 0.5μM NAA plus 1μM *m*TR) was subsequently evaluated by culturing them in solid (8 g L^{-1} agar) MS medium with different sucrose concentrations, alone or supplemented with 2–4μM NAA, and liquid MS medium at half strength with 0.5μM NAA. The best germination rate (98.3%) was achieved with somatic embryos developed in embryo development medium supplemented with 0.5 NAA and 1μM *m*TR and germinated in liquid MS medium at half strength with 0.5μM NAA. Therefore, the results obtained in *Mondia whitei* revealed a primary effect of *m*TR on somatic embryo development using cell suspension culture.

14.3.4 Embryo Germination

Meta-topolin has also been utilized in the germination phase. Solórzano-Cascante et al. (2018) compared the efficiency of BA and *m*T on shoot development in somatic embryos of *Carica papaya*. Somatic embryos used in this investigation had been regenerated from embryogenic cultures initiated from half-cut seeds cultured in half-strength MS medium supplemented with different 2,4-D concentrations (9.0, 18.0, 27.1, 36.2, or 45.2μM). Embryogenic cultures derived from the different 2,4-D treatments were combined and subsequently multiplied in solid (2.8 g L^{-1} phytagel) and liquid MS medium supplemented with 9μM 2,4-D. Somatic embryos obtained from both treatments were then used, without distinction, for analyzing the effect of BA and *m*T on germination. The influence of both cytokinins was tested in two independent experiments. In both cases, basal medium consisted of the MS formulation solidified with 9 g L^{-1} agar.

In the first trial, Solórzano-Cascante et al. (2018) used 0.0, 0.9, 1.8, 2.7, or 3.6μM BA to germinate somatic embryos derived from 2 months culture in solid medium with 9μM 2,4-D. However, the effect of different *m*T concentrations (0.0, 5.0, 10.0, 15.0, or 20.0μM) was tested using somatic embryos selected from 12-month-old cultures. Initial explants used in each case were slightly different too. While in the BA experiment, clusters of 5–10 somatic embryos at the cotyledonary stage were utilized; in the *m*T study, treatments were applied to callus sections, with approximately 20 somatic embryos per section. The effect of the different treatments was assessed 2 months after culture initiation, taking data of somatic embryo sprouting (epicotyl growth) and germination (radicle and epicotyl growth). Both BA and *m*T positively affected somatic embryo sprouting. Statistical analysis of this variable as a logistic regression revealed a significant second-order polynomial trend for both cytokinins. However, although similar maximum sprouting rates were achieved for BA (40%) and *m*T (44%), significant differences were observed in the hormone concentrations required for achieving these percentages, 1.8μM for BA versus 10μM for *m*T.

Although the effect of BA and *m*T on somatic embryo sprouting is analyzed in these experiments, it is not easy to draw clear conclusions about their comparative effect. As previously indicated, initial explants used for each experiment were different (clusters of somatic embryos and callus sections), and different numbers of somatic embryos were cultured per treatment replication (5–10 somatic embryos per cluster versus about 20 embryos per callus section). Additionally, the range of concentrations tested was very different (0.9, 1.8, 2.7, and 3.6μM for BA and 5.0, 10.0, 15.0, and 20.0μM for *m*T), and no coinciding treatments were included. Radicle development was rarely observed in these germination conditions. Thus, no rooting was evident in *m*T treatments, and less than 5% of the somatic embryos cultured in BA-containing media developed roots. Interestingly, significant differences were found in explant browning. Oxidation symptoms were observed in a considerable proportion of the somatic embryos germinated in BA-containing media (13–45%), while they were very scarce in those cultured in *m*T media (0–6%).

14.4 Concluding Remarks and Future Prospects

Meta-topolin and its riboside *m*TR have been used in the different phases of the SE process, i.e., induction of embryogenic cultures, proliferation, somatic embryos development and maturation, and embryo germination. The results have not been positive in all cases, with evident and important differences related to the species, culture conditions, and developmental stage.

In relation to the initiation step, no beneficial influence of *m*T or *m*TR was observed when they were compared to other cytokinins commonly used for this purpose, such as BA, KIN, *t*Z, or TDZ. However, in *Mondia whitei*, an improvement in the subsequent development of somatic embryos at cotyledonary stage was observed when embryogenic cultures induced in culture medium containing 40 g L^{-1} sucrose, 20μM picloram, and 1μM *m*TR were transferred to maturation conditions (Baskaran et al. 2015a).

In the proliferation phase, a positive role of *m*T on cyclic SE was reported in *Albizia lebbeck* (Saeed and Shahzad 2015). However, this hormone was not compared with other cytokinins, and there was no control without *m*T addition. In *Aloe pruinosa*, Baskaran et al. (2015b) found that BA and *t*Z were more effective than the other cytokinins tested, including *m*TR. Nevertheless, *m*TR showed a positive influence reducing the impact of polyphenol exudation, thus protecting tissue from oxidative stress.

Both *m*T and *m*TR have been used to promote development and maturation of somatic embryos. A positive effect of *m*T was observed in *Albizia lebbeck*, as higher maturation percentages and production of mature embryos were observed in *m*T-supplemented media, compared to control PGR-free medium. These parameters greatly improved when NAA and glutamine were added to culture medium containing 5μM *m*T (Saeed and Shahzad 2015). In relation to *m*TR, this *m*T derivative produced different effects depending on the species. In *Aloe pruinosa*, Baskaran et al. (2015b) reported negative effects when *m*TR was added to suspension cultures to promote somatic embryo development and maturation. In contrast, positive results were achieved in *Mondia whitei* using both solid and liquid culture media. Baskaran et al. (2015a) reported optimum production of plants and somatic embryos at advanced developmental stages in solid MS medium supplemented with 0.5μM *m*TR and 0.25μM IAA. Using cell suspension cultures, Baskaran et al. (2017) also found that production of somatic embryos at advanced developmental stages was significantly improved in *m*TR-containing media.

In relation to germination of somatic embryos, the only study concerning to this phase reported similar sprouting and rooting rates for the two cytokinins tested, BA and *m*T (Solórzano-Cascante et al. 2018).

As previously indicated (Sect. 14.1), a corrective role of some physiological disorders has been attributed to *m*T (Aremu et al. 2012). In different phases of SE, *m*T and *m*TR exhibited an important role in mitigating browning and necrosis, harmful processes in in vitro culture (Baskaran et al. 2015b; Solórzano-Cascante et al. 2018). According to Baskaran et al. (2015b), this quality could be used to improve future in vitro programs.

Therefore, *m*T and their derivatives can be considered suitable substitutes of the cytokinins traditionally used in SE protocols. Although its utilization in this developmental process is relatively recent, some of the results are promising, and an increasing use of this PGR can be expected. Nevertheless, negative results have also been obtained. Hence, as usually occurring in plant tissue culture, until a more complete knowledge of the structure-activity relationships of these PGRs is available, their adequacy should be investigated for each specific circumstance, by trial and error experimentation.

References

Altamura MM, Della Rovere F, Fattorini L, D'Angeli S, Falasca G (2016) Recent advances on genetic and physiological bases of *in vitro* somatic embryo formation. In: Germanà MA, Lambardi M (eds) *In vitro* embryogenesis in higher plants. Humana Press, New York, pp 47–86

Aremu AO, Bairu MW, Doležal K, Finnie JF, Van Staden J (2012) Topolins: a panacea to plant tissue culture challenges? Plant Cell Tissue Organ Cult 108:1–16

Baskaran P, Van Staden J (2012) Somatic embryogenesis of *Merwilla plumbea* (Lindl.) Speta. Plant Cell Tissue Organ Cult 109:517–524

Baskaran P, Kumari A, Van Staden J (2015a) Embryogenesis and synthetic seed production in *Mondia whitei*. Plant Cell Tissue Organ Cult 121:205–214

Baskaran P, Kumari A, Naidoo D, Van Staden J (2015b) *In vitro* propagation and biochemical changes in *Aloe pruinosa*. Ind Crop Prod 77:51–58

Baskaran P, Kumari A, Van Staden J (2017) Rapid propagation of *Mondia whitei* by embryonic cell suspension culture *in vitro*. S Afr J Bot 108:281–286

Beyl CA (2011) PGRs and their use in micropropagation. In: Trigiano RN, Gray DJ (eds) Plant tissue culture, development, and biotechnology. CRC Press, Boca Raton, pp 33–56

Bhojwani SS, Razdan MK (1996) Plant tissue culture: theory and practice, a revised edition. Elsevier, Amsterdam

Bogaert I, Van Cauter S, Werbrouck SPO, Doležal K (2006) New aromatic cytokinins can make the difference. Acta Hortic 725:265–270

Bradaï F, Almagro-Bastante J, Sánchez-Romero C (2017) Cryopreservation of olive somatic embryos using the droplet-vitrification method: the importance of explant culture conditions. Sci Hortic 218:14–22

Dunstan DI, Tautorus TE, Thorpe TE (1995) Somatic embryogenesis in woody plants. In: Thorpe TA (ed) *In vitro* embryogenesis in plant. Kluwer Academic Publishers, Dordrecht, pp 471–538

Egertsdotter U, Ahmad I, Clapham D (2019) Automation and scale up of somatic embryogenesis for commercial plant production, with emphasis on conifers. Front Plant Sci 10:109

Elhiti M, Stasolla C (2016) Somatic embryogenesis: the molecular network regulating embryo formation. In: Mujib A (ed) Somatic embryogenesis in ornamentals and its applications. Springer, New Delhi, pp 217–230

Elhiti M, Stasolla C, Wang A (2013) Molecular regulation of plant somatic embryogenesis. In Vitro Cell Dev Biol Plant 49:631–642

Engelmann F (2004) Plant cryopreservation: progress and prospects. In Vitro Cell Dev Biol Plant 40:427–433

Fehér A, Pasternak T, Ötvös K, Miskolczi P, Dudits D (2002) Induction of embryogenic competence in somatic plant cells: a review. Biologia 57:5–12

Guan Y, Li SG, Fan XF, Su ZH (2016) Application of somatic embryogenesis in woody plants. Front Plant Sci 7:938

Gupta PK, Timmis R (2005) Mass propagation of conifer trees in liquid cultures - progress towards commercialization. Plant Cell Tissue Organ Cult 81:339–346

Janick J (1993) Agricultural uses of somatic embryos. Acta Hortic 336:207–213

Jiménez VM (2005) Involvement of plant hormones and plant growth regulators on *in vitro* somatic embryogenesis. Plant Growth Regul 47:91–110

Karami O, Aghavaisi B, Mahmoudi Pour A (2009) Molecular aspects of somatic-to-embryogenic transition in plants. J Chem Biol 2(4):177–190

Lai S, Menon A, Turner S, Kodym A, Bunn E (2014) Development of an *in vitro* protocol for a difficult to-propagate endemic Australian dryland sedge species *Mesomelaena pseudostygia* (Cyperaceae). In Vitro Cell Dev Biol Plant 50:99–109

Lelu-Walter MA, Thompson D, Harvengt L, Sanchez L, Toribio M, Pâques LE (2013) Somatic embryogenesis in forestry with focus on Europe: state-of-the-art, benefits, challenges and future direction. Tree Genet Genomes 9(4):883–899

Lloyd G, Mc Cown B (1981) Commercially feasible micropropagation of mountain laurel, *Kalmia latifolia*, by use of shoot tip culture. Int Plant Prop Soc Proc 30:421–427

Loyola-Vargas VM, Ochoa-Alejo N (2016) Somatic embryogenesis. An overview. In: Loyola-Vargas VM, Ochoa-Alejo N (eds) Somatic embryogenesis: fundamental aspects and applications. Springer, Basel, pp 1–8

Murashige T (1977) Plant cell and organ cultures as horticultural practices. Acta Hortic 78:17–30

Murashige T, Skoog F (1962) A revised medium for rapid growth and bio assays with tobacco tissue cultures. Physiol Plant 15:473–497

Rai MK, Asthana P, Singh SK, Jaiswal VS, Jaiswal U (2009) The encapsulation technology in fruit plants—a review. Biotechnol Adv 27:671–679

Saeed T, Shahzad A (2015) High frequency plant regeneration in Indian Siris via cyclic somatic embryogenesis with biochemical, histological and SEM investigations. Ind Crop Prod 76:623–637

Salaj T, Blehová A, Salaj J (2007) Embryogenic suspension cultures of *Pinus nigra* Arn.: growth parameters and maturation ability. Acta Physiol Plant 29:225–231

Solórzano-Cascante P, Sánchez-Chiang N, Jiménez VM (2018) Explant type, culture system, 6-benzyladenine, meta-topolin and encapsulation affect indirect somatic embryogenesis and regeneration in *Carica papaya* L. Front Plant Sci 9:1769

Strnad M (1997) The aromatic cytokinins. Physiol Plantarum 101:674–688

Taiz L, Zeiger E (2010) Plant physiology, 5th edn. Sinauer Associates, Sunderland

Taiz L, Zeiger E, Møller IM, Murphy A (2015) Plant physiology and development, 6th edn. Sinauer Associates, Sunderland

Timmis R (1998) Bioprocessing for tree production in the forest industry: conifer somatic embryogenesis. Biotechnol Prog 14:156–166

von Arnold S (2008) Somatic embryogenesis. In: George EF, Hall MA, De Klerk GJ (eds) Plant propagation by tissue culture, 3rd edn. Springer, Dordrecht, pp 335–354

von Arnold S, Sabala I, Bozhkov P, Dyachok J, Filonova L (2002) Developmental pathways of somatic embryogenesis. Plant Cell Tissue Organ Cult 69:233–249

von Arnold S, Bozhkov P, Clapham D, Dyachok J, Filonova L, Högberg KA, Ingouff M, Wiweger M (2005) Propagation of Norway spruce via somatic embryogenesis. Plant Cell Tissue Organ Cult 81:323–329

Werbrouck SPO (2010) Merits and drawbacks of new aromatic cytokinins in plant tissue culture. Acta Hortic 865:103–107

Winkelmann T (2016) Somatic versus zygotic embryogenesis: learning from seeds. In: Germanà MA, Lambardi M (eds) *In vitro* embryogenesis in higher plants. Humana Press, New York, pp 25–46

Zimmerman JL (1993) Somatic embryogenesis: a model for early development in higher plants. Plant Cell 5(10):1411–1423

Meta-topolins: In Vitro Responses and Applications in Large-Scale Micropropagation of Horticultural Crops

15

Jean Carlos Cardoso

Contents

Abstract

Large-scale micropropagation is one of the key techniques in the production of vegetatively propagated horticultural crops, allowing the expansion of cultivated areas and high-quality, disease-free plantlets. This is due, among other things, to highly controlled environmental conditions, from culture medium to the source and type of light. In this controlled environment, cytokinins (CK) are the main plant growth regulators used in different phases of micropropagation in order to obtain the high-multiplication rates needed for the commercial production of millions of clonal plantlets of horticultural crops per year. However, the widely employed CKs, such as N^6-benzyladenine (BA), can lead to somaclonal variations and negative and undesired physiological effects in different stages of micropropagation. The more recently discovered natural BA derivatives,

J. C. Cardoso (✉)
Laboratory of Plant Physiology and Tissue Culture, Department of Biotechnology, Plant and Animal Production, Centro de Ciências Agrárias, Universidade Federal de São Carlos, Araras, SP, Brazil

N. Ahmad, M. Strnad (eds.), *Meta-topolin: A Growth Regulator for Plant Biotechnology and Agriculture*, https://doi.org/10.1007/978-981-15-9046-7_15

topolins, are promising alternatives for micropropagation of horticultural crops with reduced disadvantages of the standard cytokinins. The main effects of these BA derivatives in different micropropagated horticultural crops as well as their strengths and limitations compared with other CKs widely used in micropropagation are discussed in this book chapter.

Keywords

Meta-topolins · Horticulture · Plant tissue culture · Cytokinins · Shoot proliferation · Efficiency · Physiological responses

15.1 Horticulture, Propagation Systems, and the Importance of Commercial Micropropagation

Horticulture, the production of fruit, vegetables, and ornamental and medicinal plants, is a sector of great socioeconomic importance in world agriculture. The color, texture, size, and flavor of horticultural products are the predominant qualities needed for successful commercial activity in this area. Product nutraceutical value is also a quality that plays an important role.

Fruit, vegetables, and ornamental and medicinal plants can be propagated sexually, conventional vegetatively, and through micropropagation. In those propagated by seeds, the seedlings are produced commercially and used for direct (field) or indirect seeding, by transplanting pre-germinated seedlings under substrate and greenhouse conditions. In the seed propagated group, we can divide the cultivated species into two. The first group is self-propagating, for example, in lettuce cultivars (*Lactuca sativa*). In this case, after the development of the new cultivar, seed production is naturally obtained by natural self-fertilization. The other group is equivalent to F1 hybrid seeds, where heterozygous plants are used to obtain homozygous ones, naturally (tomato) or induced by controlled manual pollination, in order to obtain pure lineages (highly homozygous). Specific selected lines are crossed with each other to obtain the F1 hybrids, highly heterozygous, genetically uniform, and with characteristics of commercial interest (RHS 2019). In both cases, seed propagation reduces phytosanitary problems, when due care is taken to produce high-quality seeds.

The other group which uses conventional vegetative propagation refers to those species propagated by stem cutting (guava, chrysanthemum, and sweet potato), grafting (apple, citrus, mango, and grape), air layering (litchi), and the use of stolons (strawberries), rhizomes (banana, ginger, turmeric), divisions (banana, pineapple), and tubers (potatoes), among others. Although vegetative propagation maintains the genetic characteristics of new plantlets obtained, the methods of vegetative propagation have resulted in the dissemination and increase in the frequency of serious crop-specific diseases, especially those of viral or bacterial origin, resulting in a significant reduction in horticultural productivity.

A third group is micropropagated plants obtained under in vitro conditions, using plant tissue culture techniques. The need for micropropagated plants was based on issues found especially in vegetatively propagated plants. These included inefficient multiplication rate, failure to obtain a large number of shoots from a mother plant throughout the year, and the occurrence of diseases that were serious in these species. The latter resulted in the accumulation of contamination in plants and significant and gradate reduction in productivity along cycles of propagation (Cardoso et al. 2018), for example, garlic (*Allium sativum*) crop cultivation in Brazil, where conventional vegetative propagation through bulbils leads to the accumulation of viruses of the genera *Allexivirus*, *Carlavirus*, and *Potyvirus*, and significant reduction in bulb productivity with the advancement of propagation cycles and production of bulbs (Fernandes 2012). In different garlic cultivars that have undergone micropropagation, including clonal virus-free programs using shoot tip/meristem culture, the increased productivity can reach up to 141% in the first cycle of bulb production, compared to plants without clonal virus-free programs, with productivity up to 49% higher until the fifth cycle of bulb production, even after virus re-infection in this period (Melo Filho et al. 2006). This demonstrated a marked difference in micropropagation methods, in relation to those conventionally used, especially when micropropagation was the only method producing commercially virus-free clonal plantlets.

15.2 The Success of Commercial Micropropagation with the Discovery of Cytokinins

The historical success of plant tissue culture came about with advances in knowledge of plant hormones or plant growth regulators (PGRs), especially cytokinins. One of the oldest trials demonstrating the key role of cytokinins in the success of in vitro development was reported with tomato plants, where it was possible to obtain unlimited growth of roots in a culture medium containing minerals, sucrose, and vitamins, while the shoot tips maintained limited growth under the same conditions, without the addition of PGRs or even with the addition of only auxins to the culture media. Only when the adventitious root was induced, the restarting of growth on in vitro shoots was reported, demonstrating that an unknown diffusible factor from roots was capable of sustaining cell division (White 1934). Before that, Haberlandt (1913) had reported the presence of a diffusible substance in vascular tissue capable of stimulating cell division. However, the discovery was made of the first cytokinin, kinetin (6-furfurylaminopurine), a synthetic molecule found in autoclaved herring sperm DNA, which enabled cell proliferation of tobacco pulp cells (Miller et al. 1955). By means of this molecule and in search of similar molecules, the first natural cytokinin, zeatin, found in immature endosperms of maize (*Zea mays*) seeds was reported (Miller 1961; Letham 1963). The characterization of a cytokinin was due to its physiological effects in promoting cell division and other regulatory functions similar to kinetin (Skoog and Armstrong 1970), associated with molecules with similar molecular structure (Strnad 1997; Frébort et al. 2011).

Other synthetic molecules with cytokinin effect were also obtained with the ability to promote effects very similar to zeatin in vegetables. Of these,

N^6-benzyladenine (BA) is one of the most used and most effective synthetic PGRs for use in plant micropropagation, especially for the large-scale production of micropropagated horticultural plantlets (Oliveira et al. 2009; Silva et al. 2012; Ahmadian et al. 2017). It promotes cell division, dedifferentiation, adventitious regeneration, and the induction of multiple sprouts in individualized explants, by stimulating the cellular division of meristems located in the axillary buds of these shoots and breaking the apical dominance exerted by auxins (Müller and Leyser 2011).

In vitro cytokinins have made plant micropropagation a highly profitable commercial activity, in particular by promoting the formation of a large number of shoots from individual explants in a short period of time, making large-scale micropropagation an alternative for many species which have limited propagation by seed or even conventional vegetative propagation, such as low yield of new plantlets and important phytosanitary problems.

In addition to the isoprenoid-based cytokinins (ISCK) (Strnad 1997), some phenylureas also have cytokinin activity (Bruce and Zwar 1966), such as forchlorofenuron (CPPU; Kapchina-Toteva et al. 2000) and thidiazuron (TDZ; Sunagawa et al. 2007).

The discovery of the natural *meta*-hydroxylated analogues of BA called topolins (Strnad 1997), such as 6-(2-hydroxybenzylamino)purine and 6-(3-hydroxybenzylamino)purine also called *ortho*- and *meta*-topolin (*o*T, *m*T), was made more recently. These aromatic cytokinins were proven as efficient as isoprenoid (ISCK) and phenylurea types. Topolins have since been used in the multiplication/sprouting/shoot proliferation of different species, such as sugarcane (Vinayak et al. 2009), *Pelargonium* × *hortorum* and *P.* × *hederaefolium* (Wojtania 2010), the hybrid C35 of *Citrus sinensis* × *Poncirus trifoliata* (Chiancone et al. 2016), and *Prunus* spp. (Monticelli et al. 2017), among others.

15.3 The Importance of N^6-Benzyladenine in Commercial Micropropagation

The most widely used cytokinins for micropropagation are BA and TDZ, especially in the in vitro establishment phases, helping to maintain and stimulate cell division of shoot tips taken from donor plants and inoculated in vitro as reported in *Aloe vera* (Lavakumaran and Seran 2014) and even promoting adventitious organogenesis in non-meristem tissues, in order to obtain adventitious shoots from leaf segments of different commercial cultivars of *Anthurium andraeanum* (Cardoso and Habermann 2014). Cytokinins have also been reported to induce somatic embryogenesis, such as in *Dendrobium* 'Chengmai Pink' (Chung et al. 2005); however, the embryogenic response of somatic tissues is controlled by a complex mechanism involving hormones including cytokinins, auxins, abscisic acid, and ethylene and stress-related responses (Fehér 2015).

In in vitro establishment of culture medium, cytokinins, such as BA and TDZ, are used for the regeneration and multiplication phase, stimulating the morphogenetic response and development of multiple axillary or adventitious shoots in explants from different plant species, making micropropagation an efficient tool for clonal plant production.

In conclusion, especially for the cytokinin BA, its use in micropropagation protocols was a trigger for the development of commercial techniques, allowing regulation of in vitro regeneration processes, as well as the development of practically all existing commercial micropropagation protocols for different species used in horticulture and other agricultural applications. BA is the most widely used cytokinin in the micropropagation plant industry because of its efficiency, availability (Bairu et al. 2007), and effective response in different plant species, including herbaceous and woody plants used in horticulture.

For example, in *Prunus avium* cv. Lapins, BA was the only cytokinin capable of inducing shoot multiplication, resulting in a multiplication rate of up to 2.83 in culture medium containing 5 μM of this cytokinin and 0.5 μM indole-3-butyric acid (IBA), compared to other cytokinins such as TDZ {maximum multiplication index (MMI)—1.4)}, N^6-isopentenyladenine (iP) (MMI—1.17), and kinetin (KIN) (MMI—1.61) (Ruzic and Vujovic 2008).

Using the *Prunus* 'Flordaguard' rootstock, the addition of BA to the culture medium increased the percentage of explants with shoots (80%), and up to four shoots per explants were obtained using concentrations close to 4.0 mg L^{-1} (Radmann et al. 2011).

Similar results were obtained in a comparative study using BA and KIN cytokinins in the in vitro cultivation of gerbera, *Philodendron*, *Spathiphyllum*, and *Musa* cultivars (Vardja and Vardja 2001).

15.4 BA Is Not Effective for All Micropropagated Species

The question about testing these new cytokinins, the topolins, was needed. This was due to the fact that BA also leads to undesirable effects in micropropagated plantlets. Vardja and Vardja (2001) micropropagated different ornamental species and noted that for *Cordyline*, *Dracaena*, and *Dieffenbachia* genera, although the cytokinin BA increased the multiplication rate, there were undesirable effects, such as vitrification (hyperhydricity) and reduction in rooting and elongation, in the subsequent phases of in vitro cultivation. At the Laboratory of Plant Physiology and Tissue Culture (UFSCar), our research group observed different symptoms of BA in plant tissues of different horticultural species: in tea tree (*Melaleuca alternifolia*), we found marked reduction in size of shoots (Fig. 15.1a, b, 1.0–3.0 mg L^{-1}) and early leaf necrosis in explants (Fig. 15.1b, 3.0 mg L^{-1}) compared with normal development in BA-free culture medium (study of Carla Midori Iiyama); in strawberry (*Fragaria* × *ananassa*), some cultivars showed as symptoms severe reduction of petiole size, yellowing of leaves, and callus production (instead of shoots) using 0.5 mg L^{-1} BA, and reduction of concentration to 0.25 mg L^{-1} resulted in good shoot proliferation with no symptomatic shoots (Fig. 15.1c; study of Camila Y. Nishimura Saziki); in the medicinal plant *Phyllanthus amarus*, the main

Fig. 15.1 Different symptoms observed in horticultural species micropropagated in culture medium containing N^6-benzyladenine (BA): (**a**) multiple shoot induction (arrows) in explants of tea tree (*Melaleuca alternifolia*); (**b**) symptoms of reduction of size and leaf necrosis (ln) caused by BA in high concentration; (**c**) symptoms of size reduction and leaf yellowing in strawberry (*Fragaria* × *ananassa*); (**d**) hyperhydricity in shoots and callus development (cd) in basal region of shoots of the medicinal plant *Phyllanthus amarus*

symptoms observed using 0.5 mg L^{-1} BA were callus development in the basal region of shoots and hyperhydricity in leaves and stems (Fig. 15.1d, study of Maria Eduarda Barboza Souza de Oliveira).

There are published reports for different micropropagated species in which BA is a potent inhibitor of adventitious rooting on in vitro shoots, such as those observed in the medicinal plant *Achyrocline satureioides*, in which concentrations of BA above 2.5 μM completely inhibited the formation of roots in in vitro shoots cultivated in Woody Plant Medium (Guariniello et al. 2018). Vitrification or hyperhydricity is another physiological abnormality observed in plants cultured in vitro normally caused by the addition of cytokinins, such as BA and TDZ, into the culture medium. The type and concentration of these cytokinins as well as the type of gelling agent and the physical state of the culture medium affect this undesirable response in some micropropagated plants (Ivanova and Van Staden 2011).

For these reasons, in the micropropagation industry, studies using the alternative cytokinins, topolins, for micropropagation of horticultural plant species are mostly concentrated on comparisons of efficiency with the cytokinin BA, e.g., *m*T, compared to BA added to the culture medium for in vitro evaluations of multiplication or shoot proliferation efficiency and shoot/plantlet development (Table 15.1).

Table 15.1 The use of topolins on micropropagation of horticultural species and its effects on shoot proliferation, in vitro development and quality of plantlets

Species/hybrids/cultivars	Application of species	Culture medium	*m*T concentration used	Multiplication index or rate using *m*T	Other reported effects compared to other cytokinins	Reference
Aloe polyphylla	Me; Or	MS	5 μM *m*T	Total (near 12:1) and bigger shoots (near 8:1)	Bigger shoots; spontaneously rooting in multiplication medium; 91% successful acclimatization	Bairu et al. (2007)
Ansellia africana (Orchidaceae)	Me; Or	MS	10 μM *m*T + 5 μM NAA	9.3 ± 0.4a shoots/explant and 83% PLB formation	Increases protocorm-like bodies (PLB) formation; higher genetic stability	Battacharyya et al. (2017)
Barleria greenii	Or	MS	7 μM Me*m*TR	5.04 ± 0.62 shoots/explant	Higher multiplication rate and higher adventitious shoot length	Amoo et al. (2011)
Dendrobium nobile (Orchidaceae)	Me; Or	MS	1 mg L^{-1} *m*T + 0.5 mg L^{-1} NAA	9.2 shoots/explant	Higher and better frequency in PLB formation; genetic stability of derived plants	Battacharyya et al. (2016)
Huernia hystrix	Me	MS	20 μM *m*T + 10 μM NAA	Near 12 shoots/explant (only 4 shoots/explant using *m*T isolately)	Drastic reduction in rooting, compared to BA-treated; increased total phenolics and flavonoids	Amoo and van Staden (2013)

(continued)

Table 15.1 (continued)

Species/hybrids/cultivars	Application of species	Culture medium	*m*T concentration used	Multiplication index or rate using *m*T	Other reported effects compared to other cytokinins	Reference
Pelargonium × *hederaefolium* and *P.* × *hortorum*	Or	MS	1 mg L^{-1} *m*T	3–5 depending on cultivar used	Increased regeneration frequency at establishment; increased success propagation of all cultivars using *m*T; increased rooting percentage of shoots	Wojtania (2010)
Turmeric (*Curcuma longa*) *cv. Elite*	Me; Ar	MS medium with B5 vitamins	10 μM *m*T	6.3 shoots/explant	Shortest shoot and root length and very reduced number of roots per shoot, compared with other eight cytokinins, including BA	Salvi et al. (2002)
Musa spp. AAA 'Williams' and 'Grand Naine'	Fruit	Modified MS	15 μM *m*T or *m*TR	6.1 for 'Williams' and near of 7.0 for 'Grand Naine'	Root inhibition caused by *m*T and *m*TR; somaclonal variation does not differ among BA, *m*T, and *m*TR treatments	Bairu et al. (2008)
Pineapple 'Patawai'	Fruit	MS liquid under 100 rpm	2.5 μM *m*T	15.75	Non-reported; authors compared three in vitro systems,	Teklehayamanot et al. (2010)

		orbital shaker			TIB, semi-solid, and liquid shaking	
Prunus rootstocks	Fruit	QL macrosalts and MS microsalts	2.1 μM *m*T	2.4 for 'Ferdor' and 4.0 for 'Torinel', equal to BA treatments	Null hyperhydric shoots; higher shoot quality, chlorophyll contents, and rooting percentage	Gentile et al. (2014)
Plaintain (*Musa* AAB)	Fruit	MS liquid under bioreactor (TIB)	4.4 μM *m*T	10.7	Higher multiplication rate; non-morphologically aberrant (TDZ); effective for successive multiplication	Roels et al. (2005)
Corylus colurna	Nut	Basal medium described in the paper	8.2 μM *m*T	3.8 ± 0.14	Increased shoot length, shoot fresh and dry weight, and leaf width; higher percentages of root induction and increased survival in acclimatization	Gentile et al. (2017)
Pistachio (*Pistacia vera*)	Nut	MS medium with B5 vitamins	2.0 mg L^{-1} *m*T	4.5 ± 2.4 shoots/explant	Increased number of regenerated shoots and usable shoots	Benmahioul et al. (2012)

(continued)

Table 15.1 (continued)

Species/hybrids/cultivars	Application of species	Culture medium	*m*T concentration used	Multiplication index or rate using *m*T	Other reported effects compared to other cytokinins	Reference
Black walnut (*Juglans nigra*)	Nut	Modified liquid DKW	8.9 μM BA + 4.1 μM *m*T	6.4 (Genotype no. 55) and 5.1 (Genotype no. 189) nodes	Combinations of BA or Zea with *m*T improved percentage of shoot elongated and health of mature nodal explants	Stevens and Pijut (2018)
Coleonema album	Me/Or	MS, 3% sucrose	5.0 μM *m*T + (1 μM NAA or 2.0 μM IBA)	14.5 ± 1.76 adventitious shoots/explant	Increased length and fresh weight of shoots	Fajinmi et al. (2014)

In this sense, an advantage in the use of topolins, compared to BA, would be in maintaining or increasing the positive effects of BA (Werbrouck et al. 1996), especially for efficient induction of multiple shoots that enable the commercial micropropagation of different horticultural species (Cardoso 2018; Cardoso et al. 2018). *m*T application is usually associated with a reduction of undesirable effects reported in plants cultivated with BA or TDZ cytokinins in culture medium, such as hyperhydricity and inhibition of rooting (Bairu et al. 2007).

For this reason, and due to the similar effects to BA, topolins are used especially in the induction of multiple shoots in the multiplication phase of micropropagation (Table 15.1).

15.5 Topolins Have Effects Not Only on the Multiplication Phase

The use of topolins may also have other effects on later phases of multiplication and micropropagation, such as rooting and acclimatization (Werbrouck et al. 1996). In the micropropagation of potato (*Solanum tuberosum*) cv. Jaerla, an increase in the dry mass and survival percentage of acclimatized plantlets was observed (91.8–94.7% survival rate) by adding 0.005–0.01 mg L^{-1} of the *meta*-topolin riboside (*m*TR) compared to untreated plants (70.9% survival). This response was attributed to an increase in endogenous cytokinin content in shoots under acclimatization conditions, promoted by the application of *m*TR, which resulted in lower plant losses in this phase (Baroja-Fernandéz et al. 2002).

In banana 'Williams', the application of specific topolins, such as *meta*-methoxytopolin (Me*m*T) and *meta*-methoxytopolin-9-tetrahydropyran-2-yl (Me*m*TTHP), promoted an increase of in vitro shoots and root length under acclimatization, while other more common topolins such as *m*T or *m*TR did not have the same effect (Aremu et al. 2012). The studies done demonstrated the diversity of responses to different types of topolins, according to the species used. Interestingly, Aremu et al. (2012) observed a reduction in the levels of chlorophylls *a*, *b*, and *a* + *b* with the use of most of the topolins, compared to the control.

The use of *m*T improves the late phases of somatic embryogenesis, for example, in hybrid papaya, *m*T applied in the concentration of 10 μM stimulated sprouting in 40–45% of somatic embryos, with similar effects to BA at 1.8 μM (Solórzano-Cascante et al. 2018). Similar results were obtained by Bhattacharyya et al. (2018a) who reported increases in percentage of shoot emergence of encapsulated PLBs of *Ansellia africana* orchid by using 5.0–10.0 μM of *Mem*TTHP topolin. In *Dendrobium aphyllum* orchids, the use of *m*T up to a concentration of 15 μM increases the percentage of transverse-Thin Cell Layers (tTCLs) that proliferated shoots (79.43±0.12) and number of shoots per tTCL (11.22±0.39) (Bhattacharyya et al. 2018b).

Another positive effect of the use of *m*T was reported for *Corylus avellana* micrografting on *Corylus colurna* rootstocks, in which up to 70% survival of micrografts was reported with the shoots obtained using *m*T treatment and only 30% in the treatment with BA (Gentile et al. 2017).

The different types of topolins may also result in effects on metabolism, affecting, for example, in vitro secondary metabolite production. In *Aloe arborescens*, Amoo et al. (2012) observed significant effects of different types and concentrations of topolins on the production of in vitro secondary metabolites, such as iridoids, total phenolics, flavonoids, and condensed tannins. Similar results demonstrated that the use of *m*T at 20 μM alone in the culture medium promoted increase in total phenolic and flavonoid content in the medicinal plant *Huernia hystrix*, whereas the combination of this topolin with 10 μM NAA resulted in increase in shoot proliferation (3×) but had negative effects on the levels of these same secondary metabolites (Amoo and van Staden 2013).

15.6 Topolins Show Similar Negative Effects to BA Depending on Species, But Could Be an Important Tool for Commercial Micropropagation

Despite the positive responses cited in most studies using topolins as cytokinins, some studies have also reported non-significant or even negative effects of topolins, compared with BA used in culture medium. As examples, in 'Williams' and 'Grand Naine' bananas, the rate of abnormalities obtained with *m*T was similar to that obtained with the use of BA, both up to 15 μM, but treatments with topolins *m*T (3.3 roots) and *m*TR (0.4 roots) at 22.2 μM resulted in a drastic reduction in root numbers, whereas cytokinin BA maintained the number of roots (10.4 roots) compared to control (11.8 roots) (Bairu et al. 2008). In the same study, the authors also observed a dwarf mutant plant in the seventh cycle of in vitro multiplication, identified in the treatment with topolins and using a specific molecular marker. Drastic reductions in rooting percentage of micropropagated shoots were also observed in the medicinal plant *Huernia hystrix* using 20–25 μM *m*T (<10% rooting) compared to the use of BA at the same concentration (>90% rooting).

Thus, as observed in previous studies with different classes of plant hormones or different molecules of the same class, the cytokinins show different responses according to species and genotypes (genotype-dependent responses). Thus, it is expected that the responses to topolin-type cytokinins will also be genotype-dependent. This means that topolins have advantages in the micropropagation of some species over other cytokinins, such as BA and TDZ, while in other species BA, the most widely used cytokinin in micropropagation, remains the mainstay of efficient micropropagation protocols.

In this case, the most reasonable proposal is that topolin-type cytokinins are an important and additional tool in the micropropagation of species used in horticulture, especially those with in vitro cultivation issues (Werbrouck 2010; Amoo et al. 2011; Bandaralaje et al. 2015) and for which the most available cytokinins on the market do not result in efficient protocols for commercial micropropagation.

For example, in the micropropagation of the medicinal species *Harpagophytum procumbens*, excessive callus production at the base of the explants and shoot apical necrosis were reported as physiological disorders caused by BA addition to the culture medium (Bairu et al. 2011). These same symptoms were reported as a problem in the micropropagation of other species, such as in walnut *Corylus colurna* (Gentile et al. 2017). In both cases, the replacement of BA with *m*T resulted in resolution of the problem, and in the latter case, the use of *m*T also led to a significant increase in the percentage of acclimatized plants (78–84% survival in acclimatization), compared to symptomatic plants micropropagated using BA (24–30% survival only).

Similarly, in vitro micropropagation issues, such as early leaf senescence and loss of bud regeneration capacity in different *Pelargonium* cultivars, were overcome with *m*T at 0.5–1.0 mg L^{-1}. This resulted in increased multiplication rates of six different cultivars, including those showing symptoms of vitrification, shoot deformities, excessive callus formation, and death (67% of all cultivars) in the presence of cytokinin BA at 0.5 mg L^{-1} in the culture medium. Using *m*T as a cytokinin in the culture medium, the authors reported shoot proliferation during five subcultures in all six *Pelargonium* cultivars tested, with higher multiplication rates and none of the negative symptoms reported for BA treatments (Wojtania 2010).

15.7 Some Remarks About Differential Responses to Topolins and BA

The mechanisms underlying species peculiar responses to the various types of cytokinins are not yet elucidated. Possibly, the differences in species or cultivar responses to treatment with different types of cytokinins are associated with the fact that they represent a different stimulus for the natural biosynthesis of individual cytokinin types in the plants in question (Aremu et al. 2014) and thus lead to different responses regarding proliferation, undesirable symptoms, and subsequent effects on rooting and acclimatization of the sprouts and plantlets obtained.

In many studies, the lower toxicity of topolins, compared to BA and TDZ, is mentioned because the external application of cytokinins leads to natural CK biosynthesis in plants. The final metabolites generated by topolins are more easily degradable, especially *O*-glucosides, a readily storable form of cytokinins which are rapidly converted into plant-active cytokinin bases (Bairu et al. 2009; Strnad 1997). However, this explanation cannot be applied to all species, since undesirable effects such as inhibition of in vitro sprouting, callus production, and rooting inhibition have also been reported in plant species using some topolins, similar to those reported for the cytokinins BA and TDZ (Table 15.1).

One hypothesis that could explain the heterogeneity of responses to cytokinins among species, especially the differences between BA and topolins, may be related to the type of proliferation of new shoots under in vitro conditions in the production of commercially micropropagated plantlets. Although most species used in horticulture are micropropagated using the multiple shoot induction technique (Cardoso 2018; Cardoso et al. 2019), in some species this is precluded for a variety of reasons, such as negative effects of cytokinins on the in vitro culture environment (Duarte et al. 2019). Micropropagation using the micro-cutting technique, in which the nodes and internodes are segmented and induced to sprout new axillary shoots has been demonstrated is a viable and efficient alternative to multiplying shoots of some horticultural crops (Cardoso and Teixeira da Silva 2013; Duarte et al. 2019).

For example, in the propagation of Krymsk®5 rootstock (*Prunus fruticosa* × *Prunus lannesiana*), a higher number of shoots/explants (2.23) using BA instead of *m*T (0.57) are reported, but the use of *m*T resulted in a higher number of nodes per explant and shoot length than those cultivated with other cytokinins, including BA (Tsafouros and Roussos 2019). In this case, if each node were segmented and considered as a source of explants (microcuttings) to maintain sprout proliferation, it is possible that the multiplication rate using *m*T would increase significantly.

15.8 Conclusions

In conclusion, the cytokinins topolins discovered recently are being used as an important tool in the micropropagation of species of great importance for horticulture. In particular, the use of topolins in many cases has addressed the micropropagation and proliferation of high-quality shoots of species or genotypes in which other cytokinins were not efficient for shoot proliferation or resulted in undesirable effects, commonly caused by cytokinins BA and TDZ. However, it is important to note that the use of topolins is subject to the same symptoms caused by other cytokinins at high concentrations or even in species in which BA is still the most efficient for micropropagation. For this reason, studies that consider in particular evaluation of the genetic stability of micropropagated plants with these cytokinins are necessary because they should increase the use of topolins on a commercial scale, aiming at the production of horticultural species. In addition, topolins could increase efficiency of micropropagation in some species. The high costs of are among the major current challenges in large-scale micropropagation (Cardoso et al. 2018), and the increased efficiency of micropropagation is a preponderant factor for effective cost reduction (Chen 2016). The combination of techniques such as temporary immersion bioreactors, light-emitting diodes (LEDs), and photoautotrophic cultivation, among others, commercially emergent with the use of topolins added to the culture medium as well as the effects of topolins in tissue culture applied to plant breeding could also be considered and studied. As example, the use of topolins in the substitution of other cytokinins was also used for inducing gametic embryogenesis in microspores isolated from recalcitrant *Citrus* genotypes (Chiancone et al. 2015).

Acknowledgments JCC thanks to Conselho Nacional para o Desenvolvimento Científico e Tecnológico (CNPQ) for the processes number 304174/2015-7 and 311083/2018-8 and to Fundação de Amparo a Pesquisa do Estado de SP (FAPESP) for the processes number 2018/20673-3 and 2019/00243-7.

References

Ahmadian M, Babaei A, Shokri S et al (2017) Micropropagation of carnation (Dianthus caryophyllus L.) in liquid medium by temporary immersion bioreactor in comparison with solid culture. J Genet Eng Biotech 15(2):309–315

Amoo SO, Aremu AO, Van Staden J (2012) *In vitro* plant regeneration, secondary metabolite production and antioxidant activity of micropropagated *Aloe arborescens* Mill. Plant Cell Tissue Organ Cult 111(3):345–358

Amoo SO, Finnie JF, Van Staden J (2011) The role of *meta*-topolins in alleviating micropropagation problems. Plant Growth Regul 63:197–206

Amoo SO, Van Staden J (2013) Influence of plant growth regulators on shoot proliferation and secondary metabolite production in micropropagated *Huernia hystrix*. Plant Cell Tissue Organ Cult 11(2):249–256

Aremu AO, Bairu MW, Szüčová L et al (2012) Assessment of the role of *meta*-topolins on in vitro produced phenolics and acclimatization competence of micropropagated 'Williams' banana. Acta Physiol Plant 34:2265–2273

Aremu AO, Plačková L, Bairu MW et al (2014) How does exogenously applied cytokinin type affect growth and endogenous cytokinins in micropropagated *Merwilla plumbea*? Plant Cell Tissue Organ Cult 118:245

Bairu MW, Jain N, Stirk WA et al (2009) Solving the problem of shoot-tip necrosis in *Harpagophytum procumbens* by changing the cytokinin types, calcium and boron concentrations in the medium. S Afr J Bot 75:122–127

Bairu MW, Novak O, Doležal K et al (2011) Changes in endogenous cytokinin profiles in micropropagated *Harpagophytum procumbens* in relation to shoot-tip necrosis and cytokinin treatments. Plant Growth Regul 63(2):105–114

Bairu MW, Stirk WA, Dolezal K et al (2007) Optimizing the micropropagation protocol for the endangered *Aloe polyphylla*: can *meta*-topolin and its derivatives serve as replacement for benzyladenine and zeatin? Plant Cell Tissue Organ Cult 90:15–23

Bairu MW, Stirk W, Dolezal K et al (2008) The role of topolins in micropropagation and somaclonal variation of banana cultivars 'Williams' and 'Grand Naine'. Plant Cell Tissue Organ Cult 95(3):373–379

Bandaralaje JCAH, Hayward A, O'Brien C et al (2015) Gibberellin and cytokinin in synergy for a rapid nodal multiplication system of avocado. In: VIII Congresso Mundial de la Palta: genetic resources and nursery management (proceedings), Peru, pp 95–103

Baroja-Fernandéz E, Aguirreolea J, Martínkova H et al (2002) Aromatic cytokinins in micropropagated potato plants. Plant Physiol Biochem 40(3):217–224

Battacharyya P, Kumaria S, Tandon P (2016) High frequency regeneration protocol for *Dendrobium nobile*: a model tissue culture approach for propagation of medicinally important orchid species. S Afr J Bot 104:232–243

Battacharyya P, Kumar V, van Staden J (2017) Assessment of genetic stability amongst micropropagated *Ansellia Africana*, a vulnerable medicinal orchid species of Africa using SCoT markers. S Afr J Bot 108:294–302

Battacharyya P, Kumar V, van Staden J (2018) In vitro encapsulation based short term storage and assessment of genetic homogeneity in regenerated Ansellia Africana (Leopard orchid) using gene targeted molecular markers. Plant Cell Tissue Organ Cult 133(2):299–310

Bhattacharyya P, Kumar V, Van Staden J (2018a) In vitro encapsulation based short term storage and assessment of genetic homogeneity in regenerated *Ansellia africana* (Leopard orchid) using gene targeted molecular markers. Plant Cell Tissue Org Cult 133(2):299–310

Bhattacharyya P, Paul P, Kumari S, Tandon P (2018b) Transverse thin cell layer (t-TCL)-mediated improvised micropropagation protocol for endangered medicinal orchid *Dendrobium aphyllum* Roxb: an integrated phytomolecular approach. Acta Physiol Plant 40:137. https://doi.org/10.1007/s11738-018-2703-y

Benmahioul B, Dorion N, Kaid-Harche M et al (2012) Micropropagation and ex vitro rooting of pistachio (*Pistacia vera L.*). Plant Cell Tissue Organ Cult 108:353–358

Bruce MI, Zwar JA (1966) Cytokinin activity of some substituted ureas and thioureas. Proc R Soc B 165:245–265
Cardoso JC, Habermann G (2014) Adventitious shoot induction from leaf segments in Anthurium andreanum is affected by age of explant, leaf orientation and plant growth regulator. Hortic Environ Biotech 55:56–62
Cardoso JC, Teixeira da Silva JA (2013) Micropropagation of *Zeyheria montana* Mart. (Bignoniaceae), and endagered endemic medicinal species from the Brazilian Cerrado Biome. In vitro Cell Dev Biol Plant 49:710–716
Cardoso JC, Sheng Gerald LT, Teixeira da Silva JA (2018) Micropropagation in twenty-first century. Methods Mol Biol 1815:17–46
Cardoso JC (2018) *In vitro* responses of gerbera (*Gerbera jamesonii*) cultivars multiplied under different photoperiods. Adv Hortic Sci 32(4):557–561
Cardoso JC, Oliveira MEBS, Cardos FCI (2019) Advances and challenges on the in vitro production of secondary metabolites from medicinal plants. Hort Brasil 37:124–132. http://dx.doi.org/10.1590/S0102-053620190201
Chen C (2016) Cost analysis of plant micropropagation of *Phalaenopsis*. Plant Cell Tissue Organ Cult 126:167–175
Chiancone B, Karasawa MMG, Gianguzzi V et al (2015) Early embryo achievement through isolated microspore culture in *Citrus clementina* Hort. ex Tan., cvs. 'Monreal Rosso' and 'Nules'. Front Plant Sci 6:Article 413
Chiancone B, Martorana L, Casales F et al (2016) The effects of benzyladenine and meta-topolin on in vitro sprouting and regrowth after encapsulation of C35 citrange [*Citrus sinensis* (L.) Osb. × *Poncirus trifoliata* (L.) Raf] microcuttings. Acta Hortic 1113:99–106
Chung H-H, Chen J-T, Chang W-C (2005) Cytokinins induced direct somatic embryogenesis of *Dendrobium chiengmai* pink and subsequent plant regeneration. In Vitro Cell Dev Biol Plant 41 (6):765–769
Duarte WN, Zanello CA, Cardoso JC (2019) Efficient and easy micropropagation of Morus nigra and the influence of natural light on acclimatization. Adv Hortic Sci 33(3):433–439
Fajinmi OO, Amoo SO, Finnie JF, Van Staden J (2014) Optimization of in vitro propagation of *Coleonema album*, a highly utilized medicinal and ornamental plant. S Afr J Bot 94:9–13
Fehér A (2015) Somatic embryogenesis – stress induced remodeling of plant cell fate. Biochim Biophys Acta 1849(4):385–402
Fernandes FR (2012) Medidas gerais de controle de viroses em alho. Nosso Alho. Disponível em: https://ainfo.cnptia.embrapa.br/digital/bitstream/item/74810/1/Nosso-alho.pdf. Acessado em 25 de Abril de 2019 (text in portuguese)
Frébort I, Kowalska M, Hluska T et al (2011) Evolution of cytokinin biosynthesis and degradation. J Exp Bot 62:2431–2452
Gentile A, Frattarelli A, Nota P et al (2017) The aromatic cytokinin *meta*-topolin promotes in vitro propagation, shoot quality and micrografting in *Corylus colurna* L. Plant Cell Tissue Organ Cult 128:693
Gentile A, Jàquez Gutiérrez M, Martinez J et al (2014) Effect of *meta*-topolin on micropropagation and adventitious shoot regeneration in *Prunus* rootstocks. Plant Cell Tissue Organ Cult 118:373–381
Guariniello J, Iannicelli J, Peralta PA, Escandón AS (2018) In vivo and in vitro propagation of "macela": a medicinal-aromatic native plant with ornamental potential. Orn Hort 24(4):361–370
Haberlandt G (1913) Zur physiologie der zellteilung. Sitzungsber Akad Wiss Berlin Phys Math C 1:318–345
Ivanova M, Van Staden J (2011) Influence of gelling agent and cytokinins on the control of hyperhydricity in *Aloe polyphylla*. Plant Cell Tissue Organ Cult 104:13–21
Kapchina-Toteva VV, van Telgen HJ, Yakimova E (2000) Role of phenylurea cytokinin CPPU in apical dominance release in in vitro cultured *Rosa* hybrida L. J Plant Growth Regul 19:232–237
Lavakumaran L, Seran TH (2014) Effect of 6-benzylaminopurine and thidiazuron on in vitro shoot organogenesis of *Aloe vera* (L.) Burm. Chilean J Agric Res 74:497–501
Letham DS (1963) Zeatin, a factor inducing cell division isolated from zea mays. Life Sci 2:569–573

Monticelli S, Gentile A, Frattarelli A et al (2017) Effects of the natural cytokinin *meta*-topolin on in vitro shoot proliferation and acclimatization of *Prunus* spp. Acta Hortic 1155:375 380
Melo Filho PA, Resende RO, Cordeiro CMT et al (2006) Viral reinfection affecting bulb production in garlic after seven years of cultivation under field conditions. Europ J Plant Pathol 116:95–101
Miller CO, Skoog F, Von Saltza MH, Strong F (1955) Kinetin, a cell division factor from deoxyribonucleic acid. J Am Chem Soc 77:1392
Miller CO (1961) A kinetin-like compound in maize. Proc Natl Acad Sci USA 47:170–174
Müller D, Leyser O (2011) Auxin, cytokinin and the control of shoot branching. Ann Bot 107 (7):1203–1212
Oliveira ET, Crocomo OJ, Farinha TB, Gallo LA (2009) Large-scale micropropagation of *Aloe vera*. HortScience 44(6):1675–1678
Radmann EB, Bianchi VJ, Fachinello JC et al (2011) In vitro multiplication of 'Flordaguard' rootstock: cytokinin source and concentration effects, explants orientation and period of permanence in the culture medium. Braz Arch Biol Technol 54(1):25–34
RHS (Royal Horticultural Society) (2019) F_1 hybrids. Disposable. https://www.rhs.org.uk/advice/profile?pid=710. Accessed 14 Oct 2019
Roels S, Escalona M, Cejas I et al (2005) Optimization of plaintain (*Musa* AAB) micropropagation by temporary immersion system. Plant Cell Tissue Organ Cult 82:57–66
Ruzic DV, Vujovic TI (2008) The effects of cytokinin types and their concentration on in vitro multiplication of sweet cherry cv. Lapins (*Prunus avium* L.). Hort Sci (Prague) 35:12–21
Sunagawa H, Agarie S, Umemoto M et al (2007) Effect of urea-type cytokinins on the adventitious shoots regeneration from cotyledonary node explant in the common ice plant, *Mesembryanthemum crystallinum*. Plant Prod Sci 10:47–56
Salvi ND, George L, Eapen S (2002) Micropropagation and field evaluation of micropropagated plants of turmeric. Plant Cell Tissue Organ Cult 68:143–151
Silva RC, Luis ZG, Scherwinski-Pereira JE (2012) Short-term storage in vitro and large-scale propagation of grapevine genotypes. Pesq Agropec Bras 47(3):344–350
Skoog F, Armstrong DJ (1970) Cytokinins. Annu Rev Plant Physiol 21:359–384
Solórzano-Cascante P, Sánchez-Chiang N, Jiménez VM (2018) Explant type, culture system, 6-benzyladenine, meta-topolin and encapsulation affect indirect somatic embryogenesis and regeneration in *Carica papaya* L. Front Plant Sci 9:1769
Stevens ME, Pijut PM (2018) Rapid in vitro shoot multiplication of the recalcitrant species *Juglans nigra* L. In Vitro Cell Dev Biol Plant 54:309–317
Strnad M (1997) The aromatic cytokinins. Physiol Plant 101:674–688
Teklehayamanot T, Wannakrairoj S, Pipattanawong N (2010) Meta-topolin for pineapple shoot multiplication under three in vitro systems. Am Eur J Agric Environ Sci 7(2):157–162
Tsafouros A, Roussos PA (2019) First report of Krymsk®5 (cv. VSL2) cherry rootstock in vitro propagation: studying the effect of cytokinins, auxins and endogenous sugars. Notulae Bot Hort Agrobot Cluj Napoca 47(1):152–161
Vardja R, Vardja T (2001) The effect of cytokinin type and concentration and the number of subcultures on the multiplication rate of some decorative plants. Proc Estonian Acad Sci Biol Ecol 50(1):22–32
Vinayak V, Dhavan A, Gupta VK (2009) Efficacy of non-purine and purine cytokinins on shoot regeneration in vitro in sugarcane. Indian J Biotechnol 8:227–231
Werbrouck SPO (2010) Merits and drawbacks of new aromatic cytokinins in plant tissue culture. Acta Hortic 865:103–107
Werbrouck SPO, Strnad M, van Onckelen HA et al (1996) Meta-topolin, an alternative to benzyladenine in tissue culture? Physiol Plant 98(2):291–297
Wojtania A (2010) Effect of meta-topolin on in vitro propagation of *Pelargonium x Hortorum* and *Pelargonium x Hederaefolium* cultivars. Acta Soc Bot Pol 79(2):101–106
White PR (1934) Potentially unlimited growth of excised tomato root tips in a liquid medium. Plant Physiol 9:585–600

Optimization of Micropropagation Protocols in Some Woody Plants Using *Meta*-topolin

16

M. C. San José, M. J. Cernadas, and L. V. Janeiro

Contents

Abstract

The type and concentration of growth regulators used in micropropagation vary widely, depending on the species that are being studied and they have to be optimized depending on growth parameters: multiplication rates, shoot quality, rooting, and acclimatization. Cytokinins are an essential class of growth regulators that along with auxins, control numerous physiological and plant development processes. The choice of cytokinin is one of the most critical factors in the development of protocols for the in vitro culture of plants.

N^6-benzyladenine (BA) is currently one of the most widely used cytokinins, due to its efficacy and relatively low cost. However, the use of this cytokinin in micropropagation procedures can produce physiological disorders such as hyperhydricity, apical necrosis, senescence, rooting inhibition, and

M. C. San José (✉) · M. J. Cernadas · L. V. Janeiro
Biotechnology and Forestry Improvement Group, Instituto de Investigaciones Agrobiológicas de Galicia, CSIC, Santiago de Compostela, Spain

Diputación de Lugo, Lugo, Spain
e-mail: sanjose@iiag.csic.es

N. Ahmad, M. Strnad (eds.), *Meta-topolin: A Growth Regulator for Plant Biotechnology and Agriculture*, https://doi.org/10.1007/978-981-15-9046-7_16

acclimatization issues. These have necessitated search for alternatives to BA in order to maintain reasonable proliferation rates and acceptable plant quality.

The discovery of a new group of aromatic cytokinins, the topolins, has led to new possibilities for their use in the micropropagation of numerous species. These compounds can be considered an alternative to N^6-benzyladenine (BA), *trans*-zeatin (*t*Z), N^6-isopentenyladenine (iP), and kinetin (KIN). In the last few years, the use of *meta*-topolin (*m*T) and its derivatives has increased rapidly, especially in initiating new cultures, optimizing protocols, and inhibiting the negative features of some other cytokinins. This chapter is a review of the role of *meta*-topolin in the optimizing micropropagation in woody plant tissue cultures.

Keywords

Apical necrosis · N^6-benzyladenine · Early senescence · Hyperhydricity · *Meta*-topolin · Micropropagation · Topolins · Woody plants

16.1 Introduction

Optimizing micropropagation is a complex process involving different developmental phases which are dependent on numerous endogenous and exogenous factors (Mohamed and Alsadon 2011). Growth regulators are considered one of the most important factors in in vitro culture, since the processes of differentiation, dedifferentiation, and redifferentiation are dependent on the quality and quantity of these compounds in the culture medium (Bohra et al. 2016). The regeneration of shoots and their multiplication relies, to a great extent, on the type and concentration of growth regulators added to the culture medium, with cytokinins (CKs) as especially important (Howell et al. 2003; Aremu et al. 2012). Since their discovery, it has been demonstrated that CKs are an essential class of growth regulators that, along with the auxins, regulate numerous physiological and plant development processes, including leaf senescence, apical dominance, formation and activity of apical meristems, gametophyte development, promotion of sink activity, vascular development, and dormancy rupture. Cytokinins also play a vital role in the interaction of plants with biotic and abiotic factors, including stress due to salinity and drought, macronutrients (including nitrate, phosphorous, iron, and sulfate symbiotic nitrogen-fixing bacteria and arbuscular mycorrhizal fungi, as well as pathogenic bacteria, nematodes, and viruses) (Taiz et al. 2015).

The discovery of cytokinins originated in the 1930s when scientists were searching for chemical compounds that would enable cultivation of plant tissue in synthetic media. In the 1950s, a substance was identified and crystallized from herring sperm DNA extracts and found to stimulate the proliferation of cells in tissue culture of *Nicotiana tabacum*. The compound, N^6-furfuryladenine, was labelled kinetin (KIN; Fig. 16.1) owing to its cytokinesis promoting activity (Hall and De Ropp 1955; Miller et al. 1955). Kinetin has not been found in plants, and it is believed that it is an artificial by-product of DNA rupture. The first natural cytokinin

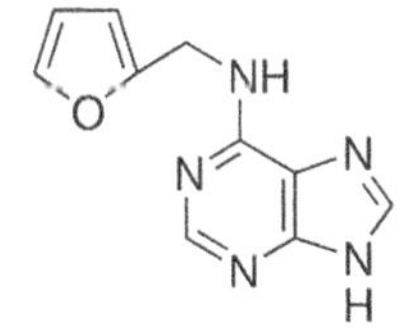

Fig. 16.1 Structure of kinetin

was isolated by Letham (1963) from immature maize seeds. This substance was identified as 6-(4-hydroxy-3-methylbut-2-enylamino)purine and given the name *trans*-zeatin. The term, cytokinins, was defined as a generic name for substances that promoted cell division and exercised other growth regulatory functions in the same way as kinetin. From these first studies, more than a hundred natural and synthetic products have been identified with similar physiological effects as those of KIN.

The natural cytokinins are N^6-substituted adenine derivatives. The chemical nature of the substitution at N^6 of the adenine enables the cytokinins to be classified into two large groups: isoprenoid and aromatic cytokinins (Figs. 16.2 and 16.3). The isoprenoid cytokinins have an isoprene-derived side chain, and the active free bases include N^6-isopentenyladenine (iP), zeatin, and the dihydrozeatin. The side chain of zeatin has a double bond, and its hydroxy group can therefore be oriented in a *trans*- or *cis*-configuration, representing *trans*- or *cis*-zeatin (*t*Z, *c*Z), respectively (Zürcher and Müller 2016).There are differences in their activity, abundancy, and stability in vivo, although there is no accepted consensus on the relevance of these variants to plant physiology. The aromatic cytokinins, which have an aromatic derivative side chains, include kinetin (KIN), N^6-benzyladenine (BA), *meta*-topolin (*m*T), and *ortho*-topolin (*o*T) (Taiz et al. 2015). Urea-based synthetic cytokinins, 6-(2-chloro-4-pyridyl)-N'-phenylurea (CPPU), and thidiazuron (TDZ) are frequently considered as a third group of cytokinins (Schmülling 2004).

Both types of cytokinins (isoprenoid and aromatic) have specific functions in the life cycle of plants. The isoprenoid cytokinins are found most frequently and in greater quantity than the aromatic ones and participate mainly in the continuation of the cell cycle, while the aromatics have a greater influence on developmental processes, such as morphogenesis and senescence (Zalabák et al. 2013). However, there is evidence of an overlap in their functions (Aremu et al. 2015).

The existence of zeatin, dihydrozeatin, and N^6-isopentenyladenine (and their conjugated forms) as natural cytokinins was accepted since their discovery, but there had been arguments about the presence of the aromatic cytokinins in plants, being considered synthetic until quite recently. However, the development of analytical techniques in recent years has led to the isolation and identification of three aromatic cytokinin families in plants. At the beginning of the 1970s, for the first time, Horgan et al. (1973, 1975) isolated the aromatic cytokinin *ortho*-topolin riboside [6-(2-hydroxybenzylamino)purine riboside], from the leaves of *Populus* × *robusta*. Jones et al. (1996) detected the presence of the aromatic cytokinins BA, *m*T and *o*T in different tissues of *Elaeis guineensis*. It was the work of Strnad et al. (1997) however that had greatest impact with the isolation of

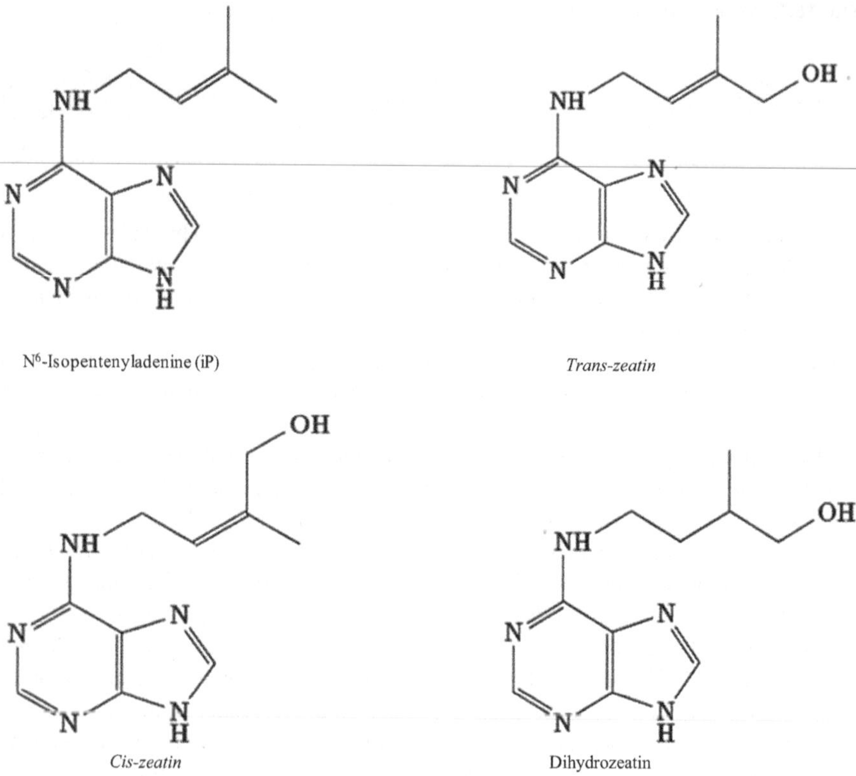

Fig. 16.2 Structure of isoprenoid cytokinins

6-(3-hydroxybenzylamino)purine from mature leaves of *Populus* × *canadensis* and proposed the name of *meta*-topolin. Given that the *ortho*- and the *meta*-topolins were isolated for the first time in black poplar leaves, they were labelled topolins (a name derived from the Czech word for poplar: topol). More recently, four new methoxy derivatives of the topolins have been identified as *ortho*-methoxytopolin (6-(2-methoxybenzylamino)purine; Me*o*T), *meta*-methoxytopolin (6-(3-methoxybencilamino)purine; Me*m*T), and their 9-β-D-ribofuranosyl derivatives (Me*o*TR and Me*m*TR). These aromatic cytokinins (ARCKs) were identified in *Arabidopsis thaliana* and *Populus* × *canadensis* (Tarkowská et al. 2003). In a subsequent review, Subbaraj (2011) listed several topolins isolated from other plant species.

BA is currently the most widely used in the micropropagation processes, mainly due to its efficacy and its relatively low cost, although in some cases it has negative effects on growth, rooting, and acclimatization, including hyperhydricity, senescence, and apical necrosis (Werbrouck et al. 1996; Aremu et al. 2012).

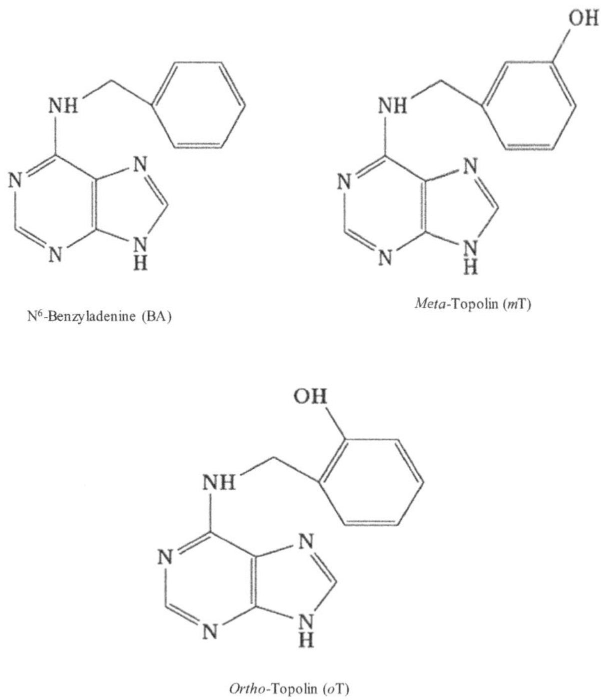

Fig. 16.3 Structure of aromatic cytokinins

The best results for the propagation of shoots, reduction in negative factors, rooting, and acclimatization of in vitro derived plants have been obtained using topolins, and in particular *m*T and its derivatives, which have made it possible to enhance micropropagation efficiency. The last 20 years have been witness to the greatly increased use topolins in plant tissue culture (PTC), in many cases as an alternative to other cytokinins, such as BA, iP, *t*Z, and KIN.

Several studies support this increase in their popularity, utilization, and the advantages they have over other cytokinins. Werbrouck (2010) reviewed the merits and drawbacks of the new aromatic cytokinins in the culture of plant tissues. Their chemistry, synthesis, natural existence, and functions were also examined by Zürcher and Müller (2016) and Subbaraj (2011). The latter also reviewed the relationship between the chemical structure and biological activities of different applications of topolins in agriculture and in the control of abiotic stress. Aremu et al. (2012) studied their practical applications, testing these compounds in a wide range of species, from food crops to medicinal plants and plantation trees, by comparing them with other cytokinins.

16.2 Optimization of Micropropagation

16.2.1 Shoot Induction and Multiplication Rates

Factors such as genotype, medium, and cytokinin concentration are vital and need to be taken into account in order to optimize in vitro propagation protocols. Several authors have reported a more vigorous growth and better quality of the shoots on using media supplemented with *m*T (Werbrouck et al. 1996; Bairu et al. 2007; Valero-Arcama et al. 2010; Wotjania 2010). Reference is made below to different works that have studied the effect of topolins, with the aim of improving shoot induction and multiplication rates in some woody species, as well as to alleviate some of the physiological disorders caused by BA (Table 16.1).

In 2016, Chiancone et al. evaluated the effect of BA and *m*T on in vitro sprouting and regrowth in citrange rootstock C35 (*Citrus sinensis* × *Poncirus trifoliata*). The use of 4.2 μM *m*T helped to obtain the largest number of shoots per explant, as well as the longest shoots, improving the results obtained with BA. In a subsequent work with Troyer citrange, they found no significant differences in the organogenesis from epicotyl cuttings between these two cytokinins (BA and *m*T). However, the authors claim that *m*T may be an alternative in the culture of *Citrus* tissues (Chiancone et al. 2017). Similar results were obtained by Niedz and Evens (2011) in their study on the regeneration in epicotyls of the sweet orange (*Citrus sinensis*), where they pointed out that *m*T showed a positive response over a wider range of concentrations, while in the case of BA, the negative effects increased in a concentration dependent manner.

Monticelli et al. (2017) also studied the effect of *m*T and BA on the micropropagation of shoots from different rootstocks of *Prunus*: "Penta" (*P. domestica*) and Myrobalan "29C" (*P. cerasifera*). In this case, the presence of *m*T (1.78 μM) had a negative effect on multiplication rates compared to the medium with BA, although the length of the shoots was greater. However, they did mention good results with 2.1 μM *m*T in the multiplication of another two rootstocks, "Ferdor" (*P. insititia* × *P. domestica*) and "Torinel" (*P. domestica*). According to the authors, these data demonstrate that the response of *Prunus* spp. to *m*T is highly dependent on the species, and even within the same species, such as in *P. domestica*, the genotype plays a very important role. In 2014, Gentile et al. working with the two latter rootstocks ("Torinel" and "Ferdor"), had already shown the potentiality of *m*T for optimizing the micropropagation protocols and for the differentiation of adventitious shoots. Although *m*T did not improve shoot proliferation compared to BA, they did show that it had a positive effect on the growth and quality of the shoots. The leaves of the shoots treated with *m*T also had a higher concentration in chlorophyll. In both genotypes, only the leaves from shoots treated with *m*T during the multiplication phase showed differentiation of adventitious buds. The best response was obtained in explants from shoots multiplied with 2.1 μM *m*T and induced to regenerate in a medium supplemented with TDZ. According to the authors, these results provide evidence of the critical role of *m*T as a pre-treatment in the growth of mother plants and are in agreement with the work of Dobránszki et al. (2002) in

Table 16.1 Use of topolins in the optimization of micropropagation protocols in some woody species

Species	Cytokinins tested	Parameters	Reference
Actinidia deliciosa	BA, TDZ, *m*T	*m*T (2 μM) improves the multiplication rates	Prado et al. (2005)
Amelanchier alnifolia	BA, iP, *m*T, *m*TTHP	Better proliferation with *m*T 10 μM. In the rooting, improves results with *m*TTHP	Moyo et al. (2018)
Citrus spp 'Troyer'	*m*T, BA	No differences in organogenesis in epicotyls	Chiancone et al. (2017)
Citrus sinensis × *Poncirus trifoliata*	BA, *m*T	*m*T (4.2 μM) improves proliferation	Chiancone et al. (2016)
Citrus sinensis 'Hamlin'	BA, *m*T	Better response of *m*T over a wide range of concentrations	Niedz and Evens (2011)
Coleonema album	BA, KIN, *m*T, *m*TR, *Mem*TR, M*em*TTHP, TDZ	5 μM *m*T + 2 μM IBA gave better results in shoot production	Fajinmi et al. (2014)
Cornus kousa × *C. hongkongensis*	BA, zeatin, KIN, *m*T	Better results with BA, although increases the amount of phenols and hyperhydricity	Lattier et al. (2014)
Crataegus aronia	BA, TDZ, *m*T	BA was the only one capable of promoting the proliferation and lengthening of the shoots	Nas et al. (2012)
Eucalyptus globulus	*m*T	*m*T (0.4 μM) improves the quality of the shoots	Vieitez (personal communication)
Ficus carica	BA, *m*T, zeatin	Better response with BA and *m*T (2.1 μM) added to artificial seed endosperm	Yahyaoui et al. (2017)
Fraxinus excelsior	*m*T	*m*T 2.1 μM for the propagation of adult material	Šedivá et al. (2017)
Ginkgo biloba	BA, iP, *m*T	*m*T (5.5 μM) improves the proliferation and quality of thc shoots	Nacheva et al. (2017)
Juglans nigra	BA, *m*T	BA + *m*T (4.1–6.2 μM) and liquid medium improve proliferation	Stevens and Pijut (2018)
Magnolia 'Ann'	BA, *m*T, iP	BA better than *m*T or 2iP	Parris et al. (2012)
Malus 'Royal Gala' and 'M26'	BA, *m*T	Pre-treatment with *m*T (2.1–6.3 μM) improves the regeneration and reduces hyperhydricity	Dobránszki et al. (2002, 2004, 2005)
Malus 'Jonagold'	BA, BAR, *m*T	Better multiplication rates with BAR and *m*T (21 μM)	Magyar-Tábori et al. (2002)

(continued)

Table 16.1 (continued)

Species	Cytokinins tested	Parameters	Reference
Paulownia elongata, P. fortunei, P. elongata × P. fortunei	BA, *m*T	*m*T (2.1–4.2 μM) improves the quality of the shoots	Clapa et al. (2014)
Pinus pinaster	BA, TDZ, iP, *m*T, zeatin, KIN	Differentiation of adventitious buds with different cytokinins. BA gives the best results	Alvárez et al. (2009)
Pinus pinaster	BA, *m*T	Better values obtained with 2.2 μM *m*T	De Diego et al. (2011)
Pinus pinea	BA, BPA, *m*T, TDZ	Better TDZ (1 μM)	Cortizo et al. (2009)
Pinus sylvestris	BA, *m*T, TDZ, zeatin	Increased organogenesis response with 25 μM *m*T	De Diego et al. (2010)
Pistacia vera	BA, *m*T, KIN, zeatin, TDZ	The best number of good quality shoots with 8.4 μM *m*T. It improves rooting and reduces apical senescence	Benmahioul et al. (2012, 2016)
Pistacia vera	BA, *m*T, KIN	*m*T 5 μM + KIN 2 μM improve the quality and length of the shoots, delaying senescence. Apical necrosis decreases	Marin et al. (2016)
Prunus domestica, P. insititia × domestica	BA, *m*T	*m*T 2.1 μM improves the length and quality of the shoots. Longer roots. Formation of buds in leaves of shoots multiplied with *m*T Hyperhydricity decreases	Gentile et al. (2014)
Prunus domestica, P. insititia × domestica, P. cerasifera	BA, *m*T	*m*T (1.78–2.1 μM) lower multiplication rate; longer shoots. Longer plants acclimatized	Monticelli et al. (2017)
Prunus microcarpa	BA, *m*T, TDZ	Greater regeneration rates with BA	Nas et al. (2010)
Pterocarpus marsupium	TDZ, *m*T	Pre-treatment of nodal segments with TDZ, transfer to *m*T 5 μM + NAA 1 μM Also improves rooting and acclimatization	Ahmad et al. (2018)
Pyrus communis	*m*T	*m*T (6–9 μM) improves proliferation, leaf gas exchange, content of phenols, and antioxidant activity	Dimitrova et al. (2016)
Rosa hybrida	MemT, Me*m*TR , FmT, FmTR, BA	F-compounds improve the proliferation rates and Me*m*TR (2.5 μM) has anti-senescence activity	Bogaert et al. (2006)
Sclerocarya birrea	BA, *m*T, *m*TR, *Mem*TR	Better adventitious buds induction in hypocotyls with 8 μM *m*T	Moyo et al. (2011)

(continued)

Table 16.1 (continued)

Species	Cytokinins tested	Parameters	Reference
Sorbus torminalis	BA, mT, MeOBAPR	Better multiplication with BA; better response to rooting with MeOBAPR	Malá et al. (2009)
Ulmus glabra	BA, *m*T	*m*T (2.1 μM) increases proliferation, delays senescence, and improves acclimatization	Malá et al. (2013)
Ulmus spp., Sorbus torminalis	*m*T, Meo*m*TR	Delay senescence, improve rooting	Doležal et al. (2011)
Vaccinium corymbosum, V. vitis-idaea	Zeatin, TDZ, *m*T	Zeatin 20 μM + NAA 1 μM gave the best results in adventitious bud differentiation	Meiners et al. (2007)

Malus. They pre-treated the mother-shoots of "Royal Gala" for 3 weeks in media with different cytokinins (none, BA, or topolins), with the subsequent regeneration induced in media with BA or TDZ. The regeneration rates were elevated independently of treatment. However, the shoots pre-treated with topolins (2.1–6.3 μM) showed an increased number of regenerated buds. According to these authors, the use of topolins in the pre-treatment of apple shoots promotes the development of a juvenile-like structure in the leaves, increasing the organogenetic potential. The favorable carry-over effect of the topolins was also observed in the cultures of other cultivars: "M26" and "Idared" (Hudák et al. 2003; Dobránszki et al. 2004). In both cases, it decreased the hyperhydricity, and in "Idared," it increased the number of regenerated buds. The results of the study conducted by Dobránszki et al. (2005) show that the type and concentration of cytokinin(s) used before the regeneration phase can modify the organogenetic potential of the leaf tissues and thus the regeneration process. In the Jonagold apple cultivar, Magyar-Tábori et al. (2002) studied the effect of N^6-benzyladenine riboside (BAR) and *m*T in an attempt to overcome the problems (difficulty in rooting or toxicity) that the use of BA has on this species. The highest multiplication rates were obtained with BAR and *m*T (21 μM).

Moyo et al. (2011) used equimolar concentrations of BA, *m*T, *meta*-topolin riboside [6-(3-hydroxybenzylamino)-9-β-D-ribofuranosylpurine; *m*TR], and Me*m*TR in the induction of adventitious buds in hypocotyl segments of *Sclerocarya birrea*. The best results, as regards percentages of explants that responded and the length of the shoots, were achieved with *m*T at 8 μM. Moyo et al. (2018) later evaluated the influence of different cytokinins applied exogenously (BA, iP, *m*T, and *m*TTHP [6-(3-hydroxybenzylamino)-9-(tetrahydropyran-2-yl)purine]) on the organogenetic response in *Amelanchier alnifolia*. Overall, the topolins (*m*T y *m*TTHP) gave the highest proliferation rates and better-quality shoots compared with BA- or iP-derived regenerants. The best results, in terms of number of shoots, were achieved with 20 μM *m*T, although the shoot quality was better with 10 μM *m*T concentration, while the higher number of roots per shoot was obtained with 1 μM *m*TTHP. Amoo et al. (2014) concluded that this compound could be another viable topolin with the added advantage of inducing rooting at low concentrations.

Ahmad et al. (2018) developed an efficient micropropagation system in *Pterocarpus marsupium* by means of pre-treatment of the nodal segments with TDZ (10 μM) for 8 days. The explants are then transferred to a medium with *m*T (5 μM) + α-naphthalene acetic acid (NAA, 1 μM). This combination of regulators succeeded in obtaining the highest number of shoots, the greatest length, and the maximum regeneration frequency. Nacheva et al. (2017) also observed enhanced proliferation of lateral buds in *Ginkgo biloba* with 5.5 μM *m*T. In the propagation of adult material of the common ash (*Fraxinus excelsior*), Šedivá et al. (2017) obtained the best results supplemented the medium with 2.1 μM *m*T.

In *Coleonema album*, a native bush species of South Africa, Fajinmi et al. (2014) compared the efficacy of four topolins (*m*T, Me*m*TR, *m*TR, and Me*m*TTHP [*meta*-methoxytopolin-9-tetrahydropyran-2-yl]) with BA, KIN, and TDZ. At equimolar concentrations, all the treatments with BA, *m*T, *m*TR, and Me*m*TTHP increased shoot production compared with KIN, with the best results obtained with 5 μM *m*T. The combination of *m*T with 2 μM of indole-3-butyric acid (IBA) improved the results, with a higher production of shoots per explant. According to these authors, there may be an additive or perhaps synergistic effect of auxins with *m*T in the proliferation of shoots in this species.

The combination of *m*T (4.1–6.2 μM) and BA (8.9 μM), together with the use of a liquid medium led to the development of a multiplication system in shoots of the black walnut (*Juglans nigra*) (Stevens and Pijut 2018). Previous attempts to propagate *Juglans* via tissue culture had achieved different results depending on a wide variety of endogenous and exogenous factors. According to these authors, the main factor in the micropropagation of this species is the interaction between genotype and the type and concentration of growth regulators added to the culture medium. Sub- or super-optimal concentration of the traditional cytokinins, such as BA or *t*Z, has produced numerous deleterious effects during the multiplication phase, such as stunted growth, exudation of phenols, hyperhydricity, fasciation, apical necrosis, and chlorosis (Heile-Sudholt et al. 1986; Revilla et al. 1989; Van Sambeck et al. 1997; Bosela and Michler 2008). The positive interaction between the cytokinins used in this study (*m*T and BA) may be due to the way in which these cytokinins are metabolized and perceived by actively growing black walnut shoots. *Meta*-topolin, even at concentrations higher than BA, is effective without inducing morphological abnormalities. These observations have been reported in other species, and it is believed that they are the result of rapid turnover of the *m*T and its translocation in the plant tissues (Kamínek et al. 1987; Bairu et al. 2009; Amoo et al. 2011). When both cytokinins are available, the walnut shoots probably respond rapidly to *m*T, as a result of its increased bioactivity, while the uptake of BA, metabolically more stable, is delayed but integral to continued elongation after the turnover of *m*T.

An effective micropropagation system was developed in pistachio (*Pistacia vera*) by Benmahioul et al. (2012, 2016). These authors studied the effect of different cytokinins (BA, *m*T, KIN, *t*Z, and TDZ) on the proliferation of axillary buds. The addition of 8.4 μM *m*T to the medium resulted in a greater number of shoots, more rapid growth, as well as better quality. Marin et al. (2016) used a combination of *m*T (5 μM) and KIN (2 μM) in the micropropagation of mature *P. vera*. The incorporation of KIN into the multiplication medium allowed compensate for the lower number of shoots produced by BA.

Malá et al. (2013) compared the effect of BA and *m*T on the in vitro multiplication of the wych elm (*Ulmus glabra*), and they achieved higher multiplication rates with *m*T (2.1 μM). The explants cultivated with *m*T produced six times more shoots than those cultivated with BA. The development of a higher number of adventitious buds was also observed. However, in this work, no significant differences were found in the length of the shoots developed with BA or *m*T.

In *Paulownia elongata*, *P. fortunei*, and in the *P. elongata* × *P. fortunei* hybrid, the highest proliferation rates were obtained with 2.1 μM BA. However, this cytokinin produced a large proliferation of basal calli. The use of *m*T (2.1–4.2 μM) ensures a relatively high and uniform proliferation (Clapa et al. 2014).

Although BA is considered the most effective and economic cytokinin in the micropropagation of many cultivars of the *Pyrus* genus (Bell and Reed 2002), some studies have shown that other cytokinins are more effective. In the multiplication of shoots of pear rootstock OHF-333 (*P. communis*), Dimitrova et al. (2016) obtained good multiplication rates with 6–9 μM *m*T. According to these authors, the use of *m*T improves the leaf gas exchange and the content of phenols, as well as total antioxidant activity.

Prado et al. (2005) described a simple and reliable protocol for the micropropagation of two selected varieties of male kiwifruit (*Actinidia deliciosa*). In both cases the best results were obtained after 12 subcultures with a combination of BA, NAA, and gibberellic acid. The effect of TDZ or *m*T was subsequently studied, leading to improved multiplication rates in the "Tomuri" variety. Their results are consistent with those reported by Fernández (2001) in the female kiwifruit cultivar "Hayward." This author showed that the best incubation period was 305 days with 2.2 μM *m*T.

Bogaert et al. (2006) evaluated the effect of different derivatives of *m*T: Me*m*T, Me*m*TR, N^6-(3-fluorobenzyl)adenine (3FBA), its riboside (3FBAR), and BA in the micropropagation of *Rosa hybrida*. In this case the best results, as regards the number of shoots, were obtained with the fluoro-type of compounds.

In *Ficus carica*, Yahyaoui et al. (2017) studied the influence on the viability, regrowth, and conversion of the addition of three cytokinins (BA, *m*T, and *t*Z) to the endosperm of artificial seeds. The best results in the regrowth and in the conversion were obtained with BA and *m*T (2.1 μM).

The effect of *m*T has also been studied in the micropropagation of some species of the *Pinus* genus. De Diego et al. (2011) considered that the application of an exogenous cytokinin is an essential factor in the culture of pine tissue (*Pinus pinaster*). Their results show that the addition of 2.2 μM *m*T to the culture medium improves the results obtained with BA. Previously, Alvárez et al. (2009) also observed differentiation of adventitious shoots in cotyledons of *P. pinaster* cultivated with various cytokinins (BA, *m*T, TDZ, iP, *t*Z, and KIN), getting the best results with BA. In the regeneration of adult trees of *P. sylvestris*, De Diego et al. (2010) obtained the highest organogenetic response with 25 μM *m*T. On the other hand, in the micropropagation of adult stone pine (*P. pinea*), Cortizo et al. (2009) obtained the best results with TDZ, compared to BA, N^6-benzyl-9-(2-tetrahydropyranyl)adenine (BPA), or *m*T.

However, unfavorable responses have also been found with the use of topolins in other species, such as in the regeneration of buds in cotyledons and hypocotyl segments of *Prunus microcarpa*. In this species, *m*T showed to be less suitable than BA (Nas et al. 2010). Later, Nas et al. (2012) studied the effect of different cytokinins (BA, TDZ, and *m*T) in the micropropagation of mature *Crataegus aronia*, finding that BA was the only cytokinin capable of promoting the proliferation and elongation of shoots. Better results were also obtained with BA than with iP or *m*T in the micropropagation of *Magnolia* "Ann." Although the micro-shoots of *Magnolia* "Ann" cultivated with *m*T had lower moisture content, suggesting a reduced hyperhydricity, it also decreased the number of shoots and their length (Parris et al. 2012). In the wild service tree (*Sorbus torminalis*), Malá et al. (2009) studied the influence of BA, *m*T, and Me*m*TR on multiplication and rhizogenesis. The highest multiplication rates were obtained with BA, while the best rooting microcuttings were those multiplied on a medium with Me*m*TR. Meiners et al. (2007) studied the effect of different types and concentrations of cytokinins (*t*Z, TDZ, *m*T) on the differentiation of adventitious buds in leaves of in vitro micropropagated shoots of *Vaccinium corymbosum* and *V. vitis-idaea*. The best results were obtained with 20 μM *t*Z in combination with 1 μM NAA, while in the media with *m*T, there was no formation of callus or shoot development. Lattier et al. (2014) studied the effect of different cytokinins (BA, KIN, *m*T, and *t*Z) in order to improve the proliferation of an interspecific dogwood hybrid (*Cornus kousa* × *C. hongkongensis*). These authors demonstrated that there is a significant interaction between the type and concentration of the cytokinin and the number, length, multiplication rate, and phenol content. The best results were obtained with BA, although it also produced a higher number of phenols and hyperhydricity in many shoots. Kinetin and *m*T had a negative effect on the length of the shoots and did not influence the multiplication rates.

As well as having been used in several species to increase the multiplication rates, the topolins have also been applied to prevent different physiological disorders (shoot tip necrosis, hyperhydricity, early senescence) that can lead to significant losses of material during the micropropagation phase.

16.2.2 Reduction in Shoot Tip Necrosis

Apical necrosis is a phenomenon that occurs with relative frequency during the multiplication phase in vitro. It leads to senescence and death of the tissues of the apical bud and often spreading to the base of the shoots. It is a physiological disorder in which several factors are involved, among which the cytokinins have a dominant role (Il'ina et al. 2006). Bairu et al. (2011) showed the differential effect of the role of the cytokinin-auxin interaction on endogenous cytokinin levels in relation to apical necrosis. These authors demonstrated that the presence of indole-3-acetic acid (IAA) in the culture medium increased the formation of cytokinin 9-glucosides in the explants treated with BA but reduced it in those treated with topolins.

In pistachio, Benmahioul et al. (2012) showed that the incorporation of *m*T (8.4 μM) to the multiplication media improved the quality of the shoots reducing the apical necrosis. In this same species, Marin et al. (2016) replaced the BA with a combination of *m*T (5 μM) and KIN (2 μM), achieving an improvement in the quality and length of the shoots and a decrease in the apical necrosis.

16.2.3 Regulation of Hyperhydricity

Hyperhydricity is a morphological and physiological disorder frequently observed in micropropagated plants as a result of the passive entry of water into the plant tissue or an active phenomenon caused by severe metabolic changes (Pâques 1991; Debergh et al. 1992). It is a serious problem in the propagation of plants in vitro that makes its mass production difficult. This phenomenon is attributed to diverse causes, such as growth regulators; the type and concentration of gelling agents; the presence of large quantities of ions in the medium, particularly NH_4^+ and Cl^-; and the high humidity of the culture containers (Paques and Boxus 1987; Franck et al. 2004). Although the role that the different factors play in the process is not exactly known, there are different works in the literature on the effect of the growth regulators, especially the cytokinins. In the majority of the works, it mentions the fact that high concentrations of cytokinins cause hyperhydricity, with it being more severe in treatments with BA.

16.2.4 Delaying Senescence

Senescence is a genetically programmed process that involves the death of cell structures, followed by the moving of degradation products to other parts of the plants (Woo et al. 2004). The leaves are the main organs where senescence is shown, as result of a gradual and significant change in their cell structure that included the breakdown of the chloroplasts, which contain up to 70% of the leaf protein. This process, mainly characterized by the disintegration of the plant pigments, is highly sensitive to growth regulators, in particular ethylene and cytokinins (Gan and Amasino 1997; Van Staden et al. 1988). Gan and Amasino (1996, 1997) have carried out several molecular studies that demonstrate the effect of cytokinins in the senescence process. The use of topolins, especially the *meta*-forms, has been increasing in the last few years in order to alleviate these physiological disorders (Aremu et al. 2012). *Meta*-topolin delays aging in many plant organs by inhibiting the breakdown of proteins and by the stimulation of RNA and protein synthesis (Palavan-Ünsal et al. 2002).

In *Ulmus glabra*, Malá et al. (2013) recultivated the explants that showed signs of senescence in multiplication medium with BA or *m*T. After 6 weeks of culture, the signs of senescence only persisted in the treatment with BA. They determined the maximum efficiency of photosystem II (PSII), and the results showed that the explants grown in a medium with *m*T had a greater variable fluorescence (F_v) over maximum fluorescence (F_v/F_m) coefficient than those developed with BA in accordance with their lack of signs of senescence (yellowing of the leaves). They also found a higher content of ethylene in the culture containers with BA. Similar results have been obtained in *Rosa hybrida* (Doležal et al. 2003). On the other hand, in *Sorbus torminalis*, Malá et al. (2009) detected higher ethylene levels in the containers that had been supplemented with *m*T, which was associated with an increased rooting capacity in these explants.

Bogaert et al. (2006) investigated the anti-senescence effect of BA and different topolins in micropropagated plants of *Rosa hybrida*. Although none of the treatments prevented senescence of the older leaves (6 weeks), the treatment with Me*m*TR showed anti-senescence activity. After 18 weeks, 50% of the plants treated with Me*m*TR were alive, whereas the next cytokinin to demonstrate anti-senescence activity, 3FBAR, the percentage of survival was only 14%.

Similarly, Gentile et al. (2014) observed that the rootstock shoots of "Ferdor" (*P. insititia* × *P. domestica*), cultivated with *m*T, had a higher content of chlorophyll than those cultivated with BA, suggesting that *m*T may have a positive effect on the delay of senescence in *Prunus* spp. Others authors also pointed out this delay in senescence on using derivatives of *m*T, such as Me*m*TR in *Ulmus* and *Sorbus* (Doležal et al. 2011).

16.2.5 Rooting and Acclimatization Stage

The success of acclimatization on plants regenerated in vitro, considered as crucial in micropropagation protocols, depends on obtaining well rooted plantlets. The obtaining of high acclimatization percentages is indispensable for the large scale application of in vitro propagation (Aremu et al. 2012, 2017). Unlike the ex vitro conditions, the plants grow in an environment with a constant temperature during micropropagation, as well as a high relative humidity, low radiation, sugars as a carbon source, growth regulators in the culture medium, and variable concentrations (often insufficient) of CO_2 (Pospíšilová et al. 2007). These differences produce variations in the leaf structure, water relations, and photosynthetic parameters that are crucial for the acclimatization of plants regenerated in vitro (Pospíšilová et al. 1999). The effect of different regulators, such as cytokinins, auxins, and abscisic acid must also be taken into account during the acclimatization. Several studies have mentioned that the survival of the plantlets regenerated in vitro is strongly influenced by the type and concentration of cytokinins applied during the multiplication phase (Werbrouck et al. 1995; Bairu et al. 2008; Valero-Aracama et al. 2010; Aremu et al. 2014).

In many articles it is stated that the use of BA during the multiplication phase has negative effects on the rooting and the subsequent acclimatization (Werbrouck et al. 1995; Bairu et al. 2008; Amoo et al. 2011). These negative effects are due to the formation of biologically inactive and chemically stable metabolites such as *N*-glucosides or alanine conjugates (Werbrouck et al. 1995). The superiority of the topolins, compared to BA, has been attributed to their distinct metabolic pathways. Werbrouck et al. (1996) carried out the first comparative analysis of the regulatory properties of BA and *m*T in *Spathiphyllum floribundum*. According to this study, BA was mainly converted into a stable derivative, N^6-benzyladenine-9-glucoside (BA9G), which is mainly located in the basal regions of the plant, while the principal derivative of *m*T was N^6-(3-*O*-β-D-glucopyranosyl)benzyladenine-9-riboside (OG*m*TR). This latter compound is present in all parts of the plant and is metabolized much faster than BA9G during acclimatization. The presence of a hydroxyl group in the side ring in *m*T provides a structural advantage over BA since, as such, it can contribute to O-glycosylation to form storage forms that are more readily degradable during acclimation (Bairu et al. 2011).

Malá et al. (2013) observed that explants of *Ulmus glabra* growing on two cytokinins (BA or *m*T) showed similar levels of their main metabolic forms (ribosides, nucleotides, and 9-glucosides). However, in the explants grown in a medium with *m*T, a much higher concentration of *O*-glucosides was detected, as previously reported by Werbrouck et al. (1996) in *Spathiphyllum*, whereas in the explants grown with BA, the 9-glucosides were the principal metabolites. These results may be related to the improvement of acclimatization frequently observed with *m*T.

In *Prunus*, Gentile et al. (2014) reported a great influence of the genotype. While in the shoots of "Ferdor" rootstock (*P. insititia* × *P. domestica*), the best results (rooting percentage, number, and length of root) were obtained in shoots treated with *m*T; the response in "Torinel" (*P. domestica*) was similar for both cytokinins (BA or *m*T). The acclimatization was greater than 90% for both genotypes, and the number of roots and the development of the aerial part of the plants acclimatized were not influenced by the cytokinins used in the multiplication phase, but the length of the roots was significantly greater in the plants obtained by multiplication with *m*T. Monticelli et al. (2017) confirmed that the response of *Prunus* spp. to *m*T is species-dependent. Although the two genotypes ("Ferdor" and "Torinel") responded well to rooting and acclimatization regardless of the cytokinin (BA or *m*T) used, the shoots multiplied with *m*T showed larger leaves and, in the cases of "Ferdor," a greater length of the internodes.

Likewise, Ahmad et al. (2018) showed that the use of *m*T in the multiplication medium has a positive influence on the rooting of shoots and acclimatization in *Pterocarpus marsupium*. Similarly in pistachio, the highest rooting percentages were obtained in the shoots developed with *m*T (8.4 μM) (Benmahioul et al. 2012, 2016).

Vieitez AM (personal communication) also showed that in *Eucalyptus globulus*, the incorporation of *m*T (0.4 μM) to the multiplication medium improves the development of the shoots (Fig. 16.4), making their subsequent rooting easier.

Fig. 16.4 Shoot multiplication of *Eucalyptus globulus* in medium supplemented with 0.4 μM *m*T

16.3 Conclusion

Plant micropropagation is a complex process consisting of successive development stages that depend on numerous exogenous and endogenous factors. Among these factors, the role of plant growth regulators (PGRs), especially auxins and cytokinins, should be highlighted. These substances are considered as one of the most crucial components for any plant tissue culture protocol. The cytokinins promote cell division and differentiation, also having an influence in numerous physiological processes like the accumulation of chlorophyll, leaf senescence, apical necrosis, hyperhydricity, rooting, and acclimatization. The choice of cytokinin is one of the most critical factors in the propagation of woody species. Although BA is the most widely used cytokinin in micropropagation, several physiological disorders have also been observed, which makes it necessary to look for other alternatives that enable acceptable multiplication rates to be obtained as well as good-quality plants. The use of the topolins in in vitro cultures has rapidly increased since their discovery. Among these new compounds, *meta*-topolin deserves special attention and is considered as one of the most promising members of the group. Numerous studies reported the benefits of *m*T in the micropropagation of diverse species because differences in its metabolism ameliorate some of the adverse effects observed with other cytokinins, such as BA. The study of new growth regulators is necessary to overcome the limitations of the existing PGRs. This study, together with a deeper knowledge of the role of endogenous cytokinins and of the effects of the exogenous cytokinins will help in the development of more effective micropropagation techniques.

Acknowledgments This work has been partially funded with different projects from MINECO (Spanish Government) and the Regional Government of Galicia (Spain).

References

Ahmad A, Ahmad N, Anis M (2018) Preconditioning of nodal explants in thidiazuron-supplemented liquid media improves shoot multiplication in *Pterocarpus marsupium* (Roxb.). In: Ahmad N, Faisal M (eds) Thidiazuron: from urea derivative to plant growth regulator. Springer Nature Singapore, Singapore, pp 175–187

Alvárez JM, Majada J, Ordás RJ (2009) An improved micropropagation protocol for maritime pine (*Pinus pinaster* Ait.) isolated cotyledons. Forestry 82:175–184

Amoo SO, Finnie JF, Van Staden J (2011) The role of *meta*-topolins in alleviating micropropagation problems. Plant Growth Regul 63:197–206

Amoo SO, Aremu AO, Moyo M et al (2014) Physiological effects of a novel aromatic cytokinin analogue in micropropagated *Aloe arborescens* and *Harpagophytum procumbens*. Plant Cell Tissue Organ Cult 116:17–26

Aremu AO, Bairu MW, Doležal K et al (2012) Topolins: a panacea to plant tissue culture challenges? Plant Cell Tissue Organ Cult 108:1–6

Aremu AO, Plačkova L, Bairu MW et al (2014) How does exogenously applied cytokinin type affect growth and endogenous cytokinins in micropropagated *Merwilla plumbea*? Plant Cell Tissue Organ Cult 118:245–256

Aremu AO, Plačkova L, Gruz J et al (2015) Accumulation pattern of endogenous cytokinins and phenolics in different organs of 1-year-old cytokinin pre-incubated plants: implications for conservation. Plant Biol 17:1146–1155

Aremu AO, Doležal K, Van Staden J (2017) New cytokinin-like compounds as a tool to improve rooting and establishment of micropropagated plantlets. Acta Hortic 1155:497–503

Bairu MW, Stirk WA, Doležal K et al (2007) Optimizing the micropropagation protocol for the endangered *Aloe polyphylla*: can *meta*-topolin and its derivatives serve as replacement for benzyladenine and zeatin? Plant Cell Tissue Organ Cult 90:15–23

Bairu MW, Stirk WA, Doležal K et al (2008) The role of topolins in micropropagation and somaclonal variation of banana cultivars 'Williams' and 'Grand Naine' (*Musa* spp. AAA). Plant Cell Tissue Organ Cult 95:373–379

Bairu MW, Jain N, Stirk WA et al (2009) Solving the problem of shoot-tip necrosis in *Harpagophytum procumbens* by changing the cytokinin types, calcium and boron concentrations in the medium. S Afr J Bot 75:122–127

Bairu MW, Novák O, Doležal K et al (2011) Changes in endogenous cytokinins profiles in micropropagated *Harpagophytum procumbens* in relation to shoot-tip necrosis and cytokinin treatments. Plant Growth Regul 63:105–114

Bell R, Reed B (2002) In vitro tissue culture of pear: advances in techniques for micropropagation and germplasm conservation. Acta Hortic 596:412–418

Benmahioul B, Dorion N, Kaïd-Harche M et al (2012) Micropropagation and ex vitro rooting of pistachio (*Pistacia vera* L.). Plant Cell Tissue Organ Cult 108:353–358

Benmahioul B, Kaïd-Harche M, Daguin F (2016) In vitro regeneration of *Pistacia vera* L. from nodal explants. J For Sci 62:198–203

Bogaert I, Van Cauter S, Werbrouck SPO (2006) New aromatics cytokinins can make the difference. Acta Hortic 725:265–270

Bohra P, Eaman AA, Sathayanarayana BN et al (2016) Influence of different growth regulators on in vitro multiplication of mixed diploid banana (*Musa* AB). Proc Natl Acad Sci India Sect B Biol Sci 86:179–185

Bosela MJ, Michler CH (2008) Media effects on black walnut (*Juglans nigra* L.) shoot culture growth in vitro: evaluation of multiple nutrient formulations and cytokinin types. In Vitro Cell Dev-Plant 44:316–329

Chiancone B, Martorana L, Casales F et al (2016) The effects of benzyladenine and *meta*-topolin on in vitro sprouting and regrowth after encapsulation of C35 citrange (*Citrus sinensis* [L.] Osb. X *Poncirus trifoliata* [L.] Raf.) microcuttings. Acta Hortic 113:99–106

Chiancone B, Martorana L, Gianguzzi V et al (2017) The effects of *meta*-topolin and benzyladenine on in vitro organogenesis from epicotyl cuttings of Troyer citrange (*Citrus sinensis* [L.] Osbeck x *Poncirus trifoliata* [L.]Raf.). Acta Hortic 1155:185–192

Clapa D, Fira A, Simu M et al (2014) Improved in vitro propagation of *Paulownia elongata*, *P. fortunei* and its interspecific hybrid *P. elongate* x *P. fortunei*. Bull UASVM Hortic 71:6–14

Cortizo M, De Diego N, Moncaleán P et al (2009) Micropropagation of adult stone pine (*Pinus pinea* L.). Trees 23:835–842

De Diego N, Montálban IA, Moncaleán P (2010) In vitro regeneration of adult *Pinus sylvestris* L. trees. S Afr J Bot 76:158–162

De Diego N, Montálban IA, Moncaleán P (2011) Improved micropropagation protocol for maritime pine using zygotic embryos. Scand J For Res 26:202–211

Debergh P, Aitken-Christie J, Cohen D et al (1992) Reconsideration of the term vitrification as used in micropropagation. Plant Cell Tissue Organ Cult 30:140–165

Dimitrova N, Nacheva L, Berova M (2016) Effect of *meta*-topolin on the shoot multiplication of pear rootstock OHF-333 (*Pyrus communis* L.). Acta Sci Pol Hort 15:46–53

Dobránszki J, Magyar-Tábory K, Jámbor-Benczúr E et al (2002) Effect of conditioning apple shoots with *meta*-topolin on the morphogenic activity of in vitro leaves. Acta Agronom Hung 50:117–126

Dobránszki J, Hudák I, Magyar-Tábory K et al (2004) Effects of different cytokinins on the shoot regeneration from apple leaves of 'Royal Gala' and 'M26'. Int J Hort Sci 10:69–75

Dobránszki J, Jámbor-Benczúr E, Reményi ML et al (2005) Effects of aromatic cytokinins on structural characteristics of leaves and their post-effects on subsequent shoot regeneration from in vitro apple leaves of 'Royal Gala'. Int J Hort Sci 1:41–46

Doležal K , Popa I, Holub J et al (2003) Heterocyclic compound based on N^6-substituted adenine, methods, of their preparation, their use for preparation of drugs, cosmetic preparations and growth regulators, pharmaceutical preparations, cosmetic preparations and growth regulators containing these compounds. WO0304144

Doležal K, Malá J, Machová P et al (2011) New cytokinin derivatives—their discovery, development and use for micropropagation of endangered tree species. BMC Proc 5:O46

Fajinmi OO, Amoo SO, Finnie JF et al (2014) Optimization of in vitro propagation of *Coleonema album*, a highly utilized medicinal and ornamental plant. S Afr J Bot 94:9–13

Fernández D (2001) Utilización de *meta*-topolina en la optimización de la micropropagación de *Actinidia deliciosa*. MS thesis, University of Oviedo, Spain

Franck T, Kevers C, Gaspar T et al (2004) Hyperhydricity of *Prunus avium* shoots cultured on gelrite: a controlled stress response. Plant Physiol Biochem 42:519–527

Gan S, Amasino RM (1996) Cytokinins in plant senescence: from spray to clone and play. BioEssays 18:557–565

Gan S, Amasino RM (1997) Making sense of senescence. Molecular genetic regulation and manipulation of leaf senescence. Plant Physiol 113:313–319

Gentile A, Gutierres MJ, Martínez J et al (2014) Effect of *meta*-topolin on micropropagation and adventitious shoot regeneration in *Prunus* rootstocks. Plant Cell Tissue Organ Cult 118:373–381

Hall RH, De Ropp RS (1955) Formation of 6-furfurylaminopurine from DNA breakdown products. J Am Chem Soc 77:6400

Heile-Sudholt C, Huetteman CA, Preece JE et al (1986) In vitro embryonic axes and seedling shoot tip culture of *Juglans nigra* L. Plant Cell Tissue Organ Cult 6:189–197

Horgan R, Hewett EW, Purse JG et al (1973) A new cytokinin from *Populus robusta*. Tetrahedron Lett 14:2827–2828

Horgan R, Hewett EW, Horgan JM et al (1975) A new cytokinin from *Populus* × *robusta*. Phytochemistry 14:1005–1008

Howell SH, Lall S, Che P (2003) Cytokinins and shoot development. Trends Plant Sci 8:453–459

Hudák I, Dobránszki J, Magyar-Tábori K, et al (2003) Effects of benzyladenine analogues on the organogenic ability of apple leaves. In: Proceedings of the 5th international symposium in the series recent advances in plant biotechnology. Plant biotechnology: progress and development, Stará Lesná, September 7-13, High Tatras, Slovak Republic, pp 22

Il'ina E, Dodueva I, Ivanova N et al (2006) The effect of cytokinins on in vitro cultured inbred lines of *Raphanus sativus* var. radiculata pers. with genetically determined tumorigenesis. Russ J Plant Physiol 53:514–522

Jones L, Martinková H, Strnad M et al (1996) Occurrence of aromatic cytokinins in oil palm (*Elaeis guineensis* Jacq.). J Plant Growth Regul 15:39–49

Kamínek M, Vaněk T, Motyka V (1987) Cytokinin activities of N^6-benzyladenosine derivatives hydroxylated on the side-chain phenyl ring. J Plant Growth Regul 6:13–120

Lattier JD, Touchell DH, Ranney TG (2014) Micropropagation of an interspecific hybrid dogwood (*Cornus* 'NCCH1'). Propag Ornam Plants 14:184–190

Letham DS (1963) Zeatin, a factor inducing cell division isolated from *Zea mays*. Life Sci 2:569–573

Magyar-Tábori K, Dobránszki J, Jámbor-Benczúr E (2002) High in vitro shoot proliferation in the apple cultivar Jonagold induced by benzyladenine analogues. Acta Agron Hung 50:191–195

Malá J, Máchová P, Cvrčková H et al (2009) Micropropagation of wild service tree (*Sorbus torminalis* [L.] Crantz). The regulative role of different aromatic cytokinins during organogenesis. J Plant Growth Regul 28:341–348

Malá J, Machová P, Cvrčková H et al (2013) The role of cytokinins during micropropagation of wych elm. Biol Plant 57:174–178

Marin JA, García E, Lorente P et al (2016) A novel approach for propagation of recalcitrant pistachio cultivars that sidesteps rooting by ex vitro grafting of tissue culture shoot tips. Plant Cell Tissue Organ Cult 124:191–200

Meiners J, Schwab M, Szankowski I (2007) Efficient in vitro regeneration systems for *Vaccinium* species. Plant Cell Tissue Organ Cult 89:69–176

Miller CO, Skoog F, Von Saltza MH et al (1955) Kinetin, a cell division factor from deoxyribonucleic acid. J Am Chem Soc 77:1392–1393

Mohamed MAH, Alsadon AA (2011) Effect of vessel type and growth regulators on micropropagation of *Capsicum annuum*. Biol Plant 55:370–374

Monticelli S, Gentile A, Frattarelli A et al (2017) Effects of the natural cytokinin *meta*-topolin on in vitro shoot proliferation and acclimatization of *Prunus* spp. Acta Hortic 1155:349–353

Moyo M, Finnie JF, Van Staden J (2011) Recalcitrant effects associated with the development of basal callus-like tissue on caulogenesis and rhizogenesis in *Sclerocarya birrea*. Plant Growth Regul 63:187–195

Moyo M, Aremu AO, Plačková L et al (2018) Deciphering the growth pattern and phytohormonal content in Saskatoon berry (*Amelanchier alnifolia*) in response to in vitro cytokinin application. New Biotechnol 42:85–94

Nacheva L, Gercheva P, Ivanova V et al (2017) *Meta*-topolin improves lateral bud proliferation in micropropagation of *Ginkgo biloba* L. Acta Hortic 1155:355–359

Nas MN, Bolek Y, Sevgin N (2010) The effects of explant and cytokinin type on regeneration of *Prunus microcarpa*. Sci Hortic 126:88–94

Nas MN, Gokbunar L, Sevgin N et al (2012) Micropropagation of mature *Crataegus aronia* L., a medicinal and ornamental plant with rootstock potential for pome fruit. Plant Growth Regul 67:57–63

Niedz RP, Evens TJ (2011) The effects of benzyladenine and *meta*-topolin on in vitro shoot regeneration of sweet orange. ARPN J Agric Biol Sci 6:64–73

Palavan-Ünsal N, Çağ S, Çetin E et al (2002) Retardation of senescence by *meta*-topolin in wheat leaves. J Cell Mol Biol 1:101–108

Pâques M (1991) Vitrification and micropropagation: causes, remedies and prospects. Acta Hortic 289:283–290

Paques M, Boxus P (1987) Vitrification: review of literature. Acta Hortic 212:155–166

Parris JK, Touchell DH, Ranney TG (2012) Basal salt composition, cytokinins, and phenolic binding agents influence in vitro growth and ex vitro establishment of *Magnolia* 'Ann'. HortScience 47:1625–1629

Pospíšilová J, Synková H, Haisel D et al (1999) Effect of elevated CO_2 concentration on acclimation of tobacco plantlets to ex vitro conditions. J Exp Bot 50:119–126

Pospíšilová J, Synková H, Haisel D et al (2007) Acclimation of plantlets to ex vitro conditions: effects of humidity, irradiance, CO_2 concentration and abscisic acid (a review). Acta Hortic 748:29–38

Prado MJ, Herrera MT, Vázquez RA et al (2005) Micropropagation of two selected male kiwifruit and analysis of genetic variation with AFLP markers. HortScience 40:740–746

Revilla A, Majada J, Rodríguez R (1989) Walnut (*Juglans regia* L.) micropropagation. Ann Sci For 46(Suppl):149s–151s

Schmülling T (2004) Cytokinin. In: Lennarz WJ, Lane MD (eds) Encyclopedia of biological chemistry. Academic Press/Elsevier Science, New York, pp 1–6

Šedivá J, Havrdová L, Maršik P (2017) Micropropagation of common ash clones resistant to fungus *Hymenoscyphus fraxineus*. Acta Hortic 1155:93–99

Stevens ME, Pijut PM (2018) Rapid in vitro multiplication of the recalcitrant species *Juglans nigra* L. In Vitro Cell Dev Biol Plant 54:309–317

Strnad M, Hanuš J, Vaněk T et al (1997) *Meta*-topolin, a highly active aromatic cytokinin from poplar leaves (*Populus* x *canadensis* Moench, cv. *Robusta*). Phytochemistry 45:213–218

Subbaraj AK (2011) Topolins: current research status and applications. Int J Biores Stress Manag 2:10–25

Taiz L, Zeiger E, Møller IM et al (2015) Plant physiology and development, 6th edn. Sinauer Associates, Inc., Sunderland

Tarkowská D, Doležal K, Tarkowski P et al (2003) Identification of new aromatic cytokinins in *Arabidopsis thaliana* and *Populus* x *canadensis* leaves by LC-(+)ESI-MS and capillary liquid chromatography fri-fast atom bombardment mass spectrometry. Physiol Plant 117:579–590

Valero-Arcama C, Kane ME, Wilson SB et al (2010) Substitution of benzyladenine with *meta*-topolin during shoot multiplication increases acclimatization of difficult- and easy-to-acclimatize sea oats (*Uniola paniculata* L.) genotypes. Plant Growth Regul 60:43–49

Van Sambeck JW, Lambus L, Kahn S, et al (1997) In vitro establishment of tissues from adult black walnut. In: Van Sambeck J (ed) Knowledge for the future of black walnut. Proceedings of V black walnut symposium, 28–31 July, USDA Forest Service, North Central Forest Experiment Station, pp 78–92

Van Staden J, Cook EL, Noodén LD (1988) Cytokinins and senescence. In: Noodén LD, Leopold AC (eds) Senescence and aging in plants. Academic Press, San Diego, pp 281–328

Werbrouck SPO (2010) Merits and drawbacks of new aromatic cytokinins in plant tissue culture. Acta Hortic 865:103–108

Werbrouck SPO, Van der Jeugt B, Dewitte W et al (1995) The metabolism of benzyladenine in *Spathiphyllum floribundum* schott 'petite' in relation to acclimatization problems. Plant Cell Rep 14:662–665

Werbrouck SPO, Strnad M, Van Onckelen HA et al (1996) *Meta*-topolin, an alternative to benzyladenine in tissue culture. Physiol Plant 98:291–297

Woo HR, Kim JH, Nam HG et al (2004) The delayed leaf senescence mutants of *Arabidopsis*, *ore1*, *ore3*, and *ore9* are tolerant to oxidative stress. Plant Cell Physiol 45:923–932

Wotjania A (2010) Effect of *meta*-topolin on in vitro propagation of *Pelargonium* x *hortorum* and *Pelargonium* x *hederaefolium* cultivars. Acta Soc Bot Pol 79:101–106

Yahyaoui E, Casamento D, Frasheri D et al (2017) Evaluation of some growth regulator effects on encapsulated in vitro-derived microcuttings of three Italian *Ficus carica* L. genotypes. Acta Hortic 1187:93–100

Zalabák D, Pospíšilová H, Šmehilová M et al (2013) Genetic engineering of cytokinin metabolism: prospective way to improve agricultural traits of crop plants. Biotechnol Adv 31:97–117

Zürcher E, Müller B (2016) Cytokinin synthesis, signaling, and function- advances and new insights. Int Rev Cell Mol Biol 324:1–38

Biotechnological Application of *Meta*-topolins as Highly Active Aromatic Cytokinins in Micropropagation of Medicinal Plants

17

Pankaj Kumar, Ishani Shaunak, Geetika Gambhir, Karuna Dhiman, and Dinesh Kumar Srivastava

Contents

Abstract

Medicinal herbs and plants are the major reservoirs for new drug formulations. Owing to pharmaceutical demands and limited wild resources, it is vital to conserve genetic diversity, which is continuously under the risk of extinction due to its overexploitation which resulted in natural habitat abolition and unmonitored medicinal plant trade. To deal with this major havoc, plant biotechnological techniques and approaches such as plant tissue culture and micropropagation

Ishani Shaunak, Geetika Gambhir and Karuna Dhiman contributed equally.

P. Kumar (✉)
Science and Engineering Research Board, Department of Science & Technology, Government of India, New Delhi, India

CSIR-Institute of Himalayan Bioresource Technology (IHBT), Palampur, HP, India

Department of Biotechnology, Dr Y.S. Parmar University of Horticulture and Forestry, Solan, HP, India

I. Shaunak · K. Dhiman · D. K. Srivastava
Department of Biotechnology, Dr Y.S. Parmar University of Horticulture and Forestry, Solan, HP, India

G. Gambhir
Department of Biotechnology, Dr Y.S. Parmar University of Horticulture and Forestry, Solan, HP, India

ICAR-Indian Agricultural Research Institute (IARI), New Delhi, India

N. Ahmad, M. Strnad (eds.), *Meta-topolin: A Growth Regulator for Plant Biotechnology and Agriculture*, https://doi.org/10.1007/978-981-15-9046-7_17

hold promise for conservation and rapid mass multiplication to obtain genetically elite (true to type) populations under in vitro conditions. The most critical factor that has an effect on in vitro plant propagation success is the choice and selection of plant growth regulators (i.e., auxins and cytokinins). Plant growth regulators are often supplemented in the nutrient medium to harness the totipotent potential such as development of shoots, roots and whole plants. Most commonly used cytokinin in plant tissue culture micropropagation protocols is N^6-benzyladenine (BA), but nowadays *meta*-topolin as highly active aromatic cytokinin along with its derivatives have been largely investigated and proved as an effective alternative to BA in micropropagation protocols of medicinal plants. This chapter addresses the biotechnological application of *meta*-topolins such as their effect on in vitro plant regeneration, micropropagation and hardening, countering micropropagation troubles and its influences on secondary metabolite production.

Keywords

Plant growth regulators · Micropropagation · Secondary metabolites · Cytokinins · Topolins · Metabolism

17.1 Introduction

Secondary metabolites/bioactive compounds present in medicinal plants are currently being widely expoited commercially at large scale and play vital role in high-value product development in the pharmaceutical industry. Due to continuous demand and the over-exploitation of medicinal plants/herbs in general, the Himalayan plants/high-altitude medicinal plants are facing threat of extinction with a reduced distribution range. In addition, various biotic and abiotic stress factors, loss in seed viability, poor seed germination and survival percentage also affect the overall growth and development of these plants (Jan and Abbas 2018; Kumar et al. 2020). Biotechnological applications such as plant tissue culture (conservation and propagation of valuable, rare and endangered plants), cell culture and tissue engineering (mass production of bioactive constituents via callus or suspension cultures), genetic engineering and pathway and metabolic engineering (for genetic makeup modification for higher biomass production) is playing a significant role in high quality plant material supply and value addition to harness potential applications.

For the cultivation of medicinal plants comprehensive understanding of the species biology is required. For in vitro plant propagation, detailed standardization and optimization procedures (phytohormones, pH, temperature, aeration, agitation, light, etc.) are required for culture initiation, growth, multiplication and acclimatization (Briskin 2000; Harsh et al. 2017; Kumar et al. 2017; Jan and Abbas 2018). Once the reliable, genetically stable and highly efficient in vitro plant regeneration protocol has been optimized, the crucial step is successful acclimatization and hardening of tissue culture raised plants. Hardening of in vitro raised culture depends on optimization of humidity, light, temperature and other environmental conditions approximating the natural environment of the cultured plant species. Re-introduction of in vitro raised plants in their natural habitat is the most critical step in plant conservation strategies and dependes on a successful acclimatization procedure (Chaturvedi et al. 2007). The process however, is time-consuming, tedious and requires extensive effort. These factors are a major obstacle in large-scale quality plant production (Harsh et al. 2017; Kumar et al. 2017).

Micropropagation is an important plant biotechnological process in which excised plant parts (explant) such as cells, tissues or organs from a particular plant, are surface sterilized and cultured on an appropriate nutrient medium to produce a large number of clonal plantlets under in vitro conditions (Kumar and Srivastava 2016). This is a valuable tool for selecting, multiplying and screening promising genotypes of Himalayan medicinal plants. Development of an efficient protocol for in vitro culture establishment is a vital procedure in plant biotechnology xhich exploits the totipotent nature of plant cells. Among the various cytokinins used in plant tissue culture protocol optimization, N^6-benzyladenine (BA) is the most commonly used. However, use of topolins and their derivatives as an alternative to BA N^6-shows efficacy in standardizing high frequency shoot and root regeneration protocols and countering various micropropagation problems arising under in vitro conditions (Werbrouck et al. 1996).

Meta-topolins (*m*T) is a hydroxylated BA analogue with the following chemical formula: 6-(3-hydroxybenzylamino)purine. It is a relatively new cytokinin with an aromatic ring but similar to other cytokinins in its metabolism. The N^6-benzyl ring of *meta*-topolins have a hydroxyl group which can lead in O-glucoside metabolite formation similar to zeatins. Although topolins and their derivatives have been identified and distinguished in various plant species, their main source is poplar leave. From the discovery of cytokinins plant tissue culture grew rapidly and exponentially in the field of their relevance. Cytokinins are indispensable group of plant growth regulators (PGRs) undeviating physiological processes, which have a distinctive attribute for promoting cell division during micropropagation (Miller et al. 1955; Heyl and Schmulling 2003) during micropropagation. This group of PGRs affects various developmental and physiological processes such as the regulation of root and shoot growth, chloroplast development, stress response, pathogen resistance and leaf senescence. Cytokinins are mainly classified into two major groups that is, naturally occurring cytokinin and synthetic ones. The former are mainly adenine derivatives and further classified on the basis of adenine N^6-side chain with an aliphatic side chain (isoprenoid group) and aromatic side chain (aromatic group) (Strnad 1997). Synthetic cytokinins have a phenylurea group such as thidiazuron (N'-phenyl-N'-l, 2,3-thiadiazol-5-yl urea) and CPPU (N-(2-chloro-4-pyridyl)-N'-phenylurea) (Mok and Mok 2001). The goal of micropropagation is the mass multiplication of homogeneous, true to type nature (regenerants) and healthy plantlets development. However, a large number of physiological disorders and in vitro abnormalities have been encountered during micropropagation protocol optimization such as low shoot multiplication rate, hyperhydricity, poor root regeneration, high production cost, underdeveloped growth and epigenetic and somaclonal variation (Smulders and de Klerk 2011). These are some of the factors that challenge the micropropagation of most plants, and for these reasons, commercial application of plant tissue culture in many plant species is also reduced. These problems can be alleviated by optimizing hormonal regulation (auxin to cytokinin ratio), using appropriate plant growth regulator (type and concentration) and by standardizing various factors which affect the in vitro morphogenic potential during efficient micropropagation protocol development. To solve this, the efficacy of topolins as naturally occurring highly active aromatic cytokinins has been highlighted in plant tissue culture systems as it has emerged as a viable choice to other cytokinins used such as N^6-benzyladenine, kinetin, and *trans*-zeatin in plant tissue culture.

Some studies have also described topolins and their substitute's role either alone or in combination with auxins in comparison to BA. *Meta*-topolin (*m*T) was reported to be very effective in high-frequency shoot regeneration and secondary metabolite production compared to BA in the case of *Huernia hystrix*, a medicinal plant. *m*T also helped solve the problem of vitrification when media was supplemented with *m*T (2.5 mg/L) and NAA (0.1 mg/L) and adenine sulfate (50 mg/L) in *Withania coagulans* (Stocks) Dunal, medicinally important plant of Solanaceae family (Joshi et al. 2016). Therefore, it can be concluded that using *m*T as an alternative to other cytokinins possibly will go ahead for remarkable change in the growth and developmental control of micropropagated plants.

17.2 *Meta*-topolins' Effect on In Vitro Plant Regeneration and Hardening During Micropropagation

Cytokinins promote plant cell division and cell differentiation processes and influence physiological aspects such as leaf senescence and chlorophyll accumulation; thus shoot formation and quality of the explant obtained with in vitro culture are highly dependent on type and concentration of cytokinins used (Mok and Mok 2001; Haberer and Kieber 2002; Sakakibara 2006). Zeatin is a naturally occurring cytokinin, but BA is cheaper than others and the most effective cytokinin in promoting in vitro shoot regeneration and proliferation. For this reason, it is currently used in commercial laboratories. However, it was also discovered that this cytokinin can also induce physiological disorders in some plant species (Amoo et al. 2011; Werbrouck 2010). In order to resolve these issue, new investigations should be done to find alternative potent cytokinins, i.e., alternative to BA, most widely used cytokinin in plant tissue culture, which could be able to maintain efficient shoot multiplication rates and production of high-quality planting material. Use of *m*T is gaining interest day by day because it leads to efficient micropropagation (Bairu et al. 2007), improved biochemical and physiological traits (Mala et al. 2013), efficient vitro root regeneration, successful acclimatization and hardening (Aremu et al. 2012a).

The main prerequisite of micropropagation/in vitro plant regeneration is to have a reliable, efficient, and high frequency regeneration protocol, i.e., efficient healthy shoots and root regeneration and successful acclimatization and hardening in the field. Topolins have been used as promising alternatives to other cytokinins in in vitro regeneration and micropropagation. There are reports suggesting *meta*-topolin (*m*T) is nearly twice as effective as BA in multiple shoot induction in a number of plant species. Use of topolins in medicinal plant micropropagation has been increasing at a rapid rate (Werbrouck 2010; Aremu et al. 2012a). Werbrouck et al. (1996) observed healthy shoot and root regeneration response when medium was supplemented with equimolar concentration (10.0μM) of *m*T and BA during *Spathiphyllum floribundum* micropropagation. In earlier studies, comparable effects of *m*T cytokinins on efficient in vitro regeneration of sugar beet (Kubalaková and Strnad 1992), *Curcuma longa* (Salvi et al. 2002), *Aloe polyphylla* (Bairu et al. 2007), and banana cv. Williams (Bairu et al. 2008) have been investigated. There is a number of reports which showed that when *mT* is used in place of BA, there was remarkable improvement in the mass in vitro propagation (Kubalaková and Strnad 1992; Strnad et al. 1997). Bairu et al. (2007) reported that *meta*-topolins resulted in a greater number of shoots in *Aloe polyphylla* micropropagation compared with other cytokinins such as BA and *trans*-zeatin at different concentrations. Use of kinetin riboside (KR), N^6-isopentenyladenine (iP) or *meta*-topolins (*m*T) in plant micropropagation, i.e., *Curcuma longa* revealed a maximum in vitro shoot regeneration response compared to other cytokinins used such as BA (Salvi et al. 2002). Different concentrations of aromatic cytokinins such as *meta*-topolin riboside (7.5μM), *meta*-methoxytopolin (15μM), and *meta*-methoxytopolin riboside (30μM) showed maximum shoot proliferation rates versus BA in banana cv. Williams (Bairu et al. 2008). From the available literature it is reported that *m*T was found more effective than other cytokinins such as BA for efficient shoot regeneration and proliferation in longa (Salvi et al. 2002), *Musa* spp. (Escalona et al. 2003; Bairu et al. 2007); *Curcuma* and *Aloe* spp. (Bairu et al. 2008), *Pelargonium* cultivars (Wojtania 2010), and *Huernia hystrix* (Amoo et al. 2013). However several adverse effects on shoot multiplication were also observed in a number of plant species such as *Vaccinium corymbosum* (Meiners et al. 2007), wild service tree (*Sorbus torminalis* L. Crantz) (Malá et al. 2009), and in *Citrus* hybrid (Niedz and Evens 2010).

Bairu et al. (2007) described *m*T as a more potent cytokinin over BA for efficient, high frequency in vitro shoot as well as root regeneration. It also resulted in reduced hyperhydricity with improved ex vitro acclimatization of *Aloe polyphylla*. These results are in accordance with investigation of Nacheva et al. (2017). These authors observed that substituting BA with *m*T remarkably improves the in vitro lateral bud induction and shoot multiplication and ensured commercial propagation in *Ginkgo biloba* plants. Amoo et al. (2015) investigated a novel aromatic cytokinin (CK) derivative, i.e., *meta*-topolin-tetrahydropyran-2-yl (*m*TTHP) effects for efficient micropropagation in *Merwilla plumbea*. *Merwilla plumbea* (Lindl.) Speta is an important medicinal bulbous plant belonging to the family Asparagaceae and in much demand in the medicinal plant market of South Africa (Williams et al. 2013).

Meta-topolin derivatives, i.e., *meta*-topolin riboside and *m*TTHP treatments resulted in improved shoot regeneration with highest adventitious shoot response over the control and thidiazuron growth regulator used. Maximum root regeneration response was also observed in *m*TTHP treatments. It also also resulted in significantly greater photosynthetic efficiency and increase in antioxidant enzyme activities.

Lata et al. (2016) *m*T as a potent aromatic cytokinin in *Cannabis sativa* L. micropropagation. *Cannabis sativa* is a high value medicinal plant of the Cannabaceae family having bioactive compounds, namely tetrahydrocannabinol (9-THC) and cannabidiol (CBD). Cannabidiol (non-psychoactive) and tetrahydrocannabinol (psychoactive) molecules have a number of potential medicinal applications such as epileptic seizures treatments in children (Pertwee 2014). Using *meta*-topolin (*m*T) as a potent highly active aromatic natural cytokinin, an efficient, reliable, simple, and cost effective one step regeneration system for high frequency shoot and root regeneration was optimized in *Cannabis sativa* from nodal explants. Highest shoot regeneration frequency as mean number of shoots was observed on Murashige and Skoog (MS) medium containing 2μM *m*T. Shoot multiplication and proliferation was also noted on the same regeneration medium and it was also able to induce a healthy root system within 4–6 weeks. For root induction, the medium was not supplemented with any kind of plant growth regulator. Remarkable achievement with 100% survival rate was observed during in vitro regenerated plantlets hardening and acclimatization. Genetic stability studies to ensure true type regenerants were also tested and confirmed using inter simple sequence repeat (ISSR) molecular markers. Cannabinoid profiles and content were found similar to each other in in vitro regenerants and in mother plant, when analyzed qualitatively and quantitatively by using gas chromatography-flame ionization detector (GC-FID). Finally, it was concluded that for large scale production of true to type *C. sativa* plants *m*T and its derivatives are an appropriate alternative being both highly potent and effective.

17.2.1 *Meta*-topolins Effect on In Vitro Rooting of Medical Plants

Auxins play a major role in in vitro rooting, but there is some evidence that cytokinins are also important. It has been found out that *m*T raised microshoots show higher rooting frequency than those raised on BA medium. Further, *m*T raised micro shoots were longer and healthier than BA raised plantlets. Aremu et al. (2012a, b) and Naaz et al. (2019) showed significant improvement in micropropagation of *Syzygium cumini* and their acclimatization to ex vitro conditions when supplemented with *m*T. Increase in root regeneration response in *m*T raised microshoots was obtained in *Uniola* (Aremu et al. 2012a), *Aloe polyphylla* (Bairu et al. 2007) and *Pelargonium* × *hortorum* (Mutui et al. 2012). Similarly, Bairu et al. (2007) observed efficient root regeneration response on micropropagated shoots of *Aloe polyphylla.* These effects were observed only on multiplication medium which had either *m*T or its riboside (*m*TR). In, *Solanum* *tuberosum* cv. Jaerla plantlets, root regeneration was increased significantly at low concentrations (5–10μg.L^{-1}) of *m*TR in culture medium (Baroja Fernandez et al. 2002). There are also reports which show that *m*T raised regenerants have a higher number of healthy roots compared to BA regenerants. For this reason, these plantlets were more resistant to environmental stresses in field conditions. However, an inhibitory effect on in vitro root regeneration response was also observed in *Musa* spp. (cv. Williams) as compared to BA (Bairu et al. 2008). Similarly, it has been reported that increased concentrations of BA (1.33μM) and *m*T (22.2μM) resulted in root regeneration reduction on plantain cv. CEMSA 3/4 (Escalona et al. 2003). However, the highest number of roots was found on 1.33μM *m*T. As topolins showed both positive and detrimental results on in vitro rooting, much research investigation is still needed for the study of topolin effects in root development biology.

17.3 *Meta*-topolins Influences on Secondary Metabolite Production

Secondary metabolites are organic compounds that help plants to cope with different environmental stresses, i.e., biotic and abiotic. Secondary metabolites are an economically very important class of compounds because they can be utilized as insecticides, colourants, antimicrobials, fragrances and therapeutics. The production of secondary metabolites is triggered in only specific stages of growth and development or during various stresses (biotic and abiotic) or the constraint of supplements. The extraction and purification of secondary metabolites are complex and arduous their production only occurs under certain environmental conditions and depends upon an individual species and genus. For these reasons, commercially available secondary metabolites, for example, pharmaceuticals, flavours, colours, antimicrobials, fragrances and pesticides are of higher value than primary metabolites (Nielsen et al. 2019).

Biotechnology emerges up as a new boom for secondary metabolite production by the use of in vitro regenerating cells, tissues, organs or entire organisms and use of advanced genetic engineering tools for desired modification and desired bioactive ingredients (Rao and Ravishankar 2002). Thus, in vitro regeneration has enormous potential in the control of plant bioactive molecule production. Different plant tissue culture approaches including cell and tissue culture, organ, callus, suspension, and hairy root cultures are used for secondary metabolite production. The application of an in vitro cell culture systems for the production of therapeutic compounds has made conceivable the manufacturing of a wide range of pharmaceuticals, such as amino acids, phenolics, flavonoids, terpenoids, saponins, alkaloids and steroids (Paric et al. 2017).

The biosynthesis of secondary metabolites through plant tissue culture depends upon various factors such as plant growth regulators, type of cell line and explants, environmental factors (temperature, humidity) and elicitors. Out of all these, plant growth regulators play vital role in growth and secondary metabolite biosynthesis. The optimum concentrations and combinations of plant growth regulators like auxins and cytokinins must be used in nutrient media to enhance the development as well as regulation of cell metabolism (Filova 2014). The exogenous application of different types and concentrations of aromatic cytokinins during plant tissue culture especially impacts the in vitro generation of bioactive secondary compounds. Cytokinins act as enhancers for the metabolism of secondary compounds and play a vital role in cyto-differentiation and sub-cellular differentiation (Jain et al. 2012). They also assist cell division and initiation of callus growth and development. Cytokinins have diverse impact on secondary metabolite production and this depends on type of metabolites and species used. For example, kinetin triggered anthocyanin production in *Haplopappus gracilis* whereas it inhibits anthocyanin synthesis in *Populus* cell cultures. Total phenolic and flavonoid content was increased in *Thymus vulgaris* and *Origanum vulgare* using BA and indole-3-butyric acid (IBA), however, it was decreased in *Ocimum basilicum* (Karalija et al. 2016). Similarly, in *Mentha piperita* us of BA alone resulted in increased essential oil yields as well as its components as menthone, menthol, pulegone and menthofuran, whereas BA in combination with IBA had no significant effect on production of essential oils (Santoro et al. 2013). Cytokinins also stimulate alkaloid biosynthesis in some tumor cell line (Rhodes et al. 1994).

In recent decades, topolins, especially *m*T have emerged as one of the important cytokinins in various micropropagation protocols. They have demonstrated promising outcomes regarding growth and development and for overcoming various physiological disorders that have been encountered under in vitro system. Normally, *m*T invigorated higher secondary metabolite production compared to other cytokinins. *m*T even at very low concentration resulted in higher production of secondary metabolites in some plant species. The use of *m*T resulted in higher proliferation of callus and shoots which was further linked with increased bioactive compound accumulation. The foremost reason for the prevalence of *m*T over other cytokinins such as BA has been due to its confined accumulation in plant tissues and due to its faster transformation rate (Kaminek et al. 1997). This aside, the metabolic final product is effectually degradable and *O*-glucoside metabolites are considered as the cytokinin storage form, which is stable under specific conditions. However, this is rapidly converted into active cytokinin bases when needed as a result of hydrolyzation of glucose in the *m*T side chain (Bairu et al. 2009). The reversible sequestration of the *O*-glucosides allows for unlimited accessibility of cytokinins over a longer period at a physiologically active level resulting in a higher rate of shoot multiplication in plant tissue cultures (Strnad 1997).

In some plant species, a higher callus growth rate and adventitious shoot regeneration within subsequently increased secondary compound production were observed using mT over other cytokinins. For example, the topolins, especially *m*T and *meta*-methoxytopolin 9-tetrahydropyran-2-yl increased the total phenolics, total flavonoids and proanthocyanidin content in subsurface and aerial parts of micropropagated banana plantlets compared to BA (Adeyemi et al. 2012). In general, secondary metabolites such as phenolics and flavonoids play a fundamental role during plant micropropagation. The increased flavonoids content revealed improved ex vitro root regeneration in differentiated shoots. A significantly higher level of total phenolics, flavonoid, iridoid and antioxidant activity was observed in *Aloe arborescens* shoots regenerated using *m*TR containing medium compared to media containing other growth regulators (Amoo et al. 2013). Thus, the utilization of topolins may be a viable approach for stimulating the biosynthesis of important secondary metabolites of value to the pharmaceutical, food and agrochemical industry.

Acknowledgments The senior author Pankaj Kumar thankfully acknowledges National Post Doctoral Fellowship Award, Science and Engineering Research Board, Department of Science & Technology, Government of India, New Delhi, India.

References

Adeyemi OA, Bairu MW, Szüčová L, Doležal K, Finnie JF, Staden JV (2012) Assessment of the role of *meta*-topolins on *in vitro* produced phenolics and acclimatization competence of micropropagated 'Williams' banana. Acta Physiol Plant 34(6):2265–2273

Amoo S, Finnie JF, Staden VJ (2011) The role of *meta*-topolins in alleviating micropropagation problems. Plant Growth Regul 63:197–206

Amoo SO, Aremu AO, Staden JV (2013) Shoot proliferation and rooting treatments influence secondary metabolite production and antioxidant activity in tissue culture-derived *Aloe arborescens* grown *ex vitro*. Plant Growth Regul 70:115–122

Amoo SO, Aremu AO, Moyo M, Sunmonu TO, Plíhalová L, Doležal K, Staden JV (2015) Physiological and biochemical effects of a tetrahydropyranyl substituted *meta*-topolin in micropropagated *Merwilla plumbea*. Plant Cell Tiss Organ Cult 121:579–590

Aremu AO, Bairu MW, Doležal K, Finnie JF, Staden JV (2012a) Topolins: a panacea to plant tissue culture challenges. Plant Cell Tiss Organ Cult 108:1–16

Aremu AO, Bairu MW, Szucova L, Doležal K, Finnie JF (2012b) Assessment of the role of *meta*-topolins on produced phenolics and acclimatization competence of micropropagated 'Williams' banana. Acta Physiol Plant 34:2265–2273

Bairu M, Stirk W, Doležal K, Staden VJ (2007) Optimizing the micropropagation protocol for the endangered *Aloe polyphylla*: can *meta*-topolin and its derivatives serve as replacement for benzyladenine and zeatin. Plant Cell Tiss Organ Cult 90:15–23

Bairu M, Stirk W, Doležal K, Staden VJ (2008) The role of topolins in micropropagation and somaclonal variation of Banana cultivars "Williams" and "Grand Naine" (*Musa spp.*). Plant Cell Tiss Organ Cult 95:373–379

Bairu MW, Jain N, Stirk WA, Doležal K, Staden VJ (2009) Solving the problem of shoot-tip necrosis in *Harpagophytum procumbens* by changing the cytokinin types, calcium and boron concentrations in the medium. South Afr J Bot 75:122–127

Baroja-Fernandez E, Aguirreolea J, Martinková H, Hanuš J, Strnad M (2002) Aromatic cytokinins in micropropagated potato plants. Plant Physiol Biochem 40:217–224

Briskin DP (2000) Medicinal plants and phytomedicines. Linking plant biochemistry and physiology to human health. Plant Physiol 124:507–514

Chaturvedi HC, Jain M, Kidwai NR (2007) Cloning of medicinal plants through tissue culture-a review. Indian J Exp Biol. 45:937–948

Escalona M, Cejas I, Gonza Lez-Olmedo J, Capote I, Roels S, Canal MJ, Rodnguez R, Sandoval J, Debergh P (2003) The effect of meta-topolin on plantain propagation using a temporary immersion bioreactor. InfoMusa 12:28–30

Filova A (2014) Production of secondary metabolites in plant tissue cultures. Res J Agric Sci 46 (1):1–10

Haberer G, Kieber JJ (2002) Cytokinins. New insights into a classic phytohormone. Plant Physiol 128(2):354–362

Harsh KC, Anil KB, Indra DB (2017) Impediments in the tissue culture of high altitude medicinal plants. Adv Biotechnol Microbiol 4:555–631

Heyl A, Schmülling T (2003) Cytokinin signal perception and transduction. Curr Opin Plant Biol 6:480–488

Jain SC, Pancholi B, Jain R (2012) *In vitro* callus propagation and secondary metabolite quantification in *Sericostoma pauciflorum*. Iran J Pharma Res 11:1103–1109

Jan S, Abbas N (2018) Himalayan phytochemicals sustainable options for sourcing and developing bioactive compounds. Elsevier Ltd, Amsterdam. ISBN:978-0-08-102227-6

Joshi H, Nekkala S, Kher MM, Nataraj M (2016) *In vitro* shoot multiplication of *Withania coagulans* (Stocks) Dunal. Plant Tiss Cult Biotech 26(2):187–195

Kamínek M, Motyka V, Vanková R (1997) Regulation of cytokinin content in plant cells. Physiol Plant 101:689–700

Karalija E, Neimarlija D, Cakar J, Paric A (2016) Elicitation of biomass and secondary metabolite production, antioxidative and antimicrobial potential of basil and oregano induced by BA and IBA application. EJMP 14(4):1–11

Kubaláková M, Strnad M (1992) The effect of aromatic cytokinins (populins) on micropropagation and regeneration *in vitro*. Biol Plant 34:578–579

Kumar P, Srivastava DK (2016) Biotechnological advancement in genetic improvement of broccoli (*Brassica oleracea* L. var. *italica*), an important vegetable crop: a review. Biotechnol Lett 38:1049–1063

Kumar P, Shaunak I, Thakur AK, Srivastava DK (2017) Health promising medicinal molecules in vegetable crops. J Genet Genomes 1:1–4

Kumar P, Shaunak I, Verma ML (2020) Biotechnological application of health promising bioactive molecules. In: Verma ML, Chandel AK (Ed.) Biotechnological production of bioactive compounds. Elsevier B.V., pp 165–189. Amsterdam. https://doi.org/10.1016/B978-0-444-64323-0.00006-0

Lata H, Chandraa S, Techena N, Khan IA, ElSohlya MA (2016) *In vitro* mass propagation of *Cannabis sativa* L.: a protocol refinement using novel aromatic cytokinin *meta*-topolin and the assessment of eco-physiological, biochemical and genetic fidelity of micropropagated plants. J Appli Res Med Arom Plants 3(1):18–26

Malá J, Machová P, Cvrčková H, Karády M, Novák O, Mikulík J, Hauserová E, Greplová J, Strnad M, Doležal K (2009) Micropropagation of wild service tree (*Sorbus torminalis* [L.] Crantz): the regulative role of different aromatic cytokinins during organogenesis. J Plant Growth Regul 28:341–348

Malá J, Machová P, Cvrčková H, Karády M, Novák O, Mikulík J, Dostál J, Strnad M, Doležal K (2013) The role of cytokinins during micropropagation of wych elm. Biol Plant 57:174–178

Meiners J, Schwab M, Szankowski I (2007) Efficient *in vitro* regeneration systems for Vaccinium species. Plant Cell Tissue Organ Cult 89:169–176

Miller CO, Skoog F, Von Saltza MH, Strong FM (1955) Kinetin, a cell division factor from deoxyribonucleic acid. J Am Chem Soc 77:1392–1393

Mok DWS, Mok MC (2001) Cytokinin metabolism and action. Annu Rev Plant Physiol Plant Mol Biol 52:89–118

Mutui TM, Mibus H, Serek M (2012) Effect of meta-topolin on leaf senescence and rooting in *Pelargonium hortorum* cuttings. Postharvest Biol Technol 63:107–110

Naaz A, Hussain SA, Anis M, Alatar AA (2019) Meta-topolin improved micropropagation in *Syzygium cumini* and acclimatization to *ex vitro* conditions. Biol Plant 63:174–182

Nacheva L, Gercheva P, Ivanova V, Ibrahim O (2017) Meta-topolin improves lateral bud proliferation in micropropagation of *Ginkgo biloba* L. ISHS Acta Hortic 1155:355–360

Niedz RP, Evens TJ (2010) The effects of benzyladenine and *meta*-topolin on *in vitro* shoot regeneration of a *Citrus citrandarin* rootstock. Res J Agric Biol Sci 6:45–53

Nielsen E, Temporiti MEE, Cella R (2019) Improvement of phytochemical production by plant cells and organ culture and by genetic engineering. Plant Cell Rep 38(10):1199–1215

Paric A, Karalija E, Cakar J (2017) Growth, secondary metabolites production, antioxidative and antimicrobial activity of mint under the influence of plant growth regulators. Acta Biol Szegediensis 61(2):189–195

Rao RS, Ravishankar GA (2002) Plant cell cultures: chemical factories of secondary metabolites. Biotechnol Adv 20:101–153

Rhodes MJC, Parr AJ, Giulietti A, Aird ELH (1994) Influence of exogenous hormones on the growth and secondary metabolite formation in transformed root cultures. Plant Cell Tiss Organ Cult 38:143–151

Sakakibara H (2006) Cytokinins: activity, biosynthesis, and translocation. Ann Rev Plant Biol 57:431–449

Salvi ND, George L, Eapen S (2002) Micropropagation and field evaluation of micropropagated plants of turmeric. Plant Cell Tissue Organ Cult 68:143–151

Santoro MV, Nievas F, Zygadlo J, Giordano W, Banchio E (2013) Effects of growth regulators on biomass and the production of secondary metabolites in peppermint (*Mentha piperita* L.) *in vitro*. AJPS 4:49–55

Smulders M, de Klerk G (2011) Epigenetics in plant tissue culture. Plant Growth Regul 63:137–146

Strnad M (1997) The aromatic cytokinins. Physiol Plant 101:674–688

Strnad M, Hanuš J, Vaněk T, Kamínek M, Ballantine J, Fussell B, Hanke D (1997) *Meta*-topolin, a highly active aromatic cytokinin from poplar leaves (*Populus canadensis* Moench, cv. Robusta). Phytochemistry 45:213–218

Subbaraj A, Funnell K, Woolley D (2010) Dormancy and flowering are regulated by the reciprocal interaction between cytokinin and gibberellin in *Zante deschia*. J Plant Growth Regul 29:487–499

Werbrouck SPO (2010) Merits and drawbacks of new aromatic cytokinins in plant tissue culture. Acta Hortic 865:103–108

Werbrouck S, Strnad M, VanOnckelen H, Debergh P (1996) Meta-topolin, an alternative to benzyladenine in tissue culture. Physiol Plant 98:291–297

Williams VL, Victor JE, Crouch NR (2013) Red Listed medicinal plants of South Africa: Status, trends, and assessment challenges. South Afr J Bot 86:23-35

Wojtania A (2010) Effect of meta-topolin on *in vitro* propagation of *Pelargonium hortorum* and *Pelargonium hederaefolium* cultivars. Acta Soc Bot Pol 79:101–106

The Use of *Meta*-topolin in Cell and Tissue Cultures for Increasing Production of Secondary Metabolites

18

Bengu Turkyilmaz Unal

Contents

Abstract

Nowadays, increasing the amount of secondary metabolite, which is an active drug and food additive source, gradually gains importance in addition to its important role in the adaptation of plants to the environment. Plant cell and tissue culture technologies have been used to produce and increase plant secondary metabolites since the late 1960s. Strain improvement, medium optimization, selection of high production cell lines, and the use of elicitors such as plant growth regulators may lead to an increase in secondary metabolite production. Cytokinins are an important group of plant growth regulators that can affect plant growth in almost all stages and regulate various biotechnological processes. *Meta*-topolin (6-(3-hydroxybenzylamino)purine) is an aromatic natural cytokinin. It is thought that the production of secondary metabolites will increase and contribute to industrial production by adding *meta*-topolin growth regulator as an elicitor into the growth medium.

Keywords

Cytokinins · Elicitors · Plant growth regulators · Secondary metabolites · Topolins

B. Turkyilmaz Unal (✉)
Biotechnology Department, Arts and Sciences Faculty, Niğde Ömer Halisdemir University, Niğde, Turkey
e-mail: bturkyilmaz@ohu.edu.tr

N. Ahmad, M. Strnad (eds.), *Meta-topolin: A Growth Regulator for Plant Biotechnology and Agriculture*, https://doi.org/10.1007/978-981-15-9046-7_18

Plants can synthesize various amounts of components or metabolites, with different structures by their versatile synthesis reactions. The plant metabolites were grouped as primary metabolites (PMs) and secondary metabolites (SMs) by Kossel (1891). PMs are present in all cells and involve in the vital activities of the organism. Secondary ones are not found in all cells and have roles in adaptation to various stress conditions, defense against microorganisms and herbivores, attracting animals, etc. SMs are known to play an important role in the environmental compatibility of plants; they are also a significant source of active drugs (Mammadov 2014). PMs are produced by plants using inorganic compounds for photosynthesis, and then they are converted into SM through various pathways. As a result of the change of environmental conditions and increasing stress factors, plants stabilize the amount of PMs and increase the amount of SMs. The SMs are generally species-specific; plants in different genera and species produce different metabolites (Seckin 2014).

Modern biology and chemistry have clarified the PMs' role in the vital life activities of living things. SMs have begun to be studied with the analytical techniques in the middle of the twentieth century; nowadays the evidences were found by intensive using of biochemical techniques and molecular biology which support their significant role in the adaptation of plants to their environment (Bourgaud et al. 2001). It has been known that SMs protect plants from pathogens (phytoalexins) due to their antibiotic, antifungal, and antiviral properties and have allelopathic interaction with other plants by inhibiting germination or secreting toxic substances too. It has also been stated that they prevent serious leaf damage with UV light protection compounds (Li et al. 1993) and increase the pollination by attracting insects to plants (Harborne 1999; Kurt 1997).

SMs are separated into three main families as phenolics, terpenes, and steroids and alkaloids depending on their biosynthetic pathways (Harborne 1999). Most of higher plants contain members of these families which are economically important chemicals. Human beings are addicted to plants as a nutritional source of carbohydrates, fat, and protein (Rao and Ravishankar 2002). Two hundred thousand species and the various chemicals in their contents provide various opportunities for human nutrition and health protection (Seckin 2014). More than 80% of the known 30,000 natural products are of herbal origin (Balandrin and Klocke 1988; Fowler and Scragg 1988). Since ancient times, humans have used secondary metabolite-rich plants as foods, herbs, and poisons. Nowadays, these plants are particularly important in the industries of food, medicine, chemical, cosmetic, and dye and in agricultural pest control.

The SM biosynthesis is difficult, and their synthesis in the laboratory is limited. Since the intracellular environment is required for the reactions, they cannot be synthesized except for a few simple structured SMs (Mammadov 2014). The collection of plants in the natural flora is labored and costly. If plants in nature are collected continuously, some species may be extinct. In the case of cultivation of plants, they are synthesized in very small amounts in certain seasons; a large amount of plants and large agricultural areas are needed for sufficient production (Erkoyuncu and Yorgancılar 2016). The world population rapidly increase, and therefore optimizing land use and obtaining high yield in plant production or secondary compound production are important. In vitro production technologies are highly advantageous compared to traditional agriculture because they are independent from geographic and seasonal changes, various environmental factors, and external parasites, provide continuous production in one type and quality and high production speed, and allow the production of compounds not found in the plant. It is also of capital importance for the protection of medicinal plant species, especially endemic and/or endangered species (Moyo et al. 2011).

Since plant cells are biosynthetically totipotent, they have the genetic information to produce all the chemicals in the plant (Rao and Ravishankar 2002). Plant biotechnology, which has the potential to use this situation, provides genetic manipulation to grow cell, tissue, organ, or whole organism under in vitro conditions and to obtain the desired compounds and creates an alternative source of secondary metabolite (SM) production and enhancement (Alfermann et al. 1980; Dornenburg and Knorr 1997; Ravishankar et al. 1999; Scragg 1997; Stöckigt et al. 1995). Advances in molecular biology and biotechnology have allowed to improve plant, to increase yield and quality, to produce more than one product, or to produce new products from genetically modified plants (Rao and Ravishankar 2002). In in vitro techniques, plant cells, tissues, or organs are grown under aseptic conditions and in a suitable media. In the history of these techniques, it has been discovered quite late that SMs can be produced by plant cell and tissue cultures. For a very long time, differentiated cells or specialized organs have been thought to be unable to produce secondary compounds (Krikorian and Steward 2012). The studies on the production of SM and enhancement with cell and tissue culture began in 1956, with Routier and Nickel's US patent application and gathered speed worldwide in the late 1960s (Bourgaud et al. 2001). One of the main reasons for this is conducting organ, tissue, and cell cultures of more than 2000 plant species; the other reason is conducting the studies for various chemical production using plant cell cultures (Sokmen and Gurel 2001).

Some problems are encountered in the production and increase studies of SMs with plant cell and tissue cultures. The problems are caused by low efficiency in the production of certain genuine SMs and imbalance of cell lines, and scale-up problems are the major ones (Ravishankar and Venkataraman 1993). To increase the natural products production and to solve these problems, different strategies have been needed in industrial production in recent years. Obtaining efficient cell lines, mutating cells, obtaining optimum media, cell immobilization, primer addition, and elicitor use are some of the strategies that can be used.

Healthy plants produce various SMs against pathogens in nature. It has been found that externally applied pathogen-origin compounds cause the same reactions. Such compounds that induce SM and cause the production of defense mechanism chemicals called phytoalexin are called elicitors. It is known that secondary pathways are activated, and a response to stress is generated by the signals. The elicitors are used to shorten the time to induce the secondary metabolite formation, to increase the culture volume, and to reach high production quantities (Barz et al. 1988; DiCosmo and Tallevi 1985; Rao and Ravishankar 2002). Due to these advantages, the enhancement of SMs with elicitor use has been one of several strategies used in commercial production (Savitha et al. 2006).

Elicitors can be classified as either endogenous or exogenous according to their origin and biotic or abiotic according to their nature (West 1981). Heavy metal salts, UV rays, and chemicals that disrupt the membrane structure are abiotic elicitors, while microorganism (chitin, chitosan, or glucans) or plant cell wall (pectin or cellulose) polysaccharides, low molecular weight organic acids, and glycoproteins are biotic elicitors (Dörnenburg and Knorr 1995). By using these, higher concentrations can be produced in less time (Barz et al. 1988). While abiotic and biotic elicitors are used for stimulation of various SMs, especially phenolic compounds, the degree of arousal, the mechanism, and types of stimulated metabolites have not been determined yet (Jeong and Park 2005). Various elicitors have been used in studies to increase efficiency in plant cell and tissue culture techniques (Palazón et al. 2003; Park et al. 1995). Fungi- or bacteria-originated elicitors, glycoproteins, inactivated enzymes, polysaccharides, purified compounds, chitosan, xanthan, cellulase, thidiazuron (TDZ), methyl jasmonate, salicylic acid, and heavy metal salts are some of the elicitors used to enhance varied SMs (Awad et al. 2014; Aydın et al. 2017; Çag et al. 2003; DiCosmo et al. 1987; Funk et al. 1987; Islek and Unal 2015; Islek et al. 2014, 2016; Matkowski 2008; Rajendran et al. 1994; Ramachandra Rao et al. 1996; Rao et al. 1996; Turkyilmaz Unal and Islek 2019).

Numerous biochemical and physiological processes in micropropagation are regulated by plant growth regulators (PGRs) (George et al. 2008). Plant hormones produced in a tissue, transported to target tissues, and effective in very low concentrations, such as indole-3-acetic acid (IAA), gibberellic acid (GA_3), zeatin (Z), abscisic acid (ABA), ethylene, and most of the others produced synthetically are used in cell and tissue culture. The exogenous treatment of PGRs can alter their synthesis, degradation, activation, or translocation as well as their susceptibility to the same or different types of endogenous PGRs (Coenen and Lomax 1997; Kamınek et al. 1997).

Cytokinins (CKs) have strong plant growth regulating activity as N^6 substituted adenine (McGaw 1987). It is an important group that can modulate various biotechnological processes by affecting nearly all stages of plant growth and development (Plihalova et al. 2016). They regulate cell division in shoots and roots, synthesize specific components of cell cycle, regulate morphogenesis, stimulate the growth of side shoots, delay leaf senescence, and enhance cell expansion in leaves and cotyledons (Davey et al. 2005; Petit-Paly et al. 1999; Synkova et al. 2006; Taiz and Zeiger 1998; Werner and Schmülling 2009). Some CKs such as kinetin (KIN) and Z have also significant anti-aging, anticarcinogenic, and anti-thrombotic effects (Rattan and Sodagam 2005; Vermeulen et al. 2002). They are also used effectively as stimulants in the studies of SM enhancement. It is known that exogenously administered CKs and auxins contribute to the increase of SMs in plants (Rao and Ravishankar 2002). The separate and combined effects of cytokinins such as N^6-benzyladenine (BA), KIN, TDZ, and N^6-isopentenyladenine (iP) and auxins like α-naphthalene acetic acid (NAA), 2,4-dichlorophenoxy acetic acid (2,4-D), and IAA on the in vitro production of SMs have been reported by many researchers (Coste et al. 2011; Luczkiewicz et al. 2001; Meyer and Van Staden 1995).

Physiological disorders like epigenetic, stunted growth and somaclonal variation are frequently observed during micro-multiplication of many plant species (Bairu et al. 2011; Smulders and De Klerk 2011). Many researchers have searched for new and more suitable CKs to be used in micropropagation for identifying new compounds in order to alleviate negative effects in vitro and to increase metabolites because such problems restrict the commercial production of many plants (Tarkowská et al. 2003). Especially in recent years, biotechnological developments in phytohormones have increased the number of this kind of studies (Strnad 1997; Tarkowski et al. 2010).

Although BA or its derivatives are present in some plant species (das Neves and Pais 1980; Ernst et al. 1983; Horgan et al. 1973, 1975; Strnad et al. 1992, 1994), they are accepted as synthetic today. Many types of aromatic CKs obtained from 6-(2-hydroxybenzylamino)purine and 6-(2-hydroxybenzylamino)purine have been identified (Strnad et al. 1992, 1994, 1997). The names of topolins as a new group of aromatic CKs, including *meta*-topolin (*m*T), are derived from the word of topol which means poplar in Czech language (Strnad et al. 1997). Topolins discovered as natural aromatic CKs have become an alternative to CKs such as BA, Z, KIN, etc. that have been used in plant cell and tissue cultures for many years (Aremu et al. 2012a). For example, it was found that *m*T yield better results in shoot proliferation, rooting, and ex vitro air conditioning of micropropagated *Corylus colurna* compared to the most commonly used cytokinin (CK) as BA in *Corylus* spp. tissue culture (Gentile et al. 2017). Topolins are effective in increasing shoot growth and rooting, preserving histogen stability and alleviating varied physiological disorders in micropropagation (Aremu et al. 2012a, b). The effects of exogenous treatment of several topolins on plant species such as cucumber (Çag et al. 2003), sugar beet (Čatský et al. 1996), and wheat (Palavan-Ünsal et al. 2002¸ 2004; Vlčková et al. 2006) were also investigated.

The use of topolins, especially their *meta*-forms, gradually increases due to their capability to increase plant growth and development and to eliminate various physiological disorders (Aremu et al. 2012a). In the production of *Coleonema album*, a medicinal and aromatic plant that is highly demanded, from the shoot tip explants directly by organogenesis, the effects of four different topolins [*m*T, *meta*-methoxytopolin riboside (Me*m*TR), *meta*-topolin riboside (*m*TR), *meta*-methoxytopolin tetrahydropyran-2-yl (Me*m*TTHP)] were investigated and compared with other CKs (KIN, BA, TDZ). At equimolar concentrations, BA, *mT* (except 10 μM), *m*TR, and *m*TTHP significantly enhanced shoot growth production compared to KIN. The most successful CK concentration in shoot production was found to be 5.0 μM *m*T (Fajinmi et al. 2014). *m*T and its derivatives have been especially used in the initiation of culture, protocol optimization, and the prevention of many types physiological disorders from in vitro (Jones et al. 1996).

Firstly, *m*T isolated from leaves of poplar differs from isoprenoid CKs such as Z and iP in terms of their biochemical and biological activity (Strnad 1997). *m*T has been used successfully in numerous plant tissue culture production protocols

(Aremu et al. 2012a, b; Fajinmi et al. 2014; Mutui et al. 2012). Lata et al. (2016) reported that *m*T treatment (2.0 μM) in *Cannabis sativa* is a single-step, simple, and cost-effective protocol that is effective in increasing new shoot formation and root stimulation. In other similar studies, it was emphasized that the use of *m*T enhances shoot proliferation and rooting efficiency, provides tissue stability, and reduces production costs in in vitro systems (Bogaert et al. 2004; Chauhan and Taylor 2018; Werbrouck et al. 1996). The same scientists also stated that *m*T may be an alternative to BAP in micropropagation (Bogaert et al. 2004; Werbrouck et al. 1996). Amoo et al. (2015) found that meta-topolin tetrahydropyran-2-yl (*m*TTHP) and *m*TR enhanced shoot formation in a bulbous medicinal plant, *Merwilla plumbea*. *m*TTHP has been also reported to be effective in *Aloe arborescens* and *Harpagophytum procumbens* even at low concentrations in rooting (Amoo et al. 2014). In a study of Mutui et al. (2012) conducted with *Pelargonium*, *m*T has active roles in ease of rooting in cut roots, as well as slowing down senescence; it may be an alternative to TDZ in ornamental plants for slowing down senescence. Some researchers have also found that exogenous treatment of *m*T and its derivatives are effective in inhibiting shoot tip necrosis and slowing down aging in many species (Aremu et al. 2012a; Malá et al. 2013).

Despite its extensive usage potential in plant tissue culture, the number of studies showing the effects of topolins on SM production is low (Aremu et al. 2012b; Moyo et al. 2011). Topolins stimulated the accumulation of phenolic compounds in micropropagated plants and increased the chance of survival of these plants in ex vitro conditions (Aremu et al. 2012b). In the study of Baskaran et al. (2012) conducted with endangered *M. plumbea* used in traditional medicine in South Africa, it was found that the use of the elicitors such as adenine sulfate, BA, iP, *m*T, *m*TTHP, TDZ, KIN and NAA, casein, glutamine, hemoglobin, mebendazole, trimethoprim, yeast extract, and yeast malt water increases SM production. In another study on the effects of only topolins and the combination of topolins with auxin on the SM production and shoot multiplication in *Huernia hystrix*, a plant from the South Africa endemic flora, it was found that *m*T is more effective in enhancing shoot multiplication and production of phenolics compared to BA (Amoo and Van Staden 2013). In a study analyzing the effects of *m*TTHP on shoot multiplication, antioxidant activity, and phytochemical and photosynthetic pigment content in *A. arborescens* and *H. procumbens* medicinal plants, it was found that the use of *m*TTHP and topolins at low concentrations increases phenolic and flavonoid production (Amoo et al. 2014).

In a study (Amoo et al. 2013) on *A. arborescens*, the shoots produced in the *meta*-methoxytopolin (Me*m*T) medium have been proved to contain more total phenolic and flavonoid compounds compared to the shoots produced in the PGR-free medium. In the study of Moyo et al. (2014), the biochemical accumulation rates of hydroxybenzoic and hydroxycinnamic acid derivatives were investigated with the CK application in the *Hypoxis hemerocallidea* micropropagation. It was stated that chlorogenic, gallic, gentisic, *m*-hydroxybenzoic, *p*-hydroxybenzoic, protocatechuic, salicylic, and *trans*-cinnamic acids were produced at high concentrations, especially in the *m*T applied cell and tissue cultures.

Secondary metabolites come into prominence in terms of human nutrition, health, and environmental protection issues with their uses as pharmaceuticals, food additives, agricultural chemicals, etc. It is very important to continuously obtain maximum yield from minimum area in minimum time in the studies of the production of the SMs and the enhancement of the production. Recent advances in biotechnology provide us great opportunities to increase the SM production. However, the development of different strategies is needed due to some problems encountered in the in vitro production. Nowadays, the use of abiotic and biotic elicitors in industrial production becomes gradually widespread. Although there are few studies in which *m*T and its derivatives are used as elicitors to increase SM production, it is known that they are more effective than many cytokinins such as BA, Z, and KIN. Additional studies on this subject are thought to contribute to the secondary metabolite production on an industrial scale in addition to creating scientific references; the use of *m*T as an elicitor is recommended.

References

Alfermann AW, Schuller I, Reinhard E (1980) Biotransformation of cardiac glycosides by immobilized cells of Digitalis lanata1. Planta Med 40:218–223

Amoo SO, Van Staden J (2013) Influence of plant growth regulators on shoot proliferation and secondary metabolite production in micropropagated *Huernia hystrix*. Plant Cell Tissue Organ Cult 112:249–256

Amoo S, Aremu A, Van Staden J (2013) Shoot proliferation and rooting treatments influence secondary metabolite production and antioxidant activity in tissue culture-derived Aloe arborescens grown ex vitro. Plant Growth Regul 70:115–122

Amoo SO, Aremu AO, Moyo M, Szüčová L, Doležal K, Van Staden J (2014) Physiological effects of a novel aromatic cytokinin analogue in micropropagated *Aloe arborescens* and *Harpagophytum procunmbes*. Plant Cell Tissue Organ Cult 116:17–26

Amoo SO, Aremu AO, Moyo M, Sunmonu TO, Plíhalová L, Doležal K, Van Staden J (2015) Physiological and biochemical effects of a tetrahydropyranyl-substituted meta-topolin in micropropagated *Merwilla plumbea*. Plant Cell Tissue Organ Cult 121:579–590

Aremu AO, Bairu MW, Doležal K, Finnie JF, Van Staden J (2012a) Topolins: a panacea to plant tissue culture challenges? Plant Cell Tissue Organ Cult 108:1–16

Aremu AO, Bairu MW, Szüčová L, Doležal K, Finnie JF, Van Staden J (2012b) Assessment of the role of meta-topolins on in vitro produced phenolics and acclimatization competence of micropropagated 'Williams' banana. Acta Physiol Plant 34:2265–2273

Awad V, Kuvalekar A, Harsulkar A (2014) Microbial elicitation in root cultures of *Taverniera cuneifolia* (Roth) Arn. for elevated glycyrrhizic acid production. Ind Crop Prod 54:13–16

Aydın S, İşlek C, Ünal BT (2017) Effect of silver nitrate solvent on total protein, total phenolic and some antioxidant enzyme activities in cell suspension culture of *Capsicum annuum* L. Turk J Agric-Food Sci Technol 5:1544–1549

Bairu MW, Aremu AO, Van Staden J (2011) Somaclonal variation in plants: causes and detection methods. Plant Growth Regul 63:147–173

Balandrin M, Klocke J (1988) Medicinal, aromatic, and industrial materials from plants. In: Bajaj YPS (ed) Medicinal and aromatic plants I. Springer, New York, pp 3–36

Barz W, Daniel S, Hinderer W, Jaques U, Kessmann H, Köster J, Tiemann K (1988) Elicitation and metabolism of phytoalexins in plant cell cultures. In: Pais MSS (ed) Plant cell biotechnology. Springer, New York, pp 211–230

Baskaran P, Ncube B, Van Staden J (2012) In vitro propagation and secondary product production by *Merwilla plumbea* (Lindl.) Speta. Plant Growth Regul 67:235–245

Bogaert I, Van Cauter S, Werbrouck S, Dolezal K (2004) New aromatic cytokinins can make the difference. In: V international symposium on in vitro culture and horticultural breeding 725, pp 265–270

Bourgaud F, Gravot A, Milesi S, Gontier E (2001) Production of plant secondary metabolites: a historical perspective. Plant Sci 161:839–851

Çag S, Palavan-Ünsal N, Büyüktuncer D (2003) Comparison of the effects of meta-topolin and other cytokinins on chlorophyll and protein contents and peroxidase activity in cucumber cotyledons. Israel J Plant Sci 51:261–265

Čatský J, Pospíšilová J, Kamínek M, Gaudinová A, Pulkrábek J, Zahradníček J (1996) Seasonal changes in sugar beet photosynthesis as affected by exogenous cytokininN 6-(m-hydroxybenzyl) adenosine. Biol Plant 38:511

Chauhan RD, Taylor NJ (2018) Meta-topolin stimulates de novo shoot organogenesis and plant regeneration in cassava. Plant Cell Tissue Organ Cult 132:219–224

Coenen C, Lomax TL (1997) Auxin—cytokinin interactions in higher plants: old problems and new tools. Trends Plant Sci 2:351–356

Coste A, Vlase L, Halmagyi A, Deliu C, Coldea G (2011) Effects of plant growth regulators and elicitors on production of secondary metabolites in shoot cultures of *Hypericum hirsutum* and *Hypericum maculatum*. Plant Cell Tissue Organ Cult 106:279–288

das Neves HJC, Pais MSS (1980) Identification of a spathe regreening factor in Zantedeschia *aethiopica*. Biochem Biophys Res Commun 95:1387–1392s

Davey M, Stals E, Panis B, Keulemans J, Swennen R (2005) High-throughput determination of malondialdehyde in plant tissues. Anal Biochem 347:201–207

DiCosmo F, Tallevi SG (1985) Plant cell cultures and microbial insult: interactions with biotechnological potential. Trends Biotechnol 3:110–111

DiCosmo F, Quesnel A, Misawa M, Tallevi S (1987) Increased synthesis of ajmalicine and catharanthine by cell suspension cultures of *Catharanthus roseus* in response to fungal culture-filtrates. Appl Biochem Biotechnol 14:101–106

Dörnenburg H, Knorr D (1995) Strategies for the improvement of secondary metabolite production in plant cell cultures. Enzym Microb Technol 17:674–684

Dornenburg H, Knorr D (1997) Challenges and opportunities for metabolite production from plant cell and tissue cultures. Food Technology (USA)

Erkoyuncu MT, Yorgancılar M (2016) Production of secondary metabolites by plant tissue culture methods Selcuk. J Agric Sci 2:66–76

Ernst D, Schäfer W, Oesterhelt D (1983) Isolation and identification of a new, naturally occurring cytokinin (6-benzylaminopurineriboside) from an anise cell culture (*Pimpinella anisum* L.). Planta 159:222–225

Fajinmi O, Amoo S, Finnie J, Van Staden J (2014) Optimization of in vitro propagation of *Coleonema album*, a highly utilized medicinal and ornamental plant South African. J Bot 94:9–13

Fowler MW, Scragg AH (1988) Natural products from higher plants and plant cell culture. In: Plant cell biotechnology. Springer, New York, pp 165–177

Funk C, Gügler K, Brodelius P (1987) Increased secondary product formation in plant cell suspension cultures after treatment with a yeast carbohydrate preparation (elicitor). Phytochemistry 26:401–405

Gentile A, Frattarelli A, Nota P, Condello E, Caboni E (2017) The aromatic cytokinin meta-topolin promotes in vitro propagation, shoot quality and micrografting in *Corylus colurna* L. Plant Cell Tissue Organ Cult 128:693–703

George EF, Hall MA, De Klerk G-J (2008) Plant growth regulators II: cytokinins, their analogues and antagonists. In: Plant propagation by tissue culture. Springer, New York, pp 205–226

Harborne JB (1999) Classes and functions of secondary products from plants. In: Chemicals from plants. World Scientific, Singapore, pp 1–25

Horgan R, Hewett E, Purse J, Wareing P (1973) New cytokinin from *Populus robusta*. Tetrahedron Lett 30:2827–2828

Horgan R, Hewett E, Horgan J, Purse J, Wareing P (1975) A new cytokinin from *Populus x robusta*. Phytochemistry 14:1005–1008

Islek C, Unal BT (2015) Copper toxicity in *Capsicum annuum*: superoxide dismutase and catalase activities, phenolic and protein amounts of in-vitro-grown plants. Pol J Environ Stud 24:2441–2445

Islek C, Ustun A, Koc E (2014) The effects of cellulase on capsaicin production in freely suspended cells and immobilized cell cultures of *Capsicum annuum* L. Pak J Bot 46:1883–1887

Islek C, Unal BT, Koc E, Kaya D (2016) Effect of thidiazuron used as an elicitor in the production of capsaicin on total protein and phenolic amounts, antioxidant enzyme activities of pepper plants. Int J Environ Agric Biotechnol 1:603–609

Jeong G-T, Park D-H (2005) Enhancement of growth and secondary metabolite biosynthesis: effect of elicitors derived from plants and insects. Biotechnol Bioprocess Eng 10:73

Jones L, Martinkova H, Strnad M, Hanke D (1996) Occurrence of aromatic cytokinins in oil palm (Elaeis guineensis Jacq.). J Plant Growth Regul 15:39

Kamınek M, Kománek D, Motyka V, Gaudinová A (1997) Brezinová A is cytokinin oxidase regulated by cytokinin levels in plant cells. In: XVIII congress of SPPS. The Scandinavian Society for Plant Physiology, Uppsala, Sweden, 1997

Kossel A (1891) Ueber die chemische Zusammensetzung der Zelle Du Bois-Reymond's Archiv/ Arch. Anat Physiol Physiol Abt:181–186

Krikorian A, Steward F (2012) Biochemical differentiation: the biosynthetic potentialities of growing and quiescent tissue. In: Steward FC (ed) Plant physiology—a treatise, vol 5. Academic, New York, pp 227–326

Kurt BGT (1997) Natural product chemistry: a mechanistic, biosynthetic and ecological approach. Swedish Pharmaceutical Press, Stockholm

Lata H, Chandra S, Techen N, Khan IA, ElSohly MA (2016) In vitro mass propagation of Cannabis *sativa* L.: a protocol refinement using novel aromatic cytokinin meta-topolin and the assessment of eco-physiological, biochemical and genetic fidelity of micropropagated plants. J Appl Res Med Aromat Plants 3:18–26

Li J, Ou-Lee T-M, Raba R, Amundson RG, Last RL (1993) *Arabidopsis* flavonoid mutants are hypersensitive to UV-B irradiation. Plant Cell 5:171–179

Luczkiewicz M, Cisowski W, Kaiser P, Ochockaz R, Piotrowski A (2001) Comparative analysis of phenolic acids in mistletoe plants from various hosts system. Acta Pol Pharm 58(5):373–391

Malá J et al (2013) The role of cytokinins during micropropagation of wych elm. Biol Plant 57:174–178

Mammadov R (2014) Secondary metabolites in seed plants. Nobel Academic Publishing, Ankara

Matkowski A (2008) Plant in vitro culture for the production of antioxidants—a review. Biotechnol Adv 26:548–560

McGaw BA (1987) Cytokinin biosynthesis and metabolism. In: Plant hormones and their role in plant growth and development. Springer, New York, pp 76–93

Meyer H, Van Staden J (1995) The in vitro production of an anthocyanin from callus cultures of Oxalis *linearis*. Plant Cell Tissue Organ Cult 40:55–58

Moyo M, Bairu M, Amoo S, Van Staden J (2011) Plant biotechnology in South Africa: micropropagation research endeavours, prospects and challenges. S Afr J Bot 77:996–1011

Moyo M, Amoo SO, Aremu AO, Gruz J, Šubrtová M, Doležal K, Van Staden J (2014) Plant regeneration and biochemical accumulation of hydroxybenzoic and hydroxycinnamic acid derivatives in *Hypoxis hemerocallidea* organ and callus cultures. Plant Sci 227:157–164

Mutui TM, Mibus H, Serek M (2012) Effect of meta-topolin on leaf senescence and rooting in *Pelargonium*× *hortorum* cuttings. Postharvest Biol Technol 63:107–110

Palavan-Ünsal N, Çağ S, Çetin E, Büyüktunçer D (2002) Retardation of senescence by meta-topolin in wheat leaves. J Cell Mol Biol 1:101–108

Palavan-Ünsal N, Cag S, Çetin E (2004) The role of meta-topolin in senescence of wheat leaf segments. J Cell Mol Biol 3:23–31

Palazón J, Cusidó RM, Bonfill M, Mallol A, Moyano E, Morales C, Piñol MT (2003) Elicitation of different *Panax ginseng* transformed root phenotypes for an improved ginsenoside production. Plant Physiol Biochem 41:1019–1025

Park H-H, Hakamatsuka T, Sankawa U, Ebizuka Y (1995) Rapid metabolism of isoflavonoids in elicitor-treated cell suspension cultures of Pueraria lobata. Phytochemistry 38:373–380

Petit-Paly G, Franck T, Brisson L, Kevers C, Chénieux J-C, Rideau M (1999) Cytokinin modulates catalase activity and coumarin accumulation in in vitro cultures of tobacco. J Plant Physiol 155:9–15

Plihalova L, Vylíčilová H, Doležal K, Zahajska L, Zatloukal M, Strnad M (2016) Synthesis of aromatic cytokinins for plant biotechnology. New Biotechnol 33:614–624

Rajendran L, Suvarnalatha G, Ravishankar G, Venkataraman L (1994) Enhancement of anthocyanin production in callus cultures of *Daucus carota* L. under the influence of fungal elicitors. Appl Microbiol Biotechnol 42:227–231

Ramachandra Rao S, Ravishankar G, Venkataraman L (1996) An improved process for the preparation of vanillin. Indian Patent 1022

Rao SR, Ravishankar G (2002) Plant cell cultures: chemical factories of secondary metabolites. Biotechnol Adv 20:101–153

Rao SR, Sarada R, Ravishankar G (1996) Phycocyanin, a new elicitor for capsaicin and anthocyanin accumulation in plant cell cultures. Appl Microbiol Biotechnol 46:619–621

Rattan SI, Sodagam L (2005) Gerontomodulatory and youth-preserving effects of zeatin on human skin fibroblasts undergoing aging in vitro. Rejuvenation Res 8:46–57

Ravishankar G, Venkataraman L (1993) Role of plant cell culture in food biotechnology: current trends, limitations and future prospects. In: Prakah J, Pierik RLM (eds) Plant biotechnology: commercial prospects and problems. Oxford IBH Press, New Delhi

Ravishankar G, Bhyalakshmi N, Ramachandra Rao S (1999) Production of food additives. In: Ramawat KG, Mérillon JM (eds) Biotechnology: secondary metabolites. Oxford IBH, New Delhi, pp 89–110

Savitha BC, Thimmaraju R, Bhagyalakshmi N, Ravishankar G (2006) Different biotic and abiotic elicitors influence betalain production in hairy root cultures of *Beta vulgaris* in shake-flask and bioreactor. Process Biochem 41:50–60

Scragg A (1997) The production of aromas by plant cell cultures. In: Biotechnology of aroma compounds. Springer, New York, pp 239–263

Seckin T (2014) Functional plant chemistry. Nobel Academic, Ankara

Smulders M, De Klerk G (2011) Epigenetics in plant tissue culture. Plant Growth Regul 63:137–146

Sokmen A, Gurel E (2001) Secondary metabolite production. In: Babaoglu M, Gurel E, Ozcan S (eds) Plant biotechnology, tissue culture and applications. Selcuk University Press, Konya, pp 211–261

Stöckigt J, Obitz P, Falkenhagen H, Lutterbach R, Endress S (1995) Natural products and enzymes from plant cell cultures. Plant Cell Tissue Organ Cult 43:97–109

Strnad M (1997) The aromatic cytokinins. Physiol Plant 101:674–688

Strnad M, Peters W, Beck E, Kaminek M (1992) Immunodetection and Identification of N6-(o-hydroxybenzylamino) purine as a naturally occurring cytokinin in *populus× canadensis* moench cv robusta leaves. Plant Physiol 99:74–80

Strnad M, Peters W, Hanuš J, Beck E (1994) Ortho-topolin-9-glucoside, an aromatic cytokinin from *Populus x canadensis* cv robusta leaves. Phytochemistry 37:1059–1062

Strnad M, Hanuš J, Vaněk T, Kamínek M, Ballantine JA, Fussell B, Hanke DE (1997) Meta-topolin, a highly active aromatic cytokinin from poplar leaves (*Populus× canadensis* Moench., cv. Robusta). Phytochemistry 45:213–218

Synkova H, Semoradova S, Schnablová R, Witters E, Husak M, Valcke R (2006) Cytokinin-induced activity of antioxidant enzymes in transgenic Pssu-ipt tobacco during plant ontogeny. Biol Plant 50:31–41

Taiz L, Zeiger E (1998) Plant physiology. Sinanuer Associates, Sunderland

Tarkowská D et al (2003) Identification of new aromatic cytokinins in *Arabidopsis thaliana* and *Populus*× *canadensis* leaves by LC-(+) ESI-MS and capillary liquid chromatography/frit–fast atom bombardment mass spectrometry. Physiol Plant 117:579–590

Tarkowski P et al (2010) An improved in vivo deuterium labeling method for measuring the biosynthetic rate of cytokinins. Molecules 15:9214–9229

Turkyilmaz Unal B, Islek C (2019) Capsaicin production in cell suspension cultures of *Capsicum annuum* L. In: Münir Öztürk KRH, Ashraf M, Ahmad MSA (eds) Crop production technologies for sustainable use and conservation physiological and molecular advances. Apple Academic Press, Waretown, p 541

Vermeulen K et al (2002) Antiproliferative effect of plant cytokinin analogues with an inhibitory activity on cyclin-dependent kinases. Leukemia 16:299

Vlčková A, Špundová M, Kotabová E, Novotný R, Doležal K, Nauš J (2006) Protective cytokinin action switches to damaging during senescence of detached wheat leaves in continuous light. Physiol Plant 126:257–267

Werbrouck SP, Strnad M, Van Onckelen HA, Debergh PC (1996) Meta-topolin, an alternative to benzyladenine in tissue culture? Physiol Plant 98:291–297

Werner T, Schmülling T (2009) Cytokinin action in plant development. Curr Opin Plant Biol 12:527–538

West CA (1981) Fungal elicitors of the phytoalexin response in higher plants. Naturwissenschaften 68:447–457

Effects of *Meta*-topolin on the Growth, Physiological and Biochemical Parameters in Plant Tissue Culture

19

Esra Koç

Contents

Abstract

Cytokinins (CKs) are essential plant growth regulators that control virtually all physiological activities in plant tissue culture (PTC). CKs are adenine derivatives carrying either an isoprene derivative or an aromatic side chain at the N^6 terminus. Aromatic CKs have greater in vivo stability than isoprenoid CKs due to their resistance against cytokinin oxidase/dehydrogenase (CKX). Although N^6-benzyladenine (BA) is the most commonly used aromatic cytokinin in PTC due to its effectiveness and low cost, hydroxylated derivatives, especially those that are hydroxylated in the *meta*-position, are emerging as viable alternatives. *Meta*-topolin (*m*T, 6-(3-hydroxybenzylamino)purine) is the most promising example in the place of BA due to its ability to obviate some of the drawbacks of BA use during the micropropagation of some medicinal and ornamental plant species and also improve shoot proliferation. New findings have been revealed in studies aimed at increasing the morphogenetic activities of this new cytokinin. A series of

E. Koç (✉)
Faculty of Sciences, Department of Biology, Ankara University, Ankara, Turkey
e-mail: ekoc@science.ankara.edu.tr

N. Ahmad, M. Strnad (eds.), *Meta-topolin: A Growth Regulator for Plant Biotechnology and Agriculture*, https://doi.org/10.1007/978-981-15-9046-7_19

substituted topolin derivatives in the N9 position was synthesized to increase the cytokinin biological activity without greater cytotoxicity. Tetrahydropyran-2-yl (THP)-substituted CKs have positive effects on rooting and the photosynthetic apparatus in micropropagated plants. Positive reports on important tissue culture parameters such as increasing shoot growth rate, alleviating physiological defects, better acclimatization and rooting and secondary metabolite production have made the topolins, especially *meta*-topolins and derivatives, popular among plant tissue culture experts. The aim of this chapter is to assess the influence of *meta*-topolin (*m*T) on the growth, senescence process and photosynthesis, antioxidants, secondary metabolite production and in vitro propagation.

Keywords

Cytokinins · Growth · *Meta*-topolin · Plant tissue culture · Secondary metabolite

19.1 Introduction

CKs, N^6-substituted adenine derivatives, are a class of plant hormones defined as factors that promote cell division. There are many other effects on plant growth and development, including the start and growth of shoots, apical dominance, senescence and photomorphogenetic development (George et al. 2008). The reason why CKs have a regulatory role in senescence is due to the retarder effect of the CKs (Zwack and Rashotte 2013). For a long time, research on CKs focused on the isoprenoid class represented by zeatins, N^6-isopentenyladenine (iP) and related compounds. Aromatic 6-benzylaminopurine (or N^6-benzyladenine) (BA) and its derivatives were considered to be totally synthetic CKs. Of these, BA is one of the most effective and affordable CK in many micropropagation systems. However, it has disadvantages in some plants. N^6-benzyladenine-9-glucoside (BA9G) is the main derivative that is deposited in the plant base in micropropagated *Spathiphyllum floribundum.* A slow or zero BA release from this stock resulted in a variety of problems during the acclimatization of the plant in a greenhouse, such as heterogeneous growth and inhibition in rooting (Werbrouck et al. 1995). Thus, one way to find an alternative to BA was to examine the BA derivatives already conjugated in the N9 position. In the N9 position, the presence of a ribose in N^6-benzyladenine-9-riboside (BAR) or a tetrahydropyran-2-yl group in N^6-benzyl-9-(2-tetrahydropyranyl) adenine (BPA) made us think that these CKs could be better protected against N9 glycosylation. The other alternative could be the use of hydroxylated BA analogues. Horgan et al. (1975) isolated the first member of this group, N^6-(2-hydroxybenzyl)adenine-9-riboside, from mature poplar leaves. In the poplar leaves, free bases of N^6-(2-hydroxybenzyl)adenine and N^6-(3-hydroxybenzyl)adenine, their ribosides and 9-glucosides were also identified (Strand 1992a, b; Strnad et al. 1997). Since the name populin had already been used for salicylic benzoate, Strnad et al. (1997) proposed the name topolins as an alternative to this class of plant hormones, derived from the Czech word "topol" (Palavan-Unsal et al. 2002).

19.1.1 The Role of *Meta*-topolin on the Senescence and Photosynthetic Apparatus

In studies on cytokinin application, this compound was found to have the potential to delay senescence in various tissues. Intense work on this subject then began. It was discovered that these hormones are involved in the preservation of the photosynthetic apparatus of plant organs. Senescence in leaves requires a change in photosynthesis apparatus. Since yellowing is very significant, chlorophyll degradation is the main parameter for the measurement of senescence in leaves (Gut et al. 1987). Approximately chlorophyll b binds to the antenna complex of PSIIs while chlorophyll a the primary electron donor in the electron transport chain. During the senescence period, the PSII reaction centre deteriorates more rapidly than the light-collecting antenna complex, and evaluation of the content and reciprocal ratios of chlorophyll a and b provides data on the physiological state of the plant. Gentile et al. (2016), in their study on *Corylus colurna*, showed that *m*T had a more positive effect than BA in preventing the degradation of chlorophyll.

One of the first symptoms of leaf senescence is the increase in protease activity (Thimann 1980). However, there is not much information about the self-regulation of proteolysis during leaf senescence. Previous studies (Richmond and Lang 1957; Wollgiehn 1967) showed that CKs delay senescence by sustaining or regulation protein and nucleic acid synthesis while inhibiting senescence. Palavan-Unsal et al. (2002) studied the main senescence parameters, chlorophyll disintegration and degradation, protease activity and soluble protein content in order to explain the relationship between senescence and a new aromatic cytokinin *m*T in excised wheat leaf. In this study, *m*T was found to be effective in preventing chlorophyll degradation. Different concentrations of *m*T were found to inhibit acid and neutral protease activity, especially neutral protease activity which decreases with increasing *m*T concentration. The fresh weight and protein content showed a linear increase with increasing *m*T concentration. It was also determined that *m*T prevents DNA degradation during senescence.

Senescence due to chlorophyll loss in ornamental plants shows itself by leaf yellowing. This is a major problem for *Pelargonium* (Mutui et al. 2005), roses (Bogaert et al. 2006), *Alstroemeria* (Mutui et al. 2006) and *Lilium longiflorum* Thunb. (Han 1997). The most characteristic sign of senescence is yellowing, starting from the veins and extending outward (Quirino et al. 2000). CKs are powerful anti-senescence hormones and play an important role in delaying the onset of leaf senescence (Thimann 1980). CKs have reduced leaf yellowing in *Pelargonium* (Mutui et al. 2004), since they play a role in the preservation of the photosynthetic apparatus of their plant organs (Thomas and Stoddart 1980). Mutui et al. (2012) investigated the effect of thidiazuron (TDZ), a substituted phenyl urea that has activity similar to *m*T and potent cytokinin in *Katinka Pelargonium* cuttings, and found that the chlorophyll content in *m*T- and TDZ-treated plant leaves was higher than in untreated control plants. These results demonstrated that CKs are effective in delaying leaf yellowing and chlorophyll degradation. CKs also activate the NADH protochlorophyllide reductase, an enzyme involved in chlorophyll biosynthesis, and reduce chlorophyll degradation (Zavaleta-Mancera et al. 1999). Some researchers have also reported that *m*T, which was exogenously applied in roses propagated in vitro and in wheat leaves, delayed chlorophyll degradation and was the best anti-senescence molecule (Tarkowská et al. 2003; Palavan-Unsal et al. 2002, 2004). Bogaert et al. (2006) determined that *m*T had no effect on rooting rate (%) and average root diameter in cuttings exposed to *m*T. Similarly, 0.05 mM *m*T had no effect on the number of roots of each cutting (Mutui et al. 2012). However, *m*T resulted in reduced root length, root surface area and total volume of roots slightly. On the other hand, TDZ significantly inhibited adventive root formation and decreased all root parameters including root formation. These results were explained by the fact that TDZ is a very strong cytokinin which is very stable in *Pelargonium* leaves and leads to inhibition of root formation. The TDZ molecule remained intact in both free and conjugated forms in the hypocotyl tissue of *Pelargonium* × *hortorum*. Exposure to TDZ is assumed to increase the accumulation and replacement of auxins in plant tissues. The reason for this superiority of *m*T over TDZ is its rapid displacement in plant tissues and thus the prevention of local accumulation, but also due to the rapid degradation of its metabolites during acclimatization (Kamínek et al. 1987). *m*T is very active in delaying leaf senescence, and it is seen as a suitable alternative to TDZ in delaying the onset of leaf yellowing in ornamental plants since it facilitates rooting in cuttings after application (Mutui et al. 2012). Holub et al. (1998) showed that the methoxy and hydroxyl derivatives of BAP have strong cytokinin activity in various biological assays. However, in delaying chlorophyll degradation in wheat leaves, it was determined that *m*T had twice the activity of BA and also *m*T increased chlorophyll content as an exogen in wheat leaf segments (Tarkowská et al. 2003). *m*T was found to be very active in the delay of senescence (Palavan-Unsal et al. 2002, 2004). It has been suggested that *m*T is a promising plant growth regulator for physiological investigations with these study results and since *m*T responds to senescence parameters in a significant way (Palavan-Unsal et al. 2002).

19.1.2 The Use of *Meta*-topolin as an Alternative Cytokinin in Tissue Culture

CKs have a great influence on many physiological properties in plant tissue culture (PTC) such as shoot formation, explant quality obtained from in vitro culture, plant cell division, differentiation, chlorophyll accumulation and leaf senescence. As mentioned, BA is the cytokinin which is most frequently used in micropropagation techniques, effective in supporting in vitro shoot propagation, used in commercial laboratories and much cheaper than naturally occurring CKs such as iP and *t*Z (Bairu et al. 2007). In optimal and higher BA concentrations, the index of abnormality increased significantly, while the production of adventive shoots decreased. This result indicates that BA can have toxic effects. Similar results have been found in different plant species (Bogaert et al. 2006; Bairu et al. 2007, 2008). Doležal et al. (2006) found that high amounts of cytokinins in tobacco callus cause a decrease in growth. They reported that this result was due to inhibition of cytokinesis as a result of cyclin-dependent kinase inhibitory activity. These problems have encouraged research in the micropropagation industry to find alternative, more cost-effective CKs while maintaining plant quality and desired growth rates. For this reason, recent research has focused on other CKs (mainly aromatic ones) instead of using BA in the micropropagation industry. The use of *m*T and its derivatives has been justified as a potential alternative to BA in the micropropagation industry (Werbrouck et al. 1996; Bogaert et al. 2006; Bairu et al. 2007). Werbrouck et al. (1995), in their study on micropropagated *Spathiphyllum floribundum* Schott cv. Petite, compared the metabolism of *m*T and its in vitro effects with BA, BPA and BAR. In vitro, BA, BPA and BAR were transformed into BA9G, which is essentially the stable derivative found in the basal part of the plant. N^6-(3-*O*-β-D-glucopyranosyl)benzyladenine-9-riboside was identified as the major derivative of *m*T. This new *O*-glucoside of *m*T found in all plant parts was metabolized much faster than BA9G during acclimatization. The effects of BA and *m*T on in vitro shoot and root production and post in vitro rooting were compared. A good in vitro root formation as well as shoot production was obtained only with *m*T. The plants developed on a medium with 10 μM or more *m*T showed better rooting than those developed with BA at the same concentrations. This new *m*T metabolite, which rapidly deteriorated during acclimatization, caused fewer problems than BA9G, which is the main metabolite of BA, BPA and BAR. The lack of in vitro root inhibition with *m*T and the easy rooting of plants on in vitro medium confirm this result. For this reason and due to good propagation rates in in vitro medium, *m*T has been shown to be an alternative to BA for the micropropagation of *S. floribundum*. This does not apply to BPA and BAR which inhibit root formation.

*m*T and its derivatives have been used in many tissue culture studies, especially in micropropagation. *m*T has been reported to be more effective than BA in stimulating shoot proliferation and rooting in many plant species and to reduce production costs in in vitro systems (Amoo et al. 2011; Vijayakumar et al. 2017). Vijayakumar et al. (2017) suggested that *m*T is safe in protocols for plant regeneration in safflower (*Carthamus tinctorius* L. cv). *m*T was also found to be useful for stimulating shoot regeneration from somatic embryos and cotyledon explants. These shoot organogenesis techniques increase the capacity of existing embryogenic systems and present previously unreported morphogenetic pathways to enhance genetic transformation and gene regulation technologies in cassava plants (Chauhan and Taylor 2018). It was found that applied *m*T concentrations were significantly lower than that of BA in terms of the abnormality index which is the ratio of abnormal shoots (shoots formed by necrosis of shoot edge and hyperhidric shoots) to normal shoots in in vitro propagated *Barleria greenii* (Amoo et al. 2011). The abnormality index is a measure of the toxic nature effect of a cytokinin. These findings suggest that topolins are effective and less toxic than BA at higher equimolar concentrations. The localized accumulation of *m*T is prevented by faster displacement in plant tissues (Kamínek et al. 1987). The metabolites of *m*T and *meta*-topolin riboside {*m*TR, 6-(3-hydroxybenzylamino)-9-β-D-ribofuranosylpurine} were found to be easily degradable (Bairu et al. 2009). The –OH group in the side chain of *m*T is affected by the formation of *O*-glucoside molecules (Werbrouck et al. 1996). These molecules have been reported as storage forms of cytokinins. The conversion and accumulation of *O*-glucosides to CKs allows CKs to remain at a physiologically active level over a long period of time, resulting in high shoot formation (Strnad et al. 1997; Amoo et al. 2011). These metabolites are also resistant to degradation by cytokinin oxidase/dehydrogenase. In addition, Bogaert et al. (2006), in the cultures of *Petunia hybrida* and *Rosa hybrida* found that closely related *meta*-methoxytopolin riboside (Me*m*TR) had histogenetic stability and anti-senescence activity. Bairu (2008) found an increase in yield of soybean callus with the following results of topolin biological activity: Me*m*TR > *m*TR > *m*T. In that study, it was reported that the highest shoot production was obtained in culture with 7 μM Me*m*TR where adventitious shoots were greater than 10 mm in length and produced a lower abnormality index than that observed in controls. The cause of the superiority of *m*TR and Me*m*TR was postulated to be due to the presence of a ribose at position N9, which can better protect the free bases against N9 glycosylation (Werbrouck et al. 1996). On the other hand, it has been reported that this superiority of Me*m*TR may be due to the presence of the methyl group in its molecular structure (Schmitz et al. 1972). Removal of the methyl group from *cis*-norzeatin causes a significant loss of activity, while the transition of the methyl group from dihydrozeatin to dihydroisozeatin was reported to have a 70-fold decrease in activity (from the three-position to the two-position) (Schmitz et al. 1972).

19.1.3 The Effect of *Meta*-topolin on Antioxidants in In Vitro Propagation

CKs modulate the activity of antioxidant enzymes (Synková et al. 2006), stimulate cell division and control morphogenesis and senescence retardance (Werner and Schmülling 2009). CKs are a key component of the growth medium used in PTC, including micropropagation. In addition, CKs remove the reactive oxygen species (ROS) that occur during normal plant metabolism (Synková et al. 2006). Excessive accumulation of ROS in situations where scavenging mechanisms are insufficient often results in oxidative damage to biomolecules. Malondialdehyde (MDA) deposition which is the end product of lipid peroxidation is a marker of oxidative stress in plants.

Another photo-protective mechanism that plants use to distribute the excitation energy in the electron transport chain takes place through the Mehler reaction (Gong et al. 2013). As stated by the authors, this usually requires disproportionate photoreduction of oxygen to superoxide anion radicals (O_2A_2) in photosystem I (PSI), which is then converted to hydrogen peroxide (H_2O_2) and oxygen under the influence of the superoxide dismutase (SOD) enzyme. H_2O_2 is converted into water and oxygen with catalase (CAT) or peroxidase (POD) enzymes to eliminate its toxic effect (Asada 2000). CK-derived plants have been noted to have significantly higher antioxidant enzyme activity compared to control (plantlets regenerated without PGRs). An increase in antioxidant enzyme activity has been reported with increased

CK content in transgenic tobacco plants (Synková et al. 2006). Dıáz-Vivancos et al. (2011) similarly reported that CKs have an enhancing effect on SOD and POD activities in *Crocus sativus*. The first response and the highest enzyme activity were detected in SOD. This increase in the activity of SOD, which is the first effective enzyme against ROS in the scavenging mechanism, indicates the accumulation of high superoxide anion radicals due to oxygen photoreduction with electron leakage in PSI. Synková et al. (2006) found an increase in antioxidant enzyme activities after TDZ application compared with other CKs in *C. sativus*. Amoo et al. (2015) found that TDZ-regenerated plants had a significant increase in the activity of CAT and POD enzymes as well as SOD compared to *m*TR and *m*TTHP (*m*T-9-tetrahydropyran-2yl) applications. This indicates oxidative stress in TDZ-derived plants. This stress is most likely due to the accumulation of ROS during an attempt to control the high excitation energy of plants. Excessive excitation energy can increase the formation of ROS during electron transport processes (Neill et al. 2002). However, overproduction of ROS may cause oxidative stress (Batková et al. 2008). The commonly used marker of oxidative stress in plants is the production of MDA, a by-product of membrane lipid peroxidation due to oxidative damage (Dewir et al. 2006). Although ROS can act as signalling molecules, overproduction can be very destructive for biomolecules (Gaspar et al. 2002). Lipid peroxidation due to ROS may result in increased membrane permeability, followed by low NPQ (non-photochemical quenching) (Kalaji et al. 2014). Stress can affect carbohydrate storage, translocation and metabolism which can lead to a lower accumulation of biomass (Mishra and Dubey 2008; Gong et al. 2013). Compared with the *m*TR and *m*TTHP applications, it was observed that there was a decrease in the biomass of leaf due to an increase in the amount of ROS and a decrease in the activity of antioxidant enzymes in TDZ-derived plants (Amoo et al. 2015).

Some authors have proposed in vitro propagation as a mean of reducing pressure on wild types and thus determined the pharmacological activity of plants using tissue culture techniques. For example, it has been reported that micropropagated *Tulbaghia violacea* has higher antioxidant content and better antimicrobial activity than naturally grown plants (Ncube et al. 2011). In vitro regenerated *Thymus lotocephalus* has higher levels of rosmarinic acid than wild types (Costa et al. 2012). Baskaran et al. (2012) demonstrated that shoot length was significantly greater in medium containing *m*T and *m*TR compared to the control and other treatments (TDZ, BA, etc.) in *M. plumbea*. The highest root length was recorded on medium containing *m*T. The present study further indicated that the *m*T and *m*TR enhanced the gallotannin content compared to the control in in vitro roots of *M. plumbea*. Aremu et al. (2013) reported that different cytokinins (CKs) had a positive effect on the accumulation of specific phenolic acids in in vitro propagated plants. In addition, these data also showed that after the acclimatization step, a decrease was observed in biosynthesis and production of most of the phenolics in tissue-cultured plants. Similarly, though there was an increase in the antioxidant activity of *M. plumbea*'s aboveground organs after acclimatization, oxygen radical absorbance capacity values were found to be lower than those detected in subsoil parts (Aremu et al. 2013).

19.1.4 The Effects of *Meta*-topolin and Its Derivatives on In Vitro Propagation and Secondary Metabolite Production

Studies aimed at further enhancing BA activity due to researchers demonstrating the superiority of aromatic CKs, especially *meta*-hydroxylated analogues in plant tissue culture, led to the synthesis of a series of substituted *m*T derivatives at the N9 location (Aremu et al. 2012a). One of these novel derivatives, 6-(3-hydroxylbenzylamino)-9-tetrahydropyran-2-ylpurine or, in other words, *meta*-topolin tetrahydropyran-2-yl (*m*TTHP), led to rooting at low concentration during micropropagation of *Aloe arborescens* and *Harpagophytum procumbens* (Amoo et al. 2014). Researches have shown that CK treatment in in vitro micropropagated plants affects the acclimatization competence, secondary metabolite production and also pharmacological activities of medicinal plants. The effect of *m*TTHP on the production and antioxidant activity of secondary metabolites has recently been stated (Aremu et al. 2014), and its physiological and/or biochemical effect on ex vitro growth was then studied. Destructive-devastating harvesting of bulbs of *Merwilla plumbea* (Lind l.) Speta, a herbal bulbous plant that is commonly used and traded in South African traditional medicine, will soon require protection. Amoo et al. (2015) investigated the effect of *m*TTHP, which is the most active cytokinin for the micropropagation of *Merwilla plumbea*, on in vitro adventitious shoot growth, rooting and carotenoid, chlorophyll *a*, chlorophyll *b*, etc. content of regenerated plants. The effects of this CK on the bulbous medicinal plant ex vitro growth, the photosynthesis efficiency and the antioxidant enzymes were also investigated. Applications with *m*TTHP and *m*TR showed the highest number of shoot increase in comparison with TDZ treatment and control. The highest rooting was determined for *m*TTHP application. Unlike *m*TTHP applications, an increase in concentration of *m*TR or TDZ above 0.5 μM resulted in a significant decrease in amounts of all photosynthetic pigments. After 6 months of ex vitro growth, regenerated plants from 0.5 μM *m*TTHP were found to have higher total leaf area, total leaf fresh weight and bulb size diameter and bulb production compared to plants treated with *m*TR and TDZ. Another study examined the effects of various concentrations of different PGRs [α-naphthalene acetic acid (NAA), BA] and topolins (*m*T, *m*TTHP, Me*m*T, Me*m*TTHP) on the growth, phytochemical and antioxidant activity by comparing different combinations. In sum, topolins have been suggested suitable alternatives for many species in comparison to traditional/widely used BA (Masondo et al. 2014). Further, they have been proven that the interaction of the NAA with CKs was effective for morphological growth and development during micropropagation and then in ex vitro acclimatization for *Eucomis autumnalis* subspecies. PGRs were found to affect secondary metabolites and antioxidant activities in *E. autumnalis* subspecies. When in vitro regenerants were acclimatized, it was determined that there was an increase in the quantitative accumulation of phytochemicals and antioxidant activities (several folds) in 4-month-old plants. The highest iridoid content was found in NAA versus Me*m*T, the highest condensed tannin content in the combination of NAA and *m*TTHP and the highest phenolic content in *m*TTHP or BAR variants (Masondo et al. 2014).

The combination of in vitro medium and specific external factors was found to significantly affect the production of plant secondary metabolites (Collin 2001; Rao and Ravishankar 2002; Pavarini et al. 2012). CKs such as BA and *m*T were found to excite production of different types of secondary metabolite through biosynthetic path excitation (Aremu et al. 2012b; Coste et al. 2011). The topolins, in particular, the novel derivative *m*TTHP, are considered to be an alternative CK for micropropagation of medicinal plant species because they enhance the acclimatization ability and induce stimulating effect on ex vitro rooting (Aremu et al. 2012b). Plantlets regenerated using *meta*-methoxytopolin (Me*m*T), Me*m*TR and Me*m*TTHP had significantly longer roots and better shoot/root ratios, higher chlorophyll a/b ratios, total chlorophyll/carotenoid ratios and higher phenolic accumulation (especially Me*m*TTHP) than control- and BA-treated plants (Aremu et al. 2012b). On the other hand, BA application was found to excite the expression of phenylalanine ammonium lyase, the key enzyme in cinnamate and anthocyanin biosynthesis in *Arabidopsis thaliana* (Deikman and Hammer 1995). Aremu et al. (2013) determined that BA excites the accumulation of protocatechuic and vanillic acid, *m*TR of *p*-coumaric and ferulic acid and iP of cinnamic acid and benzoates such as caffeic acid. It has been suggested that the presence of high amounts of these phenolics in tissue-cultured *M. plumbea* can be a potential route for large-scale production. The fact that *M. plumbea* plantlets grown in tissue culture have better antioxidant activity was explained as the presence of high amounts of other secondary metabolites (Matkowski 2008). However, it is reported that relating the antioxidant activity observed in plant extracts to any particular phenolic acid is difficult. Aremu et al. (2013) suggested that the determination of amounts of other compound (s) than phenolic acids may contribute to the oxygen radical absorbance activity values in plants.

Aremu et al. (2013) found that most quantified phenolic acids in plantlets in tissue culture were higher than in the acclimatized plants. Bairu et al. (2011) found that the total phenolic amount of the acclimatized *plant* shoots was higher than the tissue culture. In addition, it was an interesting finding that aboveground parts had more of some phenolics and better oxygen radical absorbance values after acclimatization. This shows that in different plants, different secondary metabolites can cause varying results. The use of aboveground organs will be effective in the survival of the plant as it will reduce the use of bulbs and roots. This result which indicates the presence of higher phytochemical and pharmacological activity of *M. plumbea*'s aboveground parts will prevent the devastating harvest of the bulbs of the plant used for traditional medicine (Aremu et al. 2013). The presence of CKs (e.g., *m*T and *m*TR) in the medium improved the antioxidant activity of plantlets in tissue culture. During in vitro propagation, BA significantly increased the amounts of hydroxybenzoics, whereas topolins in general caused deposition of hydroxycinnamic acid derivatives in *M. plumbea* (Aremu et al. 2013).

Aloe species are also very valuable as ornamental plants, and they have been used for centuries in traditional medicine. Due to the exploitation of the medicinal and ornamental plant trade, there are many kinds of habitat losses in this species. *A. arborescens* is one of the most important medicinal species of aloe. The effects of different aromatic cytokinin types in different amounts on the organogenesis, in vitro secondary metabolite production and the antioxidant effect of *A. arborescens* regenerated shoots have been evaluated. In the cultures with *m*T, *m*TR, Me*m*T and BAR, it was found that there was an increase in adventive shoots produced per explant with the increase in concentration and, in general, 5 μM *m*T application gave the highest number of transplantable shoots (Amoo and Van Staden 2013). Variable amounts of secondary metabolites were determined in regenerated shoots with CK applications. It was found that the applications made with *m*T, *m*TR, Me*m*T and Me*m*TR in different levels caused the total phenolics to be produced in a lower amount compared to control groups. No significant difference was found between the control group and CK applications in terms of iridoid content. The concentrations of different CKs resulted in the accumulation of variable amounts of condensed tannins, albeit at low levels (Amoo and Van Staden 2013). In another study conducted on *Pelargonium* plant, it was determined that in all CK applications (KIN, BA, *m*T, *m*TR, Me*m*TR), a significant increase in phenolic secondary metabolites was observed compared to the control group. However, with the increase in amount of *m*T, *m*TR and Me*m*TR applied, a decrease in phenolic content was observed. Phenolic secondary metabolism increased in all CK applications, suggesting that they play a role in the in vitro regulation of the phenylpropanoid biosynthetic pathway (Moyo et al. 2012). In addition, radical scavenging activities and antioxidant activities of the shoots growing at all *m*T, *m*TR, Me*m*TR and BA concentrations were higher than control groups in *A. arborescens* plants. However, in *m*T and BA applications, the antioxidant activity in general decreased with increasing cytokinin level (Amoo et al. 2012). Conversely, antioxidant activity increased with concentration increase in *m*TR and Me*m*T applications reaching the optimum at 7.5 μM concentration. The optimized regeneration protocol is simple and low cost considering the high shoot proliferation rate at low CK concentrations, the growth of additional shoots during rooting, the easy rooting of regenerated shoots and the high survival rate of acclimatized plants. Therefore, the CK used and the concentration range significantly affect the in vitro secondary metabolite production. Thus, the data suggest that the evaluation of bioactive secondary metabolite production is also necessary while improving or optimizing the micropropagation protocols of popular medicinal plants (Amoo et al. 2012). Amoo and Van Staden (2013) examined the effect of topolins on condensed tannins, secondary metabolites in micropropagated shoots of the *Huernia hystrix* (*H. hystrix*) plant, and found a higher in vitro secondary metabolite production with *m*T application than BA (Amoo and Van Staden 2013). In general, tannins and especially condensed tannins exhibit antimicrobial, antitumor, anti-inflammatory and cardiovascular protective properties (Matkowski 2008; Zhou et al. 2011). In these plant species, the excitation role of CKs (as observed only in the cytokinin-containing medium) in the production of phenolics may be due to the suppression of certain macronutrient carriers (especially nitrates), which can lead to the expression or regulation of genes involved in the biosynthetic pathway of secondary metabolites (Sakakibara et al. 2006).

19.2 Conclusion

Since the discovery of topolins as naturally occurring aromatic CKs, they are proposed as real alternatives to classical CKs in plant tissue culture such as BA, TDZ, *t*Z and KIN. Globally, there has been an increase in the use of topolins and derivatives in studies over the past 20 years. Topolins, especially *meta*-topolins and derivatives, are used for culture onset, protocol optimization and prevention of various in vitro excited physiological defects. The findings also show the efficacy of *m*T in shoot proliferation, in secondary metabolite production and in the development of pharmacological effects compared to other aromatic CKs. The presented findings provide ample evidence of the growing importance of topolins as a potential and reliable alternative to the CKs commonly used in tissue culture.

References

Amoo SO, Van Staden J (2013) Influence of plant growth regulators on shoot proliferation and secondary metabolite production in micropropagated *Huernia hystrix*. Plant Cell Tissue Organ Cult 112(2):249–256

Amoo SO, Finnie JF, Van Staden J (2011) The role of *meta*-topolins in alleviating micropropagation problems. Plant Growth Regul 63:197–206

Amoo SO, Aremu AO, Van Staden J (2012) In vitro plant regeneration, secondary metabolite production and antioxidant activity of micropropagated *Aloe arborescens* Mill. Plant Cell Tissue Organ Cult 111:345–358

Amoo SO, Aremu AO, Moyo M et al (2014) Physiological effects of a novel aromatic cytokinin analogue in micropropagated *Aloe arborescens* and *Harpagophytum procumbens*. Plant Cell Tissue Organ Cult 116(1):17–26

Amoo SO, Aremu AO, Moyo M et al (2015) Physiological and biochemical effects of a tetrahydropyranyl substituted meta-topolin in micropropagated *Merwilla plumbea*. Plant Cell Tissue Organ Cult 121:579–590

Aremu AO, Bairu MW, Doležal K et al (2012a) Topolins: a panacea to plant tissue culture challenges? Plant Cell Tissue Organ Cult 108:1–16

Aremu AO, Bairu MW, Szučová L et al (2012b) Assessment of the role of meta-topolins on in vitro produced phenolics and acclimatization competence of micropropagated 'Williams' banana. Acta Physiol Plant 34:2265–2273

Aremu AO, Grúz J, Šubrtová M et al (2013) Antioxidant and phenolic acid profiles of tissue cultured and acclimatized *Merwilla plumbea* plantlets in relation to the applied cytokinins. J Plant Physiol 170:1303–1308

Aremu AO, Moyo M, Amoo SO et al (2014) Effect of a novel aromatic cytokinin derivative on phytochemical levels and antioxidant potential in greenhouse grown *Merwilla plumbea*. Plant Cell Tissue Organ Cult 119(3):501–509

Asada K (2000) The water-water cycle as alternative photon and electron sinks. Philos Trans R Soc Lond 355:1419–1431

Bairu MW, Stirk WA, Doležal K et al (2007) Optimizing the micropropagation protocol for the endangered *Aloe polyphylla*: can meta-topolin and its derivatives serve as replacement for benzyladenine and zeatin? Plant Cell Tissue Organ Cult 90:15–23

Bairu M, Stirk W, Doležal K et al (2008) The role of topolins in micropropagation and somaclonal variation of banana cultivars "Williams" and "Grand Naine" (Musa spp. AAA). Plant Cell Tissue Organ Cult 95:373–379

Bairu MW, Jain N, Stirk WA et al (2009) Solving the problem of shoot-tip necrosis in *Harpagophytum procumbens* by changing the cytokinin types, calcium and boron concentrations in the medium. S Afr J Bot 75:122–127

Bairu MW, Amoo SO, Van Staden J (2011) Comparative phytochemical analysis of wild and in vitro-derived greenhouse-grown tubers, in vitro shoots and callus-like basal tissues of *Harpagophytum procumbens*. S Afr J Bot 77:479–484

Baskaran P, Ncube B, Van Staden J (2012) *In vitro* propagation and secondary product production by *Merwilla plumbea* (Lindl.) Speta. Plant Growth Regul 67:235–245

Batková P, Pospíšilová J, Synková H (2008) Production of reactive oxygen species and development of antioxidative systems during in vitro growth and ex vitro transfer. Biol Plant 52:413–422

Bogaert I, Van Cauter S, Prinsen E et al (2006) New aromatic cytokinin can make the difference. Acta Hortic 725:265–270

Chauhan RD, Taylor NJ (2018) Meta-topolin stimulates de novo shoot organogenesis and plant regeneration in cassava. Plant Cell Tissue Organ Cult 132:219–224

Collin HA (2001) Secondary product formation in plant tissue cultures. Plant Growth Regul 34:119–134

Costa P, Goncalves S, Valentao P et al (2012) *Thymus lotocephalus* wild plants and in vitro cultures produce different profiles of phenolic compounds with antioxidant activity. Food Chem 135:1253–1260

Coste A, Vlase L, Halmagyi A et al (2011) Effects of plant growth regulators and elicitors on production of secondary metabolites in shoot cultures of *Hypericum hirsutum* and *Hypericum maculatum*. Plant Cell Tissue Organ Cult 106:279–288

Deikman J, Hammer PE (1995) Induction of anthocyanin accumulation by cytokinins in *Arabidopsis thaliana*. Plant Physiol 108:47–57

Dewir YH, Chakrabarty D, Ali MB et al (2006) Lipid peroxidation and antioxidant enzyme activities of *Euphorbia milii* hyperhydric shoots. Environ Exp Bot 58:93–99

Díaz-Vivancos P, Majourhat K, Fernández JA et al (2011) Study of the antioxidant enzymatic system during shoot development from cultured intercalar meristems of saffron. Plant Growth Regul 65:119–126

Doležal K, Popa I, Kryštof V et al (2006) Preparation and biological activity of 6-benzylaminopurine derivatives in plants and human cancer cells. Bioorgan Med Chem 14:875–884

Gaspar T, Franck T, Bisbis B et al (2002) Concepts in plant stress physiology. Application to plant tissue cultures. Plant Growth Regul 37:263–285

Gentile A, Frattarelli A, Nota P et al (2016) The aromatic cytokinin *meta*-topolin promotes in vitro propagation, shoot quality and micrografting in *Corylus colurna* L. Plant Cell Tissue Organ Cult 128(3):693–703

George E, Hall M, Klerk GJ (2008) Plant growth regulators II: cytokinins, their analogues and antagonists. In: George E, Hall M, Kler GJ (eds) Plant propagation by tissue culture. Springer, Dordrecht, pp 205–226

Gong B, Wen D, Van den Langenberg K et al (2013) Comparative effects of NaCl and $NaHCO_3$ stress on photosynthetic parameters, nutrient metabolism, and the antioxidant system in tomato leaves. Sci Hortic 157:1–12

Gut H, Rutz C, Matile P et al (1987) Leaf senescence in a non-yellowing mutant of *Festuca pratensis*: degradation of carotenoids. Physiol Plant 70(4):659–663

Han SS (1997) Preventing postproduction leaf yellowing in Easter lily. J Am Soc Hortic Sci 122:869–872

Holub J, Hanuš J, Hanke DE, Strnad M (1998) Biological activity of cytokinins derived from ortho- and meta-hydroxybenzyladenine. Plant Growth Regul 26:109–115

Horgan R, Hewett EW, Horgan JM et al (1975) A new cytokinin from *Populus x robusta*. Phytochemistry 14:1005–1008

Kalaji HM, Schansker G, Ladle RJ et al (2014) Frequently asked questions about in vivo chlorophyll fluorescence: practical issues. Photosynth Res 122:121–158

Kamínek M, Vaněk T, Kalendová-Kulasová A et al (1987) The effect of two cytokinins on production of stem cuttings by stock plants of *Euphorbia pulcherrima Willd.* and *Gerbera jamesonii* Hook. Sci Hortic 33:281–289

Masondo NA, Aremu ADO, Finnie JF et al (2014) Plant growth regulator induced phytochemical and antioxidant variations in micropropagated and acclimatized *Eucomis autumnalis* subspecies autumnalis (*Asparagaceae*). Acta Physiol Plant 36:2467–2479

Matkowski A (2008) Plant *in vitro* culture for the production of antioxidants-a review. Biotechnol Adv 26:548–560

Mishra P, Dubey RS (2008) Effect of aluminium on metabolism of starch and sugars in growing rice seedlings. Acta Physiol Plant 30:265–275

Moyo M, Finnie JF, Van Staden J (2012) Topolins in *Pelargonium sidoides* micropropagation: do the new brooms really sweep cleaner? Plant Cell Tissue Organ Cult 110:319–327

Mutui TM, Emongor VE, Hutchinson MJ (2004) Influence of plant growth regulators on postharvest characteristics of *Alstroemeria* cut flowers. Discov Innov 16:195–201

Mutui TM, Mibus H, Serek M (2005) Effects of thidiazuron, ethylene, abscisic acid and dark storage on leaf yellowing and rooting of *Pelargonium* cuttings. J Hortic Sci Biotechnol 80:543–550

Mutui TM, Emongor VE, Hutchinson MJ (2006) The effect of Gibberellin 4+7 on the vase life and flower quality of *Alstroemeria* cut flowers. Plant Growth Regul 48:207–214

Mutui TM, Mibus H, Serek M (2012) Effect of meta-topolin on leaf senescence and rooting in Pelargonium × hortorum cuttings. Postharvest Biol Technol 63:107–110

Ncube B, Ngunge VNP, Finnie JF et al (2011) Comparative study of the antimicrobial and phytochemical properties between outdoor grown and micropropagated *Tulbaghia violacea* Harv. plants. J Ethnopharmacol 134:775–780

Neill SJ, Desikan R, Clarke A et al (2002) Hydrogen peroxide and nitric oxide as signalling molecules in plants. J Exp Bot 53:1237–1247

Palavan-Unsal N, Çağ S, Çetin E, Büyüktunçer D (2002) Retardation of senescence by meta-topolin in wheat leaves. J Cell Mol Biol 1:101–108

Palavan-Unsal N, Çağ S, Çetin E (2004) The role of meta-topolin in senescence of wheat leaf segments. J Cell Mol Biol 3:23–31

Pavarini DP, Pavarini SP, Niehues M et al (2012) Exogenous influences on plant secondary metabolite levels. Anim Feed Sci Technol 176:5–16

Quirino BF, Noh YS, Himelblau E et al (2000) Molecular aspects of leaf senescence. Trends Plant Sci 5:278–282

Rao SR, Ravishankar GA (2002) Plant cell cultures: chemical factories of secondary metabolites. Biotechnol Adv 20:101–153

Richmond AE, Lang A (1957) Effect of kinetin on protein content and survival of detached *Xanthium* leaves. Science 125:650–651

Sakakibara H, Takei K, Hirose N (2006) Interactions between nitrogen and cytokinin in the regulation of metabolism and development. Trends Plant Sci 11:440–448

Schmitz RY, Skoog F, Playtis AJ et al (1972) Cytokinins: synthesis and biological activity of geometric and position isomers of zeatin. Plant Physiol 50:702–705

Strnad M (1992a) Discovery of a new cytokinin family. Biol Plant 34:585–586

Strnad M (1992b) Aromatic cytokinins, a new family of naturally occurring plant growth regulators. Physiol Plant 85:A170

Strnad M, Hanuš J, Vaněk T et al (1997) Meta-topolin, a highly active aromatic cytokinin from poplar leaves *Populus canadensis* Moench, cv. Robusta. Phytochemistry 45:213–218

Synková H, Semorádová S, Schnablová R et al (2006) Cytokinin-induced activity of antioxidant enzymes in transgenic *Pssu-ipt* tobacco during plant ontogeny. Biol Plant 50:31–41

Tarkowská D, Doležal K, Tarkowsky P et al (2003) Identification of new aromatic cytokinins in *Arabidopsis thaliana* and *Populus* × *canadensis* leaves by LC- (+) ESI–MS and capillary liquid chromatography/frit-fast atom bombardment mass spectrometry. Physiol Plant 117:579–590
Thimann KV (1980) The senescence of leaves. In: Thimann KV (ed) Senescence in plants. CRC Press, Boca Raton, pp 85–115
Thomas H, Stoddart JL (1980) Leaf senescence. Annu Rev Plant Physiol 31:83–111
Vijayakumar J, Ponnirul P, Samuell P et al (2017) Influence of meta-topolin on efficient plant regeneration via micropropagation and organogenesis of safflower (*Carthamus tinctorius* L.) cv. NARI-H-15. Am J Plant Sci 8:688–705
Werbrouck SPO, Van der Jeugt B, Dewitte W et al (1995) The metabolism of benzyladenine in *S. floribundum* Schott 'petite' in relation to acclimatization problems. Plant Cell Rep 14:662–665
Werbrouck SPO, Strnad M, Van Onckelen HA et al (1996) Meta-topolin, an alternative to benzyladenine in tissue culture? Physiol Plant 98:291–298
Werner T, Schmülling T (2009) Cytokinin action in plant development. Curr Opin Plant Biol 12:527–538
Wollgiehn R (1967) Nucleic acid and protein metabolism of excised leaves. Symp Soc Exp Biol 21:231–246
Zavaleta-Mancera HA, Franklin KA, Ougha HJ et al (1999) Regreening of senescent *Nicotiana* leaves. I. Reappearance of NADPH protochlorophyllide oxidoreductase and light-harvesting chlorophyll a/b binding protein. J Exp Bot 50:1677–1682
Zhou HC, Lin YM, Wei SD, Tam NFY (2011) Structural diversity and antioxidant activity of condensed tannins fractionated from mangos teen pericarp. Food Chem 129:1710–1720
Zwack PJ, Rashotte AM (2013) Cytokinin inhibition of leaf senescence. Plant Signal Behav 8(7): e24737

Establishment and Management of an In Vitro Repository of Kiwifruit (*Actinidia* spp.) Germplasm

20

M. Debenham and R. Pathirana

Contents

M. Debenham
The New Zealand Institute for Plant and Food Research Limited, Palmerston North, New Zealand

R. Pathirana (✉)
The New Zealand Institute for Plant and Food Research Limited, Palmerston North, New Zealand

Plant & Food Research Australia Pvt Ltd., Waite Campus Research Precinct, Urrbrae, SA, Australia
e-mail: Ranjith.Pathirana@plantandfood.co.nz

N. Ahmad, M. Strnad (eds.), *Meta-topolin: A Growth Regulator for Plant Biotechnology and Agriculture*, https://doi.org/10.1007/978-981-15-9046-7_20

Abstract

Incursion of *Pseudomonas syringae* pv. *actinidiae* in New Zealand in November 2010 prompted the establishment of an in vitro repository of kiwifruit (*Actinidia* spp.) germplasm at the Palmerston North site of the New Zealand Institute for Plant and Food Research Limited (PFR). This chapter describes the methods used in the initiation of cultures, emphasising the use of sodium dichloroisocyanurate as the surface sterilising agent, and the methods for multiplication of cultures using *meta*-topolin as the preferred cytokinin. Once multiplied, the material is prepared for medium-term storage at 5 °C in 35-mL vials by first acclimating the plantlets for 2 weeks at a short photoperiod (10 h), with low temperature (2 °C) during the long dark period (14 h) and 22 °C during the short light period. Under cool storage, some accessions have recorded survival of up to 50 months with the majority surviving over 2 years. The accessions in the collection can be tracked to the mother plant in the field during the culture process using a data management programme that also collects data on different steps, e.g. dates of initiation, subculture, media used, results of testing for microbial contaminants and molecular fingerprinting results. Simple sequence repeat markers are used to verify the genotypes. The repository is not static as accessions are used by researchers, and we also use the resources of this repository for export of high-health advanced selections for overseas testing. Considering the value of this collection and its vulnerability, integrated conservation strategies are utilised, and cryopreservation of a core collection is part of this strategy.

Keywords

Tissue culture · Conservation · Genetic resources · Sodium dichloroisocyanurate · *Meta*-topolin · New Zealand

20.1 Introduction

Although *Actinidia* spp. originated in East Asia, with the highest genetic diversity in eastern China, the crop was domesticated and commercialised in New Zealand as kiwifruit in the last century (Ferguson and Huang 2007). It is currently the most economically important horticultural crop in New Zealand with NZ$ 3.1 billion in annual global sales and licence release in 2018/2019 (Zespri 2020). Kiwifruit has a rich genetic diversity, with several species such as *A. chinensis* var. *chinensis*, *A. chinensis* var. *deliciosa* and *A. arguta* already in commercial production. Several other species such as *A. eriantha*, *A. latifolia*, *A. indochinensis* and *A. kolomikta* are considered economically important and are being used in breeding programmes, with various combinations producing fertile interspecific hybrids (Datson and Ferguson 2011; Wang et al. 2000; Zhong et al. 2012). New Zealand has one of the largest collections of kiwifruit genetic resources outside mainland China, comprising 17 *Actinidia* spp. (McNeilage 2015), and this collection is held under field conditions in three sites by the New Zealand Institute for Plant and Food Research

Limited (PFR) in Te Puke (Bay of Plenty, North Island), Kerikeri (Northland, North Island) and Riwaka (Tasman, South Island).

Being a dioecious genus with multiple ploidy series within species (Ferguson and Huang 2007), kiwifruit poses challenges for breeding and conservation efforts (Beatson et al. 2014; Li et al. 2014; Pathirana et al. 2016). The incursion of *Pseudomonas syringae* pv. *actinidiae*—Virulent (Psa-V)—in November 2010 into Te Puke and its fast spread across the North Island of New Zealand posed a threat not only to production (Vanneste 2017; Peacock 2014) but also to PFR's field collections of *Actinidia* genetic resources (Datson et al. 2015; Debenham et al. 2016). This prompted the ex situ conservation and establishment of an in vitro repository of *Actinidia* genetic resources at the Palmerston North site of PFR. This being the first such repository in the world for *Actinidia* (Datson and Ferguson 2011), methodology for different steps of the in vitro culture process was developed within PFR. This conservation effort not only helped rescue many Psa-V susceptible accessions but also supported deployment of associated technologies such as in vitro mutagenesis (Pathirana et al. 2016) and cryopreservation (Mathew et al. 2018, 2019; Pathirana et al. 2020). In this chapter, we give the optimised protocols for tissue culture initiation, proliferation, conservation at low temperature, rooting and greenhouse acclimation for growing on. Although zeatin has been routinely used for the proliferation stage of propagation in kiwifruit in vitro culture (Wang et al. 2006; Pathirana et al. 2016), recent research has shown that *meta*-topolin (*m*T) is a better option for difficult-to-root kiwifruit genotypes (Saeiahagh et al. 2019). Here we describe both zeatin- and *m*T-based strategies used in kiwifruit in vitro conservation work.

20.2 Materials

20.2.1 Tissue Culture Facilities and General Equipment for Plant Tissue Culture

- Precision balance
- Magnetic stirrer
- pH meter
- Autoclave
- Laminar airflow cabinet
- Glass bead steriliser
- Refrigerator and freezer
- Culture room with controlled temperature (25 °C) and photoperiod (16 h) at a photosynthetic photon flux density (PPFD) of 40 μmol m^{-2} s^{-1}
- Culture room with controlled temperature (5 °C) and photoperiod (8 h) at a PPFD of ~5 μmol m^{-2} s^{-1}
- Growth cabinet with light for providing photoperiod and temperature control with capability of maintaining 22 °C for 10 h and 2 °C for14 h
- Forceps, scalpels and scalpel blades, secateurs

- Adjustable volume single-channel pipettes with capacities of 2 μL, 10 μL, 20 μL, 200 μL, 1 mL and 5 mL and tips to suit
- Sterile syringe filters (0.2 μm) and syringes (5, 10, 20 and 30 mL)
- Sterile tissue culture vessels ~250 mL
- Sterile ~35-mL screw cap clear vials with suitable stands
- Volumetric flasks
- Sterile plastic Petri dishes (100 × 25 mm; 60 × 15 mm)
- Sterile glass containers (jars)
- Shaker
- Schott bottles of different capacities
- Plastic wrap

20.2.2 Culture Media and Solutions

Prepare all the media and solutions using distilled or reverse osmosis water, analytical grade reagents and store at 4 °C, except for vitamins which should be stored at −20 °C.

The formulation for Murashige and Skoog (MS) salts (Murashige and Skoog 1962) is shown in Table 20.1. MS medium can be purchased from suppliers as a powdered mixture containing all macro- and micro-elements, with or without ammonium as well as half-strength MS macro salt solution in combination with full-strength MS micro salt solution.

Stock Solutions

Plant Growth Regulators

All plant growth regulators (PGR) are prepared at a concentration of 0.1 mg mL^{-1}. Dissolve the quantity of PGR in a small volume of 5-M NaOH and add distilled H_2O to the desired volume. Store the PGR stock solutions at 4 °C.

Vitamins

Stock solutions of B5 vitamins (Gamborg et al. 1968) are prepared according to the formulations given in Table 20.1. Dispense vitamins into aliquots and store at −20 °C.

Shoot Initiation, Maintenance and Proliferation Media

PrM1—MS macro- and microelements, B5 vitamins and 20-g/L sucrose, solidified using 7-g/L agar. The culture medium is supplemented with the plant growth regulators indole-3-butyric acid (IBA) (0.05 mg/L) and zeatin (0.2 mg/L). Adjust pH to 6 prior to autoclaving.

PrM2—Half-strength MS macroelements, full-strength MS microelements, B5 vitamins and 30 g/L sucrose, solidified with a mixture of 4-g/L agar and 1.5-g/L Phytagel. This medium is supplemented with 0.05 g/L IBA, 0.1 mg/L gibberellic acid and 0.66 g/L *m*T. Adjust pH to 6 prior to autoclaving.

Table 20.1 Composition of Murashige and Skoog (MS) culture medium (Murashige and Skoog 1962) and B5 vitamins (Gamborg et al. 1968)

Macroelements	$mg\ L^{-1}$
NH_4NO_3	1650
$CaCl_2 \cdot 2H_2O$	440
$MgSO_4 \cdot 7H_2O$	370
KH_2PO_4	170
KNO_3	1900
Microelements	**$mg\ L^{-1}$**
$MnSO_4 \cdot H_2O$	16.9
$ZnSO_4 \cdot 7H_2O$	8.60
H_3BO_3	6.20
KI	0.83
$Na_2MoO_4 \cdot 2H_2O$	0.25
$CuSO_4 \cdot 5H_2O$	0.025
$CoCl_2 \cdot 6H_2O$	0.025
Iron source	**$mg\ L^{-1}$**
$FeSO_4 \cdot 7H_2O$	27.8
$Na_2EDTA \cdot 2H_2O$	37.3
B5 vitamins	**$mg\ L^{-1}$**
Nicotinic acid	1.0
Pyridoxine-HCl	1.0
Thiamine-HCl	10.0
Myo-inositol	100.0
Sucrose	$30\ g\ L^{-1}$
pH	6.00

Medium for Cool Storage

Half-strength MS macroelements, full-strength MS microelements, B5 vitamins and 30 g/L sucrose, solidified with 7.5 g/L agar. Adjust pH to 6 prior to autoclaving.

Bacteriological Media for Screening Established Plantlets for Endogenous Microorganisms

BactM—During subculture, sections of the basal part of the cut shoots are transferred to this medium on Petri plates. The medium consists of potato dextrose agar supplemented with 5 g/L peptone.

20.3 Methods

20.3.1 Collection and Surface Sterilisation of Explants

There are two ways to source explants for initiation: (a) by selecting healthy source material, preferably from plants maintained under optimal conditions in a growth chamber, greenhouse or fresh flushes from the field, or (b) by sprouting, in water at room temperature for 2–3 weeks, mature canes that have been held at 0 °C for a minimum of 4 weeks. We have successfully used material that has been stored this way for up to 8 months. A solution of 0.3% (v/v) Kathon LXE® biocide (Dow Chemical Company) may be incorporated in the water for sprouting the canes as this may reduce contamination of the water and potentially the shoots.

Sprouted shoots that are approximately 5–7 cm long typically comprising 3–7 nodes are used for surface sterilisation. All the leaves are removed and the shoots placed into clean Schott bottles or flasks and washed under running tap water for 1–2 min. Add water and antibacterial soap (2% v/v) and shake vigorously for 1 min, and then rinse repeatedly until soap bubbles are no longer produced. Rinse the cuttings in 70% ethanol for 30 s, then place in a solution of 5 g/L sodium dichloroisocyanurate (Sigma) with 0.2% (v/v) Tween® 20, tighten the cap and shake for 25 min on an orbital shaker (50 rpm). Under aseptic conditions, rinse the cuttings three times with sterile water, and cut into nodal sections, each 2–4 mm long, and place 6–8 sections onto Petri dishes containing PrM1 or PrM2 media. Shoot tips can also be used. Maintain in a growth room under light (35–-50 μmol m^{-2} s^{-1} at shelf level) with a 16-h photoperiod at 23–27 °C (standard culture conditions). Several plates per accession need to be established to allow for contamination during the early phase of establishment.

20.3.2 Establishment and Proliferation of In Vitro Cultures

During the first few days of culture growth, observe plates daily for contamination. If one or two explants are contaminated in a plate, the clean explants can be carefully transferred to fresh PrM1 or PrM2 plates to prevent the spread of bacteria or fungal contaminants onto clean explants.

Once the shoots have produced sufficient growth (1–4 weeks; 2- to 4-cm long), the individual shoots can be removed and plated in 250-mL tubs or in an equivalent vessel for further proliferation and growth. Subculture every 4–6 weeks to obtain a sufficient number of shoots to prepare the material for cool storage.

20.3.3 Screening Established Plants for Endogenous Microorganisms

After the first round of subculture and establishment of plantlets in tissue culture, the material can be tested for endogenous microorganisms by removing a thin section of the basal tissue of each plantlet and placing on the BactM media plates. These sections are serially numbered to allow tracking of the individual plantlet which the basal tissue came from. The corresponding shoots are transferred to PrM1 or PrM2 tubs and numbered. The shoots in BactM are maintained under standard culture conditions for 10 days, with careful examination each day, by which time any contamination should be visible. All the shoots corresponding to the infected basal tissue in BactM media need to be removed from proliferation media as these are potentially contaminated.

20.3.4 Preparing Cultures for Cool Storage and Management Under Cool Storage

Once cultures are established and sufficient plantlet numbers are produced in proliferation media, dissect 2- to 3-cm-long shoots from 5- to 8-week-old in vitro stock plants, and transfer two shoots to individual 35-mL clear vials or their equivalent containing plant growth regulator-free media for cool storage. Eight vials per accession should be established, keeping one or two tubs with stock plants as a backup. The in vitro cultures in vials are maintained at standard culture conditions for 2–4 weeks and then cold acclimated for 10 days at alternating temperatures with a short photoperiod (22 °C with 5 μmol m^{-2} s^{-1} light for 10 h/ 2 °C dark for 14 h). These vials can now be transferred to cool storage at 5 °C under low light intensity (1–3 μmol m^{-2} s^{-1}) with a 16-h photoperiod (Fig. 20.1).

All the cultures in cool storage are checked every 3 months with the plant in each vial assessed for survival. Special attention should be given to plants stored longer than 1 year. When 5–6 out of the 8 vials have between 20 and 50% of viable plant tissue (stem) visible and the other vials have either dead plantlets or <20% viable tissue, the accession requires regeneration. The viable plants are used to reestablish cultures under standard culture conditions and multiplied for another cycle of cool storage.

Fig. 20.1 The cool storage facility at the Palmerston North site of the New Zealand Institute for Plant and Food Research Limited. Each accession is held in eight 35-mL plastic screw cap vials at 5 °C

20.3.5 Regeneration After Cool Storage

Vials containing plants that require regeneration are removed from cool storage and plants carefully removed under aseptic conditions. The plants are dissected, dead tissue removed and live nodal sections are transferred to PrM1 or PrM2 in 250-mL tubs or their equivalent and maintained under standard culture conditions. The proliferating plantlets are then cultured and prepared for another round of cool storage as already described.

20.4 Discussion

20.4.1 Surface Sterilisation and Effect of Cytokinin in Proliferation Media

This collection started in response to the incursion of Psa-V in November 2010 into Te Puke in the Bay of Plenty Region where the largest kiwifruit field collection in New Zealand is held. Large numbers of accessions were introduced to tissue culture, but the success rates were low (10%) (Debenham et al. 2016) due to high levels of contamination as well as high genetic diversity of the collection, resulting in differential response to media used (Fig. 20.2). Since then, improved sterilisation procedures and testing for endogenous microorganisms were introduced. For example, changing the sterilising solution from a 20–30% solution of domestic bleach (Janola®; 5% NaOCl, 4-g/L NaOH) to sodium dichloroisocyanurate increased success rates. The domestic bleach solution is very alkaline (pH 11.2) which imparts a stress to tender explant tissue. In contrast, the sodium dichloroisocyanurate solution has a pH of 5.4 and provides effective sterilisation with less stress to plant tissues.

Another observation was that plantlets of some accessions declined in quality within the first few weeks after introduction to tissue culture. This was observed

Fig. 20.2 Response of different *Actinidia kolomikta* genotypes to the same culture medium

when we used zeatin as cytokinin in proliferation media. Recently, we have tested *m*T as an alternative cytokinin, initially for material that was difficult to root after the proliferation step (Saeiahagh et al. 2019), and found that the declining shoots can be recovered when these are transferred to *m*T media (PrM2). As we have not yet tested this in a large number of accessions, we have given two proliferation media in this protocol, one containing zeatin (PrM1) and the other containing *m*T (PrM2). We are slowly moving to PrM2 and conducting further experiments to determine if *m*T is superior to zeatin. In this protocol, we have not described the rooting and exflasking of plantlets to greenhouse. This is described in Saeiahagh et al. (2019), where we show that *m*T is the preferred cytokinin in proliferation media, particularly when the accession has difficulties in the rooting stage.

20.4.2 Testing for Endogenous Microorganisms

In order to ensure plant materials in the repository are in the best possible condition, it was necessary to develop robust methods for detecting endogenous microorganisms in our cultures, particularly Psa-V. It has been demonstrated that Psa-V can survive within plant tissue without growing out on to standard tissue culture media, thus avoiding detection (Minardi et al. 2012; Muganu et al. 2009; Tyson et al. 2018). However, supplementation of the medium with peptone allows the detection of endogenous Psa-V within 3 days (Tyson et al. 2018) and a range of other bacteria. Holding the basal part of plantlets for 2–3 weeks in BactM media during culture establishment stage provides confidence that Psa-V and other bacteria that grow on peptone-supplemented medium are unlikely to be present in plants in the culture collection. In addition, any material destined for overseas undergoes another test where all the plantlets produced after 2–3 subcultures are grown for 7 days in PrM1 or PrM2 supplemented with 3 g/L peptone. This test is conducted at the time of plating the basal part of plantlets in BactM. Any shoots showing contamination in peptone-supplemented media are discarded at this stage.

20.4.3 The Status of the Collection

To support the breeding effort in New Zealand, *Actinidia* genetic resources comprising 17 species in the form of clones and seeds were introduced during the period from 1956 to 1982, not only from China and Japan but also from nurseries and botanical gardens in Europe and Australia (McNeilage 2015). In addition to these introductions, PFR's orchard system comprises advanced selections under different stages of screening and also experimental populations used by researchers of the institute. The in vitro repository currently comprises all of the introduced species as well as some interspecific hybrids (Table 20.2), but the majority of the accessions are *A. chinensis* var. *chinensis* and *A. chinensis* var. *deliciosa*. Only the germplasm accessions (about 50% of the total collection) are held in medium-term cool storage,

Table 20.2 Current status of the in vitro repository of *Actinidia* germplasm held at the New Zealand Institute for Plant and Food Research Limited

Species or cross	Number of accessions
A. arguta	13
A. arguta var. *purpurea*	12
A. callosa var. *callosa*	2
A. chinensis var. *chinensis*	769
A. chrysantha	1
A. chinensis var. *deliciosa*	132
A. deliciosa var. *coloris*	2
A. deliciosa × *A. deliciosa* var. *coloris*	1
A. eriantha f. *eriantha*	4
A. eriantha f. *eriantha* × *A. chinensis*	1
A. glaucophylla var. *robusta*	1
A. guilinensis	2
A. hypoleuca	1
A. indochinensis	2
A. kolomikta	1
A. lanceolata	2
A. latifolia var. *latifolia*	1
A. macrosperma var. *macrosperma*	19
A. macrosperma var. *macrosperma* × *A. melanandra* var. *melanandra*	1
A. melanandra var. *melanandra*	8
A. polygama	23
A. polygama × *A. chinensis* var. *chinensis*	1
A. setosa	2
A. chinensis var. *chinensis* × *A. melanandra* var. *melanandra*	3
A. chinensis var. *deliciosa* × *A. chinensis* var. *chinensis*	3
A. chinensis var. *deliciosa* × *A. eriantha* f. *eriantha*	5
Total	1012

with advanced selections and experimental populations being maintained for shorter periods under standard culture conditions.

At any given time, about 80% of the germplasm collection is held in cool storage with the rest being rejuvenated for the storage cycle. Depending on the genotype, it takes between 2 and 6 months to return a cultivar back to cool storage after rejuvenation. Depending on the genotype, an accession can last from 4 to 50 months in cool storage, with the majority surviving over 2 years. The condition of plantlets at the time of cool storage and genotype have influence on the longevity in cool storage.

20.4.4 Molecular Markers for Quality Control of the Repository

When shoots for initiation arrive for inclusion in the repository, two leaf samples are taken in vials, labelled and stored at −80 °C for DNA extraction and fingerprinting for trueness-to-type at later dates. DNA from 100 germplasm accessions of the repository has so far been tested against the DNA from field or greenhouse plants from which the initiations were made (Wiedow et al. 2017). For this, six simple sequence repeat (SSR) markers were used, and the results provide confidence in the labelling and tracking of tissue cultures in the repository. We plan to apply the testing to all other germplasm accessions in the collection to confirm the trueness-to-type.

20.4.5 Database and Tracking of Plants in Culture

It is important to keep track of the plants from which the vines were taken for initiating tissue cultures. A database was developed to help manage the records associated with the large numbers of plants in the repository and to capture relevant data on the initiation, growth, multiplication and movements of tissue as well as details on pathogen screening. All information on the in vitro plants and their mother plants is collected in the in-house developed database called Germplasm Management System (GMS) (Seelye et al. 2016). A barcode system developed allows us to track the date of initiation, media used, name of the worker for each step with dates, location of culture in the lab, DNA fingerprinting result, any biosecurity compliance details and the location of original plant from which the material was taken for initiation. The GMS has now been linked to the E-Brida software (https://www.e-brida.com) used by the breeders, allowing us to search for parents and relationships among accessions.

20.4.6 Management and Future of the Germplasm Collection

Integration of conservation strategies is important to safeguard the collection of *Actinidia* germplasm in New Zealand. With thousands of unique genotypes, it is expensive to duplicate all accessions in different field sites. While it is more cost-effective to manage the accessions in vitro compared to field, this also has the risk of loss, for example, due to contamination. Somaclonal variation is another risk in tissue culture, and we have detected this at a low frequency in kiwifruit (Saeiahagh et al. 2019). Therefore, to secure the genetic diversity of kiwifruit, we plan to conserve a core collection of kiwifruit accessions in long-term cryostorage. A droplet-vitrification protocol was developed using *A. chinensis* var. *chinensis* as a model (Mathew et al. 2018, 2019). This protocol has since been simplified and improved and applied to five other *Actinidia* spp. with higher regeneration rates (Pathirana et al. 2019, 2020). The in vitro repository of kiwifruit germplasm and the knowledge on the response of different genotypes to tissue culture acquired during the establishment of this collection will make this task easier.

20.5 Conclusions

An in vitro germplasm repository comprising representatives of the diverse genetics of *Actinidia* has been established at PFR. This collection complements the field collections, conserves genotypes that have been lost to Psa-V in the field and supports the PFR breeding programme which focuses on developing cultivars with novel traits and resistance to Psa-V. Protocols for explant collection, initiation, contaminant testing, cool storage, regeneration and DNA fingerprinting have been developed alongside an auditable system for tracking material. While ex situ conservation through medium-term storage of germplasm under in vitro conditions is useful, at PFR we apply integrated conservation strategies. Therefore, a plan to conserve a core collection of kiwifruit accessions in long-term cryostorage is being implemented.

References

Beatson R, Datson P, Ferguson A, Montefiori M (2014) Use of kiwifruit germplasm resources for genetic improvement. Acta Hortic 1048:25–34

Datson PM, Ferguson AR (2011) Actinidia. In: Kole C (ed) Wild crop relatives: genomics and breeding resources, tropical and subtropical fruits. Springer-Verlag, Berlin, pp 1–20. https://doi.org/10.1007/978-3-642-20447-0-1

Datson P, Nardozza S, Manako K, Herrick J, Martinez-Sanchez M, Curtis C, Montefiori M (2015) Monitoring the *Actinidia* germplasm for resistance to *Pseudomonas syringae* pv. *actinidiae*. Acta Hortic 1095:181–184

Debenham MC, Seelye JF, Mullan AC (2016) An *in vitro* repository for clonal kiwifruit. Acta Hortic 1113:93–97. https://doi.org/10.17660/ActaHortic.2016.1113.13

Ferguson AR, Huang H (2007) Genetic resources of kiwifruit: domestication and breeding. Horticult Rev 33:1–121

Gamborg OL, Miller RA, Ojima K (1968) Nutrient requirements of suspension cultures of soybean root cells. Exp Cell Res 50(1):151–158. https://doi.org/10.1016/0014-4827(68)90403-5

Li D, Liu Y, Li X, Rao J, Yao X, Zhong C (2014) Genetic diversity in kiwifruit polyploid complexes: insights into cultivar evaluation, conservation, and utilization. Tree Genet Genomes 10(5):1451–1463

Mathew L, McLachlan A, Jibran R, Burritt DJ, Pathirana R (2018) Cold, antioxidant and osmotic pre-treatments maintain the structural integrity of meristematic cells and improve plant regeneration in cryopreserved kiwifruit shoot tips. Protoplasma 255(4):1065–1077. https://doi.org/10.1007/s00709-018-1215-3

Mathew L, Burritt DJ, McLachlan A, Pathirana R (2019) Combined pre-treatments enhance antioxidant metabolism and improve survival of cryopreserved kiwifruit shoot tips. Plant Cell Tissue Organ Cult 138:193–205. https://doi.org/10.1007/s11240-019-01617-3

McNeilage MA (2015) Genetics and breeding of kiwifruit. Acta Hortic 1096:185–190

Minardi P, Ardizzi S, Lucchese C, Bertaccini A (2012) Micropropagation of kiwifruit: one-year endophytic colonisation of *Pseudomonas syringae* pv. *actinidiae* in shoots of *Actinidia deliciosa* cv. Hayward. J Plant Pathol 94(4 Suppl):S4.70

Muganu M, Fratarcangeli L, Paolocci M, Balestra GM (2009) Improvement of the quality of nursery kiwifruit plants: sanitation of micropropagated shoots infected by *Pseudomonas syringae* pv. *syringae* (Van Hall). Italus Hortus 16(5):227–230

Murashige T, Skoog F (1962) A revised medium for rapid growth and bioassays with tobacco tissue cultures. Physiol Plant 15:473–497

Pathirana R, Deroles S, Hoeata K, Montefiori M, Tyson J, Wang T, Datson PM, Hellens RP (2016) Fast-tracking kiwifruit breeding through mutagenesis. Acta Hortic 1127:217–222. https://doi.org/10.17660/ActaHortic.2016.1127.34

Pathirana R, Mathew L, Jibran R, Hunter DA, Morgan ER (2019) Cryopreservation and cryotherapy research on horticultural crops in New Zealand. Acta Hortic 1234:29–36. https://doi.org/10.17660/ActaHortic2019.1234.4

Pathirana R, Mathew L, McLachlan A (2020) A simplified method for high recovery of kiwifruit (*Actinidia* spp.) shoot tips after droplet vitrification cryopreservation suitable for long-term conservation. Plant Cell Tissue Organ Culture. https://link.springer.com/article/10.1007/s11240-020-01860-z#citeas

Peacock L (2014) *Pseudomonas syringae* pv. *actinidiae*-virulent impacts on the New Zealand kiwifruit industry. Acta Hortic 1055:55–62

Saeiahagh H, Mousavi M, Wiedow C, Bassett HB, Pathirana R (2019) Effect of cytokinins and sucrose concentration on the efficiency of micropropagation of 'Zes006' *Actinidia chinensis* var. *chinensis*, a red-fleshed kiwifruit cultivar. Plant Cell Tissue Organ Cult 138:1–10. https://doi.org/10.1007/s11240-019-01597-4

Seelye JF, Corpe S, Debenham MC (2016) A management system for tracking high health in vitro kiwifruit germplasm. Acta Hortic 1113:113–118. https://doi.org/10.17660/ActaHortic.2016.1113.16

Tyson JL, Vergara MJ, Butler RC, Seelye JF, Morgan ER (2018) Survival, growth and detection of *Pseudomonas syringae* pv. *actinidiae* in *Actinidia* in vitro cultures. N Z J Crop Hortic Sci 46 (4):319–333. https://doi.org/10.1080/01140671.2017.1414064

Vanneste JL (2017) The scientific, economic, and social impacts of the New Zealand outbreak of bacterial canker of kiwifruit (*Pseudomonas syringae* pv. *actinidiae*). Annu Rev Phytopathol 55:377–399

Wang SM, Huang HW, Jiang ZW, Zhang ZH, Zhang SR (2000) Studies on *Actinidia* breeding by species hybridization between *A. chinensis* and *A. eriantha* and their hybrids' progenies. In: Huang HW (ed) Advances in *Actinidia* research. Science Press, Beijing, pp 123–127

Wang T, Ran Y, Atkinson RG, Gleave AP, Cohen D (2006) Transformation of *Actinidia eriantha*: a potential species for functional genomics studies in *Actinidia*. Plant Cell Rep 25:425

Wiedow C, Debenham M, Bassett HM, Corpe S, Seelye J, Pathirana R, Gardiner SE (2017) Quality control in a kiwifruit germplasm collection. Paper presented at the fourth international symposium on molecular markers in horticulture, Napier, New Zealand, 7–10 March, 2017

Zespri (2020) Zespri Group Limited and subsidiaries annual report for the year ended 31 March 2019. https://www.zespri.com/en-NZ/annualreports. Accessed 2 Jun 2020

Zhong C, Wang S, Jiang Z, Huang H (2012) 'Jinyan', an interspecific hybrid kiwifruit with brilliant yellow flesh and good storage quality. HortScience 47(8):1187. https://doi.org/10.21273/hortsci.47.8.1187

Matsuda M, [illegible] (2009) [illegible]

[illegible]

Pathirana R, Deroles S, Hoeata K, Montefiori M, Tyson J, Wang T, [illegible]

[illegible]

New Generation of *Arabidopsis thaliana* Cytokinin Oxidase/Dehydrogenase Inhibitors Affect Shoot/Root Growth and Seed Yield

21

Jana Bíbová, Marek Zatloukal, Radoslav Koprna, Lukáš Spíchal, and Miroslav Strnad

Contents

J. Bíbová (✉)
Laboratory of Growth Regulators, Institute of Experimental Botany of the Czech Academy of Sciences and Palacký University, Olomouc, Czech Republic

Centre of the Region Haná for Biotechnological and Agricultural Research, Department of Chemical Biology and Genetics, Faculty of Science, Palacký University, Olomouc, Czech Republic

M. Zatloukal · R. Koprna · L. Spíchal
Centre of the Region Haná for Biotechnological and Agricultural Research, Department of Chemical Biology and Genetics, Faculty of Science, Palacký University, Olomouc, Czech Republic

M. Strnad
Laboratory of Growth Regulators, Institute of Experimental Botany of the Czech Academy of Sciences and Palacký University, Olomouc, Czech Republic

N. Ahmad, M. Strnad (eds.), *Meta-topolin: A Growth Regulator for Plant Biotechnology and Agriculture*, https://doi.org/10.1007/978-981-15-9046-7_21

Abstract

This chapter describes the preparation of a new group of cytokinin oxidase/dehydrogenase (CKX) enzyme inhibitors usable for different agricultural applications. Recently developed CKX inhibitors, 2-chloro-6-(3-methoxyphenylamino)purine (INCYDE) and 2-fluoro-6-(3-methoxyphenylamino)purine (INCYDE-F) were modified by the tetrahydropyran-2-yl (THP) group at the *N9*-position to obtain 2-chloro-6-(3-methoxyanilino)-9-(tetrahydropyran-2-yl)purine (INCYDE-THP) and 2-fluoro-6-(3-methoxyanilino)-9-(tetrahydropyran-2-yl)purine (INCYDE-F-THP), respectively. These compounds were used to inhibit CKX from *Arabidopsis thaliana* (*At*CKX2). Further, their biological activity was investigated in a receptor binding assay where its ability to activate cytokinin receptors was assessed. Activation of the cytokinin signalling pathway was determined by ARR5/GUS reporter gene assay. All the assays showed that INCYDE-THP is a medium *At*CKX2 inhibitor, average cytokinin receptor agonist as well as inducer of ARR5 gene expression (a member of the A-type response regulator family classified as cytokinin primary response genes). INCYDE-F-THP was, however, found to be almost inactive. Their biological activity was also compared in three different bioassays based on the stimulation of tobacco callus growth, the retention of chlorophyll in excised wheat leaves and the dark induction of betacyanin synthesis in *Amaranthus* cotyledons. The two compounds were, however, only mildly active ($<10\%$ of BAP) in tobacco callus and *Amaranthus* bioassay and almost inactive in the senescence test. Finally, in vitro and in vivo experiments carried out with INCYDE-THP on different plant species showed positive effects on plant growth and development. In summary, new perspective inhibitors of *At*CKX2 were found. These positive effects are also reflected in increased yields of some crops, such as wheat, corn and oilseed rape, and some vegetables, especially when applied by watering. In conclusion, INCYDE-THP is a promising candidate for plant irrigation and seed coating that could contribute as an agrochemical approach to the growth of shoot/root biomass than standard molecular biology methods.

Keywords

Topolin · Aromatic cytokinin · INCYDE · Cytokinin oxidase/dehydrogenase · Plant growth and development

Cytokinins (CKs) are hormones that regulate multiple processes in plants including cell division and differentiation, regulation of seed germination, bud formation and senescence processes (Mok 1994). They also control organ development and play an important role in response to different types of biotic and abiotic stress (Werner and Schmülling 2009). Naturally occurring CKs are N^6-substituted adenine derivatives with an isoprenoid or an aromatic side chain (Mok and Mok 2001). Besides free bases, CKs form corresponding nucleosides, nucleotides and conjugates with glucose, xylose and amino acids.

Irreversible degradation of CKs is catalysed by the enzyme cytokinin oxidase/dehydrogenase (CKX, EC 1.5.99.12, Schmülling et al. 2003). The sequence analysis of the *Arabidopsis thaliana* genome has revealed the *CKX* gene family with seven homologues (*AtCKX1–AtCKX7*, Bilyeu et al. 2001). Natural CKXs are present at very low levels in plants. Greater amounts of *At*CKX2 protein were obtained by heterologous expression in *S. cerevisiae* (Werner et al. 2001) and *P. pastoris* (Frébortová et al. 2007). The new CKX inhibitors were tested in the *Pichia* system. This allowed us to obtain large amounts of homogenous, active protein from yeast culture media and to test the ability of a library of novel adenine derivatives to inhibit the activity of the enzyme. Some 2,6-disubstituted adenine derivatives have been shown to be biologically active in CKX enzyme assays. For example, strong inhibitory *meta*-methoxytopolin derivatives have been prepared by shortening the bond between the benzyl ring and the N^6-adenine by one carbon atom (Zatloukal et al. 2008) and adding a halogen atom to the C2-position of the purine ring. More specifically, 2-chloro-6-(3-methoxyphenylamino)purine and 2-fluoro-6-(3-methoxyphenylamino)purine were prepared and described as highly active inhibitors of the enzyme *At*CKX2 from *A. thaliana* (Spíchal et al. 2012; Zatloukal et al. 2008; Spíchal et al. 2004). The inhibition of enzymes like this can postpone the degradation of endogenously occurring CKs in plants (Spíchal et al. 2004). The first compound inhibitor of cytokinin degradation (INCYDE), 2-chloro-6-(3-methoxyphenylamino)purine, described above, is able to significantly affect the response of plants to saline or cadmium stress at low nanomolar concentration (10 nM) and improve the yield of a range of crops and vegetables such as tomatoes, *Bulbine natalensis* Baker (Asphodelaceae) and *Rumex crispus* L. (Polygonaceae) (Gemrotová et al. 2013). In addition, it is also used in the micropropagation of some plants, for example, *Eucomis autumnalis* (Aremu et al. 2015) and banana trees (Aremu et al. 2012). The second compound, 2-fluoro-6-(3-methoxyphenylamino) purine, was termed INCYDE-F and has been found to be even more active in field treatments of some crops.

Mok and Mok (2001) reported that if the *N9*-substitution of CKs is occupied by special protection groups, such as tetrahydropyran-2-yl (THP), tetrahydrofuran-2-yl (THF), 4-chlorobutyl or methyl, the arising 9-substituted CK derivatives show even higher biological activity than the original free bases. In the case of THP and THF groups in particular, they mimic the structure of sugars. Further, neither THP nor THF derivatives have been shown to be toxic, which is uncommon in cytokinin ribosides (Doležal et al. 2007; Szüčová et al. 2007, 2009). Bearing in mind that the majority of 9-substituted CKs still possess CK activity, blocking the *N9*-atom by suitable protecting group can serve as a barrier to the in situ formation of 9-glucosides and can prolong CK persistence in living plant systems (Bairu et al.

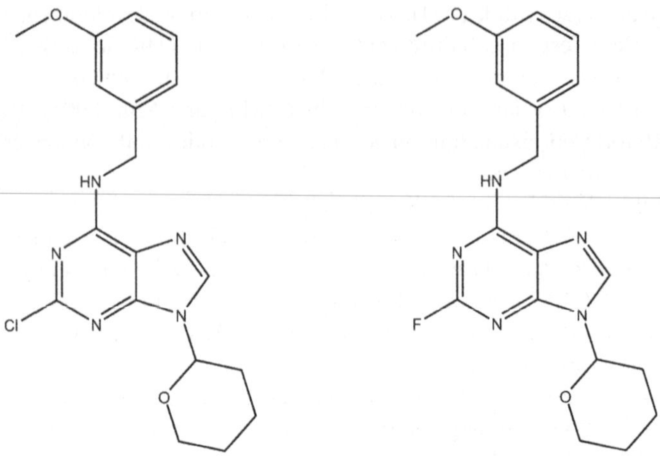

Fig. 21.1 Chemical structure of 2-chloro-6-(3-methoxyanilino)-9-(tetrahydropyran-2-yl)purine (INCYDE-THP) and 2-fluoro-6-(3-methoxyanilino)-9-(tetrahydropyran-2-yl)purine (INCYDE-F-THP)

2011; Podlešáková et al. 2012). 9-Substitution by a THP or THF group can also increase the anti-senescent activity of ARCKs (Young and Letham 1969; Szüčová et al. 2009).

The foregoing provided the motivation for preparing new 2,6,9-trisubstituted purine derivatives, 2-chloro-6-(3-methoxyanilino)-9-(tetrahydropyran-2-yl)purine (INCYDE-THP, Fig. 21.1) and 2-fluoro-6-(3-methoxyanilino)-9-(tetrahydropyran-2-yl)purine (INCYDE-F-THP, Fig. 21.1) derived from INCYDE and INCYDE-F and test their biological activities. Modulation of endogenous cytokinin levels with exogenous CKX inhibitors has already shown possible applications in agriculture and biotechnology, and new derivatives may thus lead to further improvements of their biological activities in planta.

21.1 Materials and Methods

21.1.1 General Procedures

Elemental analyses (C, H and N) were performed on an EA1108 CHN analyser (Fisons Instruments). The melting points were determined on a Büchi melting point B-540 apparatus. Analytical thin layer chromatography (TLC) was carried out using silica gel 60 WF254 plates (Merck) and mobile phase $CHCl_3$/MeOH (9:1, v/v). ES^+ mass spectra were recorded using direct probe on Waters ZMD 2000 mass spectrometer. The mass monitoring interval was 10–1500 amu. The spectra were collected by using 3.0 s cyclical scans and applying sample cone voltage 25 V at source block temperature 150 °C, desolvation temperature 80 °C and desolvation gas flow rate 200 L/h. The mass spectrometer was directly coupled to a MassLynx data system. NMR spectra were measured in a Bruker Avance AV 300 spectrometer operating at a temperature of 300 K and a frequency of 300.13 MHz (^{1}H) and 75.48 MHz (^{13}C), respectively. Samples were prepared by dissolving the compounds in DMSO-d_6. Tetramethylsilane (TMS) was used as the internal standard.

21.1.2 Chemicals

The starting material for the compound syntheses was 2,6-dichloro-9-(tetrahydropyran-2-yl)purine or 2,6-dichloro-9-(tetrahydrofuran-2yl)purine prepared from 2,6-dichloropurine and 3,4-dihydro-2*H*-pyran or 2,3-dihydrofuran in acidic conditions. Another starting material can be 2-fluoro-6-chloro-9-(tetrahydrofuran-2-yl)purine, synthesized from 2-amino-6-chloropurine by reaction with tetrafluoroboric acid in the presence of sodium nitrate aqueous solution (Kim et al. 2003) and subsequent reaction with 3,4-dihydro-2*H*-pyran or 2,3-dihydrofuran.

21.1.3 Synthesis

2-Chloro-6-(3-Methoxyanilino)-9-(Tetrahydropyran-2-yl)Purine (INCYDE-THP)

To a suspension of 2-chloro-6-(3-methoxyanilino)purine (10 g, 0.036 mol) in ethylacetate (250 mL) and 3,4-dihydropyran (10 mL, 0.1 mol), trifluoroacetic acid (10 mL, 0.13 mol) was added dropwise. The reaction mixture was then warmed up to 60 °C and stirred for 2 h at the same temperature. Then it was cooled to 10–15 °C and neutralized with 7 M methanolic ammonia (20 mL). The solvent was removed in a vacuum rotary evaporator, and the residue was partitioned between ethyl acetate (200 mL) and water (200 mL). The water phase was extracted with ethyl acetate (2 × 50 mL). The combined extracts were washed with brine (30 mL), dried with anhydrous sodium sulphate and evaporated to give a crude product (foam, 12.95 g, quant.). Purity: HPLC-MS min. 93%. The crude product was recrystallized from ethyl acetate–petrolether (1:10). Yield: 10.0 g (77.0%). HPLC purity: 97+ %. ^{1}H-NMR, DMSO-d_6, TMS, ppm: 1.58m (2H, THP); 1.72–1.78m (1H, THP); 1.98d (2H, THP); 2.18–2.31m (1H, THP); 3.65–3.75m (1H, THP); 3.76s (3H, CH_3); 4.02d (1H, THP); 5.63d (1H, THP); 6.66–6.70m (1H, Ar–H); $J_a = 8$; $J_b = 2.2$ Hz, 7.26t (1H, Ar–H); $J = 8$ Hz, 7.45d (H, Ar–H); $J = 8.0$ Hz, 7.58t (1H, Ar–H); $J = 2.2$ Hz, 8.54s (1H, C8), 10.27s (H, N6-H).

2-Fluoro-6-(3-Methoxyanilino)-9-(Tetrahydropyran-2-yl)Purine (INCYDE-F-THP)

2-Fluoro-6-(3-methoxyanilino)-9-(tetrahydropyran-2-yl)purine was synthesized using the same procedure as for INCYDE-THP (yield: 75%). This compound was prepared by the reaction of 2-fluoro-6-chloro-9-(tetrahydropyran-2-yl)purine (2.73 g, 0.01 mol), *m*-anisidine (1.23 g, 0.01 mol) and triethylamine (3.5 mL,

0.025 mol) in 2-propanol (20 mL) at 60 °C for 16 h. The reaction mixture was then cooled to room temperature. The white precipitate was filtered off, rinsed with *n*-butanol (3 × 10 mL) and water (3 × 10 mL) and dried in a drying oven to constant weight. Yield: 5.55 g (61%) of yellowish crystalline powder. TLC (chloroform-methanol, 85:15): one single spot. HPLC purity: 99+ %. ^{1}H-NMR, DMSO-d_6, TMS, ppm: 1.58m (2H, THP); 1.72–1.75m (1H, THP); 1.97d (2H, THP); 2.20–2.32m (1H, THP); 3.65–3.72m (1H, THP); 3.76s (3H, CH3); 4.01d (1H, THP); 5.60d (1H, THP); 6.69dd; J_a = 8.1 Hz; J_b = 2.2 Hz (1H, Ar–H), 7.26t; J = 8.1 Hz (1H, Ar–H), 7.45dd; J_a = 8.1 Hz; J_b = 2.2 Hz (1H, Ar–H), 7.53t; J = 2.2 Hz (1H, Ar–H), 8.51s (1H,C8H), 10.36bs (1H,N6H).

CKX Inhibition Assay

Pichia pastoris strain X-33 harbouring pGAPZα A::AtCKX2 vector grew in YPD medium for 2 days in the presence of Zeocin antibiotic, and the strain was then incubated for 3 days without any antibiotics. Cells were removed by centrifugation, and the prepared medium was used for CKX activity measurement. For measurement of the CKX activity in the presence of the tested compounds, a 96-well plate was used. 100 μL reaction solution consisted of 0.1 M KH_2PO_4 (pH 7.4), 1 mM phenazine methosulphate (PMS) and 0.2 mM 3-(4,5-dimethylthiazol-2-yl)-2,5-diphenyltetrazolium bromide (MTT) in a final concentration and tested compound in a range of 0.1–100 μM concentrations. iP (30 μM) was added to the reaction mixture as a substrate for *AtCKX2*. After pipetting all components, the plate was incubated for 30 min at 37 °C in dark, and after the incubation time, the reaction was stopped by 25 μL of 35% acetic acid per well. The absorbance was measured at 578 nm by a spectrophotometer (Synergy H4). Absorbance of the sample without iP was subtracted. The values shown are means of three replicates, and the entire test was repeated at least twice (Frébort et al. 2002).

Bacterial Receptor Assay

In the bacterial receptor assay, the *Escherichia coli* strain KMI001 was used. The strain harbours plasmids pIN-III-AHK4 and pSTV28-AHK3, which express the cytokinin receptors CRE1/AHK4 and AHK3 and fusion gene cps::lacZ, respectively. β-galactosidase activity was measured using a fluorescent substrate 4-methylumbelliferyl-β-D-galactoside. The OD_{600} of the remaining culture was determined, and β-galactosidase activity was calculated as nmol 4-methylumbelliferone × OD_{600}^{-1} × h^{-1}. The tested compounds were measured in three repetitions at least twice (Spíchal et al. 2004).

Reporter Gene Assay

We cultivated transgenic *Arabidopsis* plants harbouring ARR5::GUS reporter (D'Agostino et al. 2000) in MS medium containing cytokinin BAP, CKX inhibitors INCYDE and INCYDE-F and novel compounds INCYDE-THP and INCYDE-F-THP in concentrations 0.01, 0.1, 1, 5 and 10 μM; DMSO (0.1%) was tested as a solvent control. Seeds were surface-sterilized in 70% ethanol and then placed into wells of 6-well microtiter plate (TPP, Switzerland) containing 3 mL of MS medium in each well. After the incubation time, a plate was added on the shaker (22 °C, 450 rpm) for 3 days. The plants were incubated with the tested compounds for 16 h. Quantitative estimation of the level of ARR5/GUS gene induction was done according to the method published by Romanov et al. (2002). After extraction of proteins, the GUS activity was determined using incubation with fluorogenic substrate MUG (4-methylumbelliferyl glucuronide, 1 h, 37 °C), and then fluorescence at 365 and 460 nm (excitation and emission wavelengths) was recorded, as described in detail by Spíchal et al. (2004).

Cytokinin Bioassays

Standard bioassays based on the stimulation of cytokinin-dependent tobacco callus growth, the retention of chlorophyll in excised wheat leaves and the dark induction of betacyanin synthesis in *Amaranthus* cotyledons were carried out as previously described (Holub et al. 1998). Relative cytokinin bioassay activities of prepared derivatives at optimal concentration were compared with the activity of 6-benzylaminopurine (BAP) (100% means 10^{-6} M BAP for the tobacco callus bioassay, 10^{-5} M BAP for the *Amaranthus* betacyanin bioassay and 10^{-4} M BAP in the case of the senescence bioassay).

In Vitro Experiments

Sterile seeds of *A. thaliana*, ecotype Columbia, were added to the surface of ½ MS medium with the addition of the tested compounds. After 2 days in the dark, the plates were relocated to a growing chamber in a vertical position in regulated conditions (16 h light/8 h dark, 22 °C). Every 5 days, the length of the main root of each plant was monitored.

In Vivo Experiments

Arabidopsis thaliana, ecotype Columbia, was used for the experiments as a wild type. The plants grew in a growing chamber in regulated conditions (16 h light/8 h dark, 22 °C, 60% humidity). The seeds were planted into the growth substrate, and after germination and development of the first two real leaves, the tested compounds were applied by spraying (0.1, 1 and 10 μM) with addition of 0.02% detergent Silwet. In case of soaking (0.1, 1 and 10 μM), 1 mL of tested solution was pipetted onto the soil. DMSO was applied as a control. The treatment was usually applied until the stage of an adult plant, after which they were harvested and quantified. Finally, every plant was harvested for seeds, and 100 seed weight was monitored in the dry stage.

To test the effect on wild-type plants, *Arabidopsis* (ecotype Col-0) was grown in soil under controlled conditions in a growth chamber (16 h light/8 h dark, light intensity 120 μmol m^{-1} s^{-1}, 22 °C, 60% humidity). In the stage of approximately the middle of the principal growth stage 1 and the beginning of principal growth stage 5 (according to *Arabidopsis* growth stages described by Boyes et al. 2001), 1–10 mL of INCYDE-THP was applied regularly every third day in 0.1 and 1 μM concentration using different application rates (experiment 1, 1 mL of 1 μM INCYDE-THP water solution 12 times; experiment 2, 10 mL of 0.1 and 1 μM INCYDE-THP water solution 1, 3 and 5 times) by drenching each plant pot. DMSO (0.01%) was used as a solvent control. The plants were grown until the end of the generative stage when morphological parameters (primary inflorescence stem length, number of auxiliary branches, number of lateral branches, fresh shoot biomass, number of flowers, number of siliques) and seed yield were recorded.

21.1.4 Field Trials

Silage Maize

Based on the results, field plot experiments were performed on two different locations (Kroměříž and Olomouc, Czech Republic) in the years 2012 and 2013 to evaluate the effect of seed coating with INCYDE-THP on biomass formation of maize. In the season 2012, the maize seeds (hybrid Amadeo, KWS, FAO 230) were coated with 10 and 50 μM water solution of INCYDE-THP (8 L of solution per 1000 kg of seeds). They were dried and then sown at a rate of 90,000 seeds per hectare in 4 replications per each variant in randomized positions (each replication was 10 square meter plot). The plants were harvested when they reached the stage of 33–34% of dry mass. They were then analysed for moisture and standardized dry mass. The experiment was repeated in the season 2013 with 50 μM concentration of INCYDE-THP in the same experimental design with maize hybrid Zidane (KWS, FAO 250).

Winter Wheat

To evaluate the effect of seed coating with INCYDE-THP on morphological yield-forming traits of winter wheat, a field plot experiment was performed in 2012 in the of locality of Olomouc, Czech Republic. Seeds of winter wheat bakery variety Diadem (Selgen a.s, Czech Republic) were coated with 10 and 50 μM concentration of INCYDE-THP mixed with seed treatment agent Cruiser 350FS (Syngenta, 10 L of solution per 1000 kg of seeds); seeds coated with Cruiser 350FS only were used as a control variant. The seeds were dried and then sown at a rate of 3,500,000 seeds per hectare in 4 replications per each variant in randomized positions (each replication was 10 square meter plot). The plants were analysed for a number of productive tillers, medium tillers and weak tillers.

Freezing Tolerance of Winter Rapeseed

To evaluate the effect of seed coating with INCYDE-THP on yield-forming traits of winter rapeseed, a field, plot experiment was performed in the season of 2010/2011 at the locality of Opava, Czech Republic. Seeds of winter rapeseed variety Benefit (Selgen a.s., Czech Republic) were coated with 50 μM concentration of INCYDE-THP mixed with the seed coating treatment Cruiser OSR (Syngenta, 15 L of solution per 1000 kg of seeds); seeds coated with Cruiser OSR only were used as a control variant. The seeds were dried and then sown at the sowing rate of 700,000 seeds per hectare in 4 replications per each variant in randomized positions (each replication was 10 square meter plot). In this experiment, plant emergence, freezing tolerance, and final seed yield were scored. The frost tolerance was estimated by the method described by Janáček and Prášil (1991).

Container Experiment: Barley—Emergence Test of Treated Barley Seeds
In 2012, a container experiment was established at Holice, Olomouc, which was carried out in a cultivation chamber with controlled temperature, light and humidity regimes. INCYDE-F THP (labelled RR-B) was applied to the seeds at concentrations of 10 and 50 μM (calculated on the final concentration of the substance in the dressing, including water). The spring barley variety Amulet was also used for this experiment. The experiment was established on July 24, 2012, and the final evaluation took place after 21 days of the experiment, i.e. on August 14, 2012. Within each variant, 15 seeds were sown in a 165 mL planter; depth of sowing was 2.5 cm. The sown dry topsoil from the locality of Holice, Olomouc, and peat substrate for cultivation of nutmegs in the ratio of 70:30 were used for all the cultivations. The whole experiment was established in two variants according to the intensity of the dressing (experiments A and B). The "wet variant B" was watered 6 times during the experiment as 30 mL of water per plant at each watering; the so-called dry variant A was watered with a half-dose of water, i.e. by 15 mL at the same time (24.7, 26.7, 31.7, 3.8, 7.8 and 10.8). Plant cultivation was carried out in ClimaBoxes (Adaptis A1000, Conviron) under 16/8 h light/dark conditions, temperature 18 °C/15 °C and air humidity 60%. The height of the plants, the total weight of the individual plants and separately the weight of the shoots and roots were measured. The statistical programme PAST v. 2.08 was used for statistical evaluation of the experiment (significant differences of tested variants to the control variants). Between the individual variants and the control, the statistical significance of the monitored traits was calculated using ANOVA at the level of significance $P < 0.05$ and $P < 0.01$.

21.2 Results and Discussion

21.2.1 Chemistry

The first attempts in the preparation of 9-disubstituted THP cytokinin derivatives resulted in the mixtures of 7- and 9-derivatives, which generally contained preferentially the 9-isomer. For this reason, the synthetic approach had to be subjected to further modifications. As described here, *N9*-atom substitution can be performed in two steps using 2,6-dichloropurine (and/or 2-fluoro-6-chloropurine) as the starting material. In the first step, *N9* of purine is substituted by a selected protecting group, such as THP or THF. The second step is realized via nucleophilic substitution of the N^6-atom by the appropriate amine, similar to the preparation of free bases (Szüčová et al. 2009). A different situation arises in case of reactivity of 2-fluoro-6-chloropurine derivatives, where there is a low regioselectivity of the purine C_2/C_6 mixed halogen system (2-Cl and 2-F-purine positions have similar reactivity towards nucleophiles). To minimize reaction promiscuity, reaction temperatures have to be lower, a longer reaction time is required, and the final product yields are remarkably lower than for 2,6-dichloro-derivatives. Here we also report a basically higher yielding but not very versatile alternative approach, starting from mother INCYDE molecules that are transformed into their corresponding 9-THP or THF derivatives by reaction with 3,4-dihydro-2*H*-pyran or 2,3-dihydrofuran, resp., in strong acidic conditions, using trifluoroacetic acid as catalyst. The identity and purity of the prepared compounds were verified by ^{1}H NMR, mass spectrometry (chemical ionization), HPLC-UV and elemental analysis (Doležal et al. 2007).

21.2.2 Inhibition of AtCKX2 Enzyme

To assess the features of the compounds that underlay the inhibition of *At*CKX2, the CKX inhibition assay was used (Frébort et al. 2002). Classical cytokinin N^6-isopentenyladenine (iP) was chosen as a suitable substrate for the assay (Fig. 21.2). Substitution of the purine ring by phenyl side chain and halogen C2-substitution were found to be essential for strong *At*CKX2 inhibition (Zatloukal et al. 2008). The ability of the novel compounds to inhibit the activity of CKX was compared with that of thidiazuron, a weak inhibitor of CKX ($IC_{50} = 55$ μmol L^{-1}). More potent compounds than thidiazuron have IC_{50} values lower than thidiazuron. INCYDE and INCYDE-F are effective inhibitors of *At*CKX2 activity, which are approximately ten times stronger than thidiazuron. INCYDE-THP is only three times weaker than the structurally related compound INCYDE indicating that it can at least partially act as an original inhibitor of CKX activity in plant tissues and thus increase endogenous cytokinin levels. On the other hand, the visible difference in activity of INCYDE-F-THP ($IC_{50} = 380$ μmol L^{-1}) and its structurally related

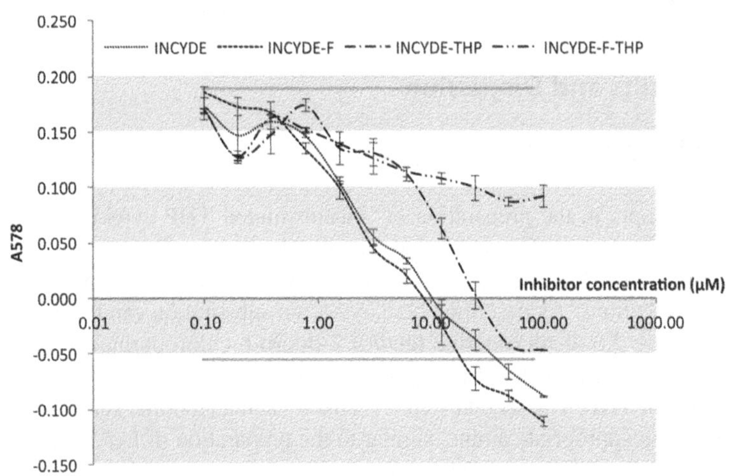

Fig. 21.2 The effect of novel compounds on inhibition of recombinant *At*CKX2. The IC50 value, the compound concentration that inhibited the enzyme activity to 50%, was calculated from the dose-response curves

CKX inhibitor INCYDE-F ($IC_{50} = 5.2\ \mu mol\ L^{-1}$) indicates that the biological activity of the compound may also be caused by other mechanisms than CKX inhibition.

21.2.3 The Receptor Binding Assay

Newly developed compounds have also been used for studying their binding to cytokinin receptors. Their affinity is compared to the naturally occurring isoprenoid cytokinin, *trans*-zeatin (*t*Z), which is recognized by the cytokinin receptors AHK3 and CRE1/AHK4 with a high affinity (EC_{50} ca. 1 μM). The curves in Fig. 21.3 show that compared to the cytokinin *trans*-zeatin (*t*Z), the novel compound INCYDE-THP has very low ability to activate the cytokinin receptors AHK3 and CRE1/AHK4, and INCYDE-F-THP does not activate any of the receptors. This indicates that if cytokinin-like activity is observed after application of the compounds, it is due to a mechanism different from direct activation of the cytokinin receptor pathway. The comparison with structurally related CKX inhibitors INCYDE and INCYDE-F also indicates that the novel compounds have original modes of action different from the CKX inhibitors. Interestingly, the AHK3 receptor bound the tested compounds with a somewhat higher affinity but much lower than that for *t*Z and lower than INCYDE and INCYDE-F (Fig. 21.3). This indicates that AHK3 has a broader specificity for very different ligands than CRE1/AHK4 (Spíchal et al. 2004; Doležal et al. 2006).

21.2.4 Reporter Gene Assay

In *Arabidopsis*, cytokinins induce transcription of gene ARR5, a member of the A-type response regulator family classified as cytokinin primary response gene and

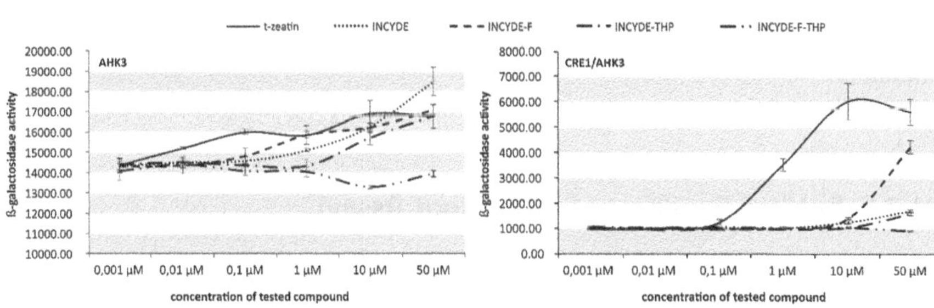

Fig. 21.3 Dose-dependent activation of cytokinin receptors AHK3 (left) and CRE1/AHK4 (right) in *Escherichia coli* receptor assay (Spíchal et al. 2004) by cytokinin *trans*-zeatin (*t*Z), inhibitors of cytokinin oxidase-dehydrogenase INCYDE (2-chloro-6-(3-methoxyanilino)purine), INCYDE-F (2-fluoro-6-(3-methoxyanilino)purine) and new compounds INCYDE-THP (2-chloro-6-(3-methoxyanilino)-9-(tetrahydropyran-2-yl)purine) and INCYDE-F-THP (2-fluoro-6-(3-methoxyanilino)-9-(tetrahydropyran-2-yl)purine). Values represent means of three repetitions; error bars represent SD

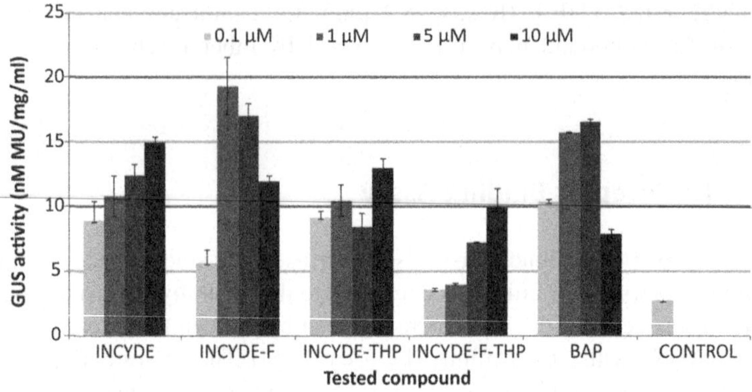

Fig. 21.4 Activation of cytokinin responsive reporter gene ARR5 GUS (Romanov et al. 2002) by cytokinin 6-benzylaminopurine (BAP) and inhibitors of cytokinin oxidase-dehydrogenase INCYDE, INCYDE-F, INCYDE-THP and INCYDE-F-THP. Bars represent means of three repetitions; error bars represent SD

expressed downstream of sensing by cytokinin receptors. We cultivated transgenic *Arabidopsis* plants harbouring ARR5::GUS reporter (D'Agostino et al. 2000) with BAP, INCYDE, INCYDE-F and novel compounds INCYDE-THP and INCYDE-F-THP, respectively. The activity of gene expression product β-glucuronidase (GUS) under the control of ARR5 promoter has been measured in transgenic *Arabidopsis* seedlings (Romanov et al. 2002). BAP was used as a positive control in this assay (Fig. 21.4). Quantitative assay showed INCYDE-THP and INCYDE-F-THP dose-dependent induction of ARR5/GUS expression. Taking into account that INCYDE-THP activated cytokinin receptors only weakly and INCYDE-F-THP fails to activate cytokinin receptors at all (see Fig. 21.3), it can be concluded that these compounds do not act as cytokinins directly, but after their application to the plants, the cytokinin pathway is activated indirectly. Comparison with activities of CKX inhibitors INCYDE and INCYDE-F also documents that INCYDE-THP and INCYDE-F-THP have different mode of action to these structurally related CKX inhibitors.

21.2.5 In Vitro Experiments (The Root Assay)

Development of the root system of *Arabidopsis* plants has been studied in root assays where different concentrations of prepared compounds are tested. For comparison of the biological activity of new compounds, classical aromatic cytokinin BAP is used as a standard. 35S:CKX1 *Arabidopsis* seedlings were *grown* in vitro on a standard MS medium containing 0.1 µM BAP, the same concentration of INCYDE-THP and the same concentration of CKX inhibitor INCYDE. In the case of BAP treatment, reduction in root length was visible in concentrations of 10 nM and higher. The optimal concentration for main root development is registered for

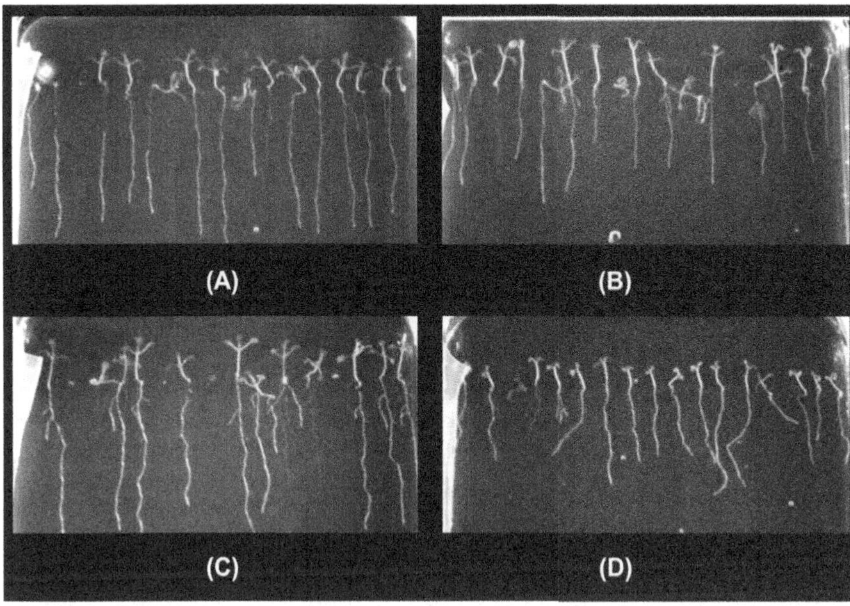

Fig. 21.5 Comparison of the effect of 0.1 μM CKX inhibitor INCYDE (b), 0.1 μM compound 12 (c, INCYDE-THP) and cytokinin BAP (d) on wild-type *Arabidopsis* root growth and development in vitro compared to control (a, DMSO)

1 and 5 nM concentrations of BAP. The presence of INCYDE-THP in the growth medium had positive effect on the root growth of cytokinin-deficient 35S:CKX1 *Arabidopsis* plants. This effect was further studied using wild-type *Arabidopsis* plants and compared with biologically and structurally relevant compounds, a cytokinin BAP and CKX inhibitor INCYDE. Figure 21.5 clearly shows that whereas cytokinin BAP and CKX inhibitor INCYDE predictably inhibited root growth, application of INCYDE-THP showed positive effect on the root length and lateral branching.

21.2.6 In Vitro Effects of INCYDE-THP on Shoot and Root Growth of Cytokinin-Deficient *Arabidopsis* Plants

To prove that INCYDE-THP has a positive effect on plant growth and development through inhibition of cytokinin CKX degradation enzyme (i.e. by increasing and stabilizing plant endogenous cytokinin levels) and that phenotype complementation is not a general effect of exogenously applied cytokinin, CKX1-overproducing (35S: CKX1) *Arabidopsis* seedlings were grown in vitro on a standard MS medium containing 0.1 μM BAP and the same concentration of INCYDE and INCYDE-THP, respectively. Figure 21.6 clearly shows that the application of INCYDE-THP as well as CKX inhibitor INCYDE released the plants from growth inhibition caused

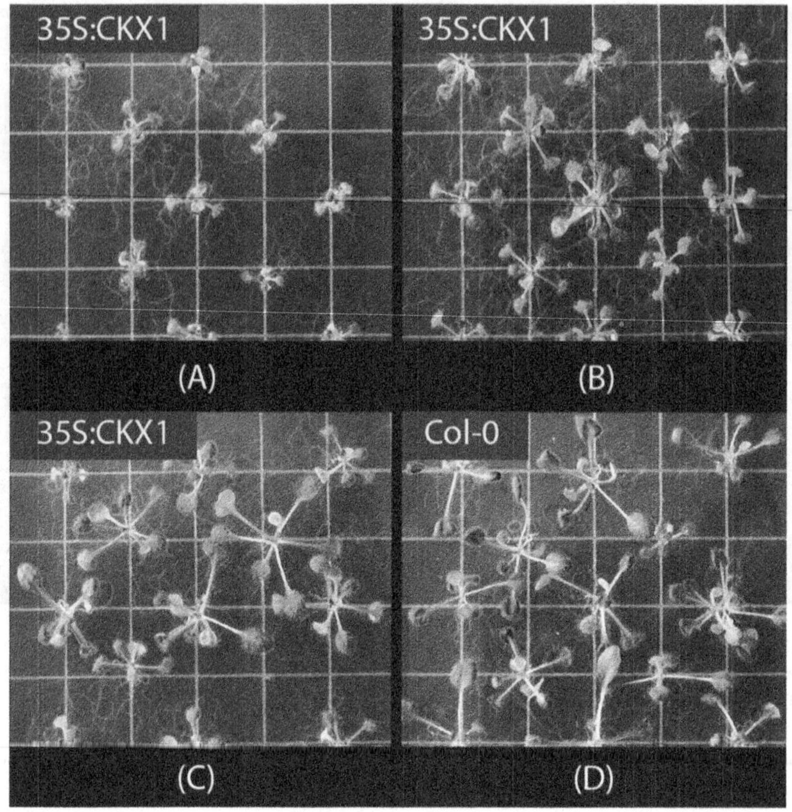

Fig. 21.6 The effect of INCYDE-THP and CKX inhibitor INCYDE application on complementation of the wild-type phenotype of AtCKX1 *Arabidopsis* seedlings. (**a**) *At*CKX1 control, (**b**) *At*CKX1 cultivated on 0.1 μM INCYDE-THP, (**c**) *At*CKX1 cultivated on 0.1 μM INCYDE, (**d**) control plants with wild-type phenotype

by decreased cytokinin level and led to restoration of wild-type shoot phenotype in a dose-dependent manner (Fig. 21.6). Such an effect cannot be achieved by the use of the cytokinin BAP.

To further prove that the action of INCYDE-THP is different from structurally similar CKX inhibitor INCYDE, the effect of the two compounds on the root growth of the 35S:CKX1 *Arabidopsis* plants was assessed. As shown in Fig. 21.7, whereas INCYDE had no influence on root biomass in the tested concentration range, INCYDE-THP had a strong dose-dependent positive effect. Constitutive overexpression of CKX in 35S:CKX1 plants leads to enhanced root growth. If a CKX inhibitor can restore WT phenotype, it is expected that inhibition of CKX activity in 35S:*At*CKX1 plants will have no effect or inhibit root growth. Such observation was made in the case of INCYDE, but not INCYDE-THP. Albeit INCYDE-THP can inhibit CKX activity (Fig. 21.7), the results shown in this

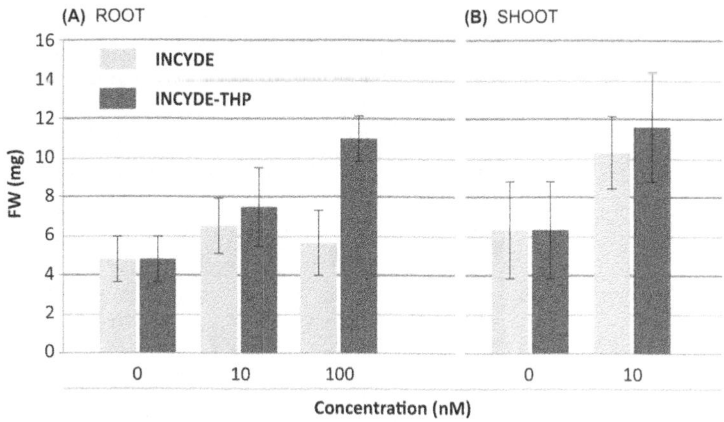

Fig. 21.7 (**a**) The effects of INCYDE-THP and INCYDE on stimulation of root biomass in *At*CKX1 *Arabidopsis* plants. (**b**) Comparison of the effect of 0.1 μM CKX inhibitor INCYDE and 0.1 μM INCYDE-THP on *At*CKX1 *Arabidopsis* shoot growth and development in vitro. Bars represent means of fresh weight of roots/shoot of treated plants; error bars indicate SD ($n = 30$)

example clearly indicate that the biological activity of INCYDE-THP is different from that of the specific CKX inhibitor INCYDE.

21.2.7 In Vivo Experiments

The tested compounds were also studied on *A. thaliana* plants in vivo. Different compounds, concentrations, application modes and doses were tested. Different phenotypic parameters such as shoot length, shoot weight, number of branches, number of flowers and siliques were analysed for each experiment. Finally, the density of the root system was also evaluated. Spraying of compounds was used at the beginning of phenotypic analysis. No significant positive effect of compound application on *Arabidopsis* shoot and root enlargement was observed. Enlargement of shoot biomass and higher number of branches of the main stem were, however, observed as a result of the positive effect of the compound applied as drenching. INCYDE-THP in the growth medium of in vivo-grown *Arabidopsis* plants showed a positive effect on both shoot and root growths (see Fig. 21.8).

The initial experiment 1 showed that regular drenching with 1 μM INCYDE-THP leads to the stimulation of the overall shoot biomass (Fig. 21.8). In experiment 2, the effect of drenching in two concentrations (0.1 and 1 μM) of INCYDE-THP was tested in three application rates (1, 3, 5 times). All had stimulatory effect on the overall shoot biomass and significantly increased the morphological parameters (Table 21.1). To summarize, higher doses increased the primary inflorescence stem length by 16–34%, number of auxiliary branches by 6–25%, number of lateral branches by 77–249%, fresh shoot biomass by 6–22%, number of flowers by

Fig. 21.8 The effect of INCYDE-THP applied by drenching on shoot growth and phenotypic development of *Arabidopsis* plants

20–89%, number of siliques by 5–66% and seed yield by 6–26%, compared to control, respectively.

The data indicate that drenching application with INCYDE-THP significantly increases plant performance leading to the increase in shoot architecture, biomass and total seed yield in *Arabidopsis* model plant. Other positive effects of the compound were large number of flowers and siliques for each *Arabidopsis* plant and larger number of seeds from each silique and greater weight per 1000 seeds compared to control plants (Fig. 21.9).

All the results presented here led to conclusions similar to those obtained using the different approaches of molecular biology described in previous *Arabidopsis* publications (Galuszka et al. 2004; Ashikari et al. 2005; Sharma et al. 2008; Bartrina et al. 2011) (Fig. 21.10).

Table 21.1 Phenotypic analysis of *Arabidopsis*-treated plants by INCYDE-THP. The experiment was repeated three times with comparable results. Thirty plants were used in each variant. Asterisks indicate statistical significance $*P < 0.05$, $*P < 0.01$, $***P < 0.001$ (Student's *t*-test, $n = 30$)

		0.1 μM			1 μM		
Phenotypic trait	Control	1×	3×	5×	1×	3×	5×
Primary stem length (cm)	33.26 ± 3.4	32.78 ± 5.5 (98%)	32.6 ± 3.8 (119%)***	38.61 ± 5.2 (116%)***	32.66 ± 3.8 (98%)	39.07 ± 3.2 (117%)***	**44.66 ± 5.9 (134%)*** **
Number of side branches	3.59 ± 0.7	4.06 ± 1.3 (113%)	3.82 ± 0.6 (106%)	3.94 ± 1.1 (109%)	4.29 ± 1.5 (119%)	4.29 ± 1.1 (119%)*	**4.50 ± 1.3 (125%)**
Number of lateral branches	4.71 ± 1.8	8.59 ± 3.1 (182%)***	8.35 ± 3.3 (177%)***	9.82 ± 4.4 (208%)***	13.35 ± 4.7 (283%)***	13.50 ± 3.9 (286%)***	**16.44 ± 1.3 (349%)*** **
Number of flowers	75.71 ± 27.8	116.12 ± 70 (153%)*	91.23 ± 50.1 (120%)	99.59 ± 59.3 (131%)	139.65 ± 51.7 (184%)***	118.53 ± 36.6 (156%)***	**143.56 ± 66.3 (189%)*** **
Number of pods	133.597 ± 27.8	133.94 ± 47 (100%)	154.29 ± 25.6 (115%)	141.24 ± 31.7 (105%)	134.82 ± 31.1 (100%)	155.29 ± 44.1 (116%)	**220.00 ± 37.4 (164%)*** **
Fresh shoot biomass (g)	1.09 ± 0.3	1.18 ± 0.4 (108%)	1.16 ± 0.2 (106%)	1.18 ± 0.3 (108%)	1.17 ± 0.3 (107%)	1.23 ± 0.3 (113%)	**1.81 ± 0.3 (166%)*** **
Seed yield (g) per plant	**0.29 ± 0.2**	**0.30 ± 0.1 (102%)**	**0.31 ± 0.1 (106%)**	**0.35 ± 0.1 (120%)**	**0.37 ± 0.1 (126%)**	**0.36 ± 0.1 (122%)**	0.33 ± 0.1 (111%)

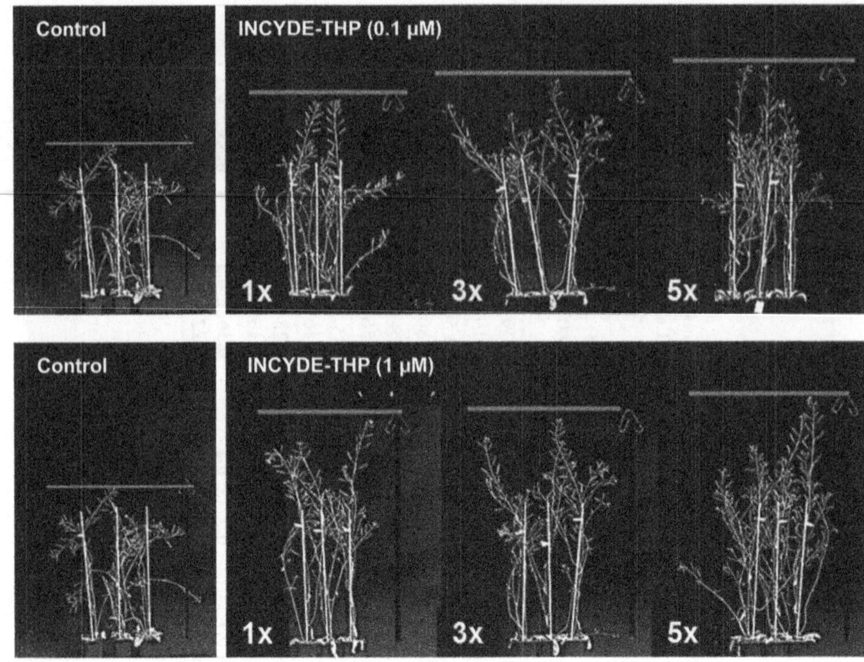

Fig. 21.9 The effect of INCYDE-THP applied by drenching in different application rates on shoot growth and biomass development of *Arabidopsis* plants

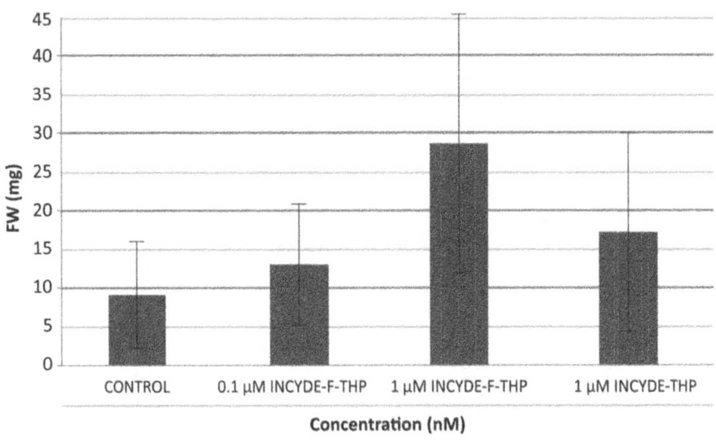

Fig. 21.10 The effect of INCYDE-THP applied by drenching in different application rates on seed yield of *Arabidopsis* plants

21.3 Field Testing of INCYDE-THP

21.3.1 Effect of Seed Coating with INCYDE-THP on Shoot Biomass of Silage Maize

The results from all field plot experiments are summarized in Table 21.2. The data show that the seed treatment with INCYDE-THP positively influenced the dry biomass of maize without any effect on the plant height. The average silage yield counted from all trials shows 110.9% increase in the case of a concentration of 10 μM (counted from two trials) and 105.3% for a concentration of 50 μM (counted from three trials), respectively. These data indicate that INCYDE-THP can be used to advantage for seed applications to improve the biomass of crops such as maize in field conditions.

21.3.2 Effect of Seed Coating with INCYDE-THP on Tillering of Winter Wheat

The data summarized in Table 21.3 show that seed treatment with INCYDE-THP led to significant dose-dependent increase in number of productive tillers and reduction of medium and weak tillers, respectively. The number of productive tillers was increased to 106–124% compared to untreated control. These data indicate that

Table 21.2 Summary of effects of on-seed application of INCYDE-THP on maize silage yield obtained in field plot experiments performed at two localities in the years 2012 and 2013 with maize hybrids Amadeo (KWS, FAO 230) and Zidane (KWS, FAO 250). Average silage yield of control was 50.95 t/ha (in Olomouc) and 49.60 t/ha (in Kroměříž), respectively. Asterisks indicate *P* value 0.01–0.05

Variant	Absolute green matter/dry matter yield	Relative dry matter yield (% of control)	Height before harvest (cm/% of control)	Location, year
INCYDE-THP (50 μM)	68.78/19.23	101.33	246.4/100.8	Olomouc, 2013
	66.47/22.33	107.82*	–	Kroměříž, 2013
INCYDE-THP (10 μM)	43.20/17.01	110.94	237.3/98.3	Olomouc, 2012
INCYDE-THP (50 μM)	44.31/17.80	116.06	235.8/97.7	Olomouc, 2012
INCYDE-THP (10 μM)	52.19/23.06	110.85	276.6/101.0	Kroměříž, 2012
INCYDE-THP (50 μM)	49.52/22.18	106.63	273.5/99.8	Kroměříž, 2012

Table 21.3 Summary of effect of seed application of INCYDE-THP on tillering of winter wheat. Asterisks indicate significant differences, $^{**}P \geq 0.01$ and $^{*}P = 0.01–0.05$

Variant	Number of productive tillers (% of control)	Number of medium tillers (% of control)	Number of weak tillers (% of control)	Location, harvest year
Control	2.33 tillers	1.60 tillers	1.15 tillers	Olomouc, 2012
INCYDE-THP (10 μM)	106.5	75.0*	102.2	Olomouc, 2012
INCYDE-THP (50 μM)	124.2**	78.1*	80.4	Olomouc, 2012

Table 21.4 Summary of the effect of seed application of INCYDE-THP on yield-influencing traits and final seed yield of winter rapeseed. Seed yield was counted as the mean of four plots; other traits were counted from 15 randomly chosen plants. Asterisks indicate significant differences, $^{**}P \geq 0.01$ and $^{*}P = 0.01–0.05$

Trait	Control	INCYDE-THP (50 μM)
Seed yield (t/ha)	5.80	5.98
Seed yield (% of control)	100	103.1
Leaf weight—autumn (October 22)	17.24	11.69**
Root length (cm)—autumn (October 22)	14.03	14.63
Number of emerged plants per sq. meter—autumn (October 11, 2010)	49.5	60.0
Plant height (cm—autumn (October 11, 2010)	6.0	6.6
Estimated frost resistance (artificial frost test—lethal temperature LT50	−11.7	−12.3*

INCYDE-THP can be used to advantage for seed applications to improve the tillering as an important yield-forming trait of cereals such as winter wheat in field conditions.

21.3.3 Effect of Seed Coating with INCYDE-THP on the Emergence and Freezing Tolerance of Winter Rapeseed

To evaluate the effect of seed coating with INCYDE-THP on the yield-forming traits of winter rapeseed, a field plot experiment was performed. The results summarized in Table 21.4 show that seed coating of INCYDE-THP stimulated plant emergence (approximately 20% more plants emerged in the same time compared to control). The treatment significantly decreased leaf biomass compared to control, whereas it had a slightly stimulatory effect on the root length, leading to improved ratio between shoot and root towards the root system. The artificially estimated frost tolerance was significantly increased by approximately 5%. The treatment in this

Table 21.5 Seedling test—emergence of treated barley seeds. Experiment A: water stress (half-water regime). Experiment B: optimal watering. Asterisks indicate significant differences, ** $P \geq 0.01$ and *$P = 0.01–0.05$

Variant	Plants length (cm) at date			Plants weight		Roots weight		Leaves weight	
	3.8.	9.8.	14.8.	Absolute value (g)	Relative (% of control)	Absolute value (g)	Relative (% of control)	Absolute value (g)	Relative (% of control)
Experiment A: control	8.13	19.40	25.81	1.32	**100.00**	0.80	**100.00**	0.52	**100.00**
INCYDE-F-THP (50 μM)	10.78	21.82	26.87	2.38**	**180.23****	1.47**	**183.65****	0.92**	**175.00****
Experiment B: control	16.42	28.27	31.79	3.03	**100.00**	1.75	**100.00**	1.28	**100.00**
INCYDE-F-THP (50 μM)	14.50	24.90*	28.55**	2.75	**90.69**	1.70	**97.14**	1.05**	**81.86****

trial increased the final seed yield by 3.1%. These data indicate that INCYDE-THP can be used to advantage for seed applications to improve the winter tolerance and final seed yield of winter rapeseed in field conditions.

21.3.4 Emergence Test of Treated Barley Seeds

A container experiment analysing the root and shoot growth of treated barley seeds was carried out in a cultivation chamber with controlled temperature, light and humidity regimes. INCYDE-F THP was applied as seed coating at 10 and 50 μM concentrations, respectively. The whole experiment was established in two variants according to the intensity of the dressing (experiments A and B). The "dry variant A" was watered with half-dose of water six times during the experiment (15 mL of water per plant), while "wet variant B" was watered with 30 mL at each watering. 10 μM was not effective, and thus only the data from the 50 μM INCIDE-F-THP experiment are included here. The substance was not effective in stimulating barley germination and its subsequent growth under normal water regime conditions; it was even slightly inhibitory in terms of the generally reduced shoot and root weight in seedlings. INCIDE-THP, in contrast, showed a strong stimulatory effect on plant germination and subsequent growth under drought stress conditions, as shown in Table 21.5. An unusually strong stimulating effect on shoot and root weight was observed (75–80% over the control).

21.4 Conclusions

In summary, we have prepared new ARCK-derived inhibitors of *At*CKX2. According to the biological data analyses, these extensively studied compounds are comparable to the CKX inhibitor thidiazuron, a synthetic cytokinin with strong biological activity. Their sensing by cytokinin receptors was, however, much weaker. Finally, from in vivo experiments, the seed yield and shoot/root biomass were increased using cytokinin analogues to *Arabidopsis* plants.

Acknowledgements The authors of this review are grateful to Alexander Oulton, PhD., for his critical reading of the chapter. This research was supported by Grant No. 18-07563S from the Czech Science Foundation and ERDF project "Plants as a tool for sustainable global development" (No. CZ.02.1.01/0.0/0.0/16_019/0000827).

References

Aremu AO, Bairu MW, Szüčová L, Finnie JF, Van Staden J (2012) The role of meta-topolins on the photosynthetic pigment profiles and foliar structures of micropropagated "Wiliams" bananas. J Plant Physiol 169:530–1541

Aremu AO, Stirk WA, Masondo NA, Plačková L, Novák O, Pěnčík A, Zatloukal M, Nisler J, Spíchal L, Doležal K, Finnie JF, van Staden J (2015) Dissecting the role of two cytokinin analogues (INCYDE and PI-55) on in vitro organogenesis, phytohormone accumulation, phytochemical content and antioxidant activity. Plant Sci 238:81–94

Ashikari M, Sakakibara H, Lin S, Yamamoto T, Takashi T, Nishimura A, Angeles ER, Qian Q, Kitano H, Matsuoka M (2005) Cytokinin oxidase regulates rice grain production. Science 309:741–745

Bairu MW, Novák O, Doležal K, van Staden J (2011) Changes in endogenous cytokinin profiles in micropropagated *Harpagophytum procumbens* in relation to shoot-tip necrosis and cytokinin treatments. Plant Growth Regul 63:105–114

Bartrina I, Otto E, Strnad M, Werner T, Schmülling T (2011) Cytokinin regulates the activity of reproductive meristems, flower organ size, ovule formation, and thus seed yield in *Arabidopsis thaliana*. Plant Cell 23:69–80

Bilyeu KD, Cole JL, Laskey JG, Riekhof WR, Esparza TJ (2001) Molecular and biochemical characterization of a cytokinin oxidase from maize. Plant Physiol 125:378–386

D'Agostino I, Deruère J, Kieber JJ (2000) Characterization of the response of the Arabidopsis response regulator gene family to cytokinin. Plant Physiol 124:1706–1717

Doležal K, Popa I, Hauserová E, Spíchal L, Chakrabarty K, Novák O, Kryštof V, Voller J, Holub J, Strnad M (2007) Preparation, biological activity and endogenous occurrence of N^6-benzyladenosines. Bioorg Med Chem 15:3737–3747

Frébortová J, Galuszka P, Werner T, Schmülling T, Frébort I (2007) Functional expression and purification of cytokinin dehydrogenase from *Arabidopsis thaliana* (AtCKX2) in *Saccharomyces cerevisiae*. Biol Plant 51:673–682

Galuszka P, Frébortová J, Werner T, Yamada M, Strnad M, Schmülling T, Frébort I (2004) Cytokinin oxidase/dehydrogenase genes in barley and wheat. Cloning and heterologous expression. Eur J Biochem 271:3990–4002

Gemrotová M, Kularni MG, Stirk WA, Strnad M, van Staden J, Spíchal L (2013) Seedlings of medicinal plants treated with either a cytokinin antagonist (PI-55) or an inhibitor of cytokinin degradation (INCYDE) are protected against the negative effects of cadmium. Plant Growth Regul 71:137–145

Janáček J, Prášil I (1991) Quantification of plant frost injury by nonlinear fitting of an s-shaped function. Cryo Lett 12(7):47–52

Kim HS, Ohno M, Xu B, Kim HO, Choi Y, Ji XD, Maddileti S, Marquez VE, Harden TK, Jacobson KA (2003) 2-Substitution of adenine nucleotide analogues containing a bicyclo[3.1.0]hexane ring system locked in a northern conformation: enhanced potency as P2Y1 receptor antagonists. J Med Chem 46:4974–4987

Mok MC (1994) Cytokinins and plant development—an overview. In: Mok DWS, Mok MC (eds) Cytokinins. Chemistry, activity and function. Boca Raton, CRC Press, pp 155–166

Mok DWS, Mok MC (2001) Cytokinin metabolism and action. Annu Rev Plant Physiol Plant Mol Biol 52:89–118

Podlešáková K, Zalabák D, Čudejková M, Plíhal O, Szüčová L, Doležal K, Spíchal L, Strnad M, Galuszka P (2012) Novel cytokinin derivatives do not show negative effects on root growth and proliferation in submicromolar range. PLoS One 7:39293

Romanov GA, Jieber JJ, Schmülling T (2002) A rapid cytokinin response assay in *Arabidopsis* indicates a role for phospholipase D in cytokinin signalling. FEBS Lett 515:39–43

Schmülling T, Werner T, Riefler M, Krupková E, Bartrina I (2003) Structure and fiction of cytokinin oxidase/dehydrogenase genes of maize, rice, *Arabidopsis* and other species. J Plant Res 116:241–252

Spíchal L (2012) Cytokinins – recent news and views of evolutionally old molecules. Funct Plant Biol 39:267–284

Spíchal L, Rakova NY, Riefler M, Mizuno M, Romanov GA, Strnad M, Schmülling T (2004) Two cytokinin receptors of *Arabidopsis thaliana*, CRE1/AHK4 and AHK3, differ in their ligand specificity in a bacterial assay. Plant Cell Physiol 45:1299–1305

Szüčová L, Spíchal L, Doležal K, Zatloukal M, Greplová J, Galuszka P, Kryštof V, Voller J, Popa I, Massino FJ, Jorgensen JE, Strnad M (2009) Synthesis, characterization and biological activity of ring-substituted 6-benzylamino-9-tetrahydropyran-2-yl and 9-tetrahydrofuran-2-ylpurine derivatives. Bioorg Med Chem 17:1938–1947

Werner T, Schmülling T (2009) Cytokinin action in plant development. Curr Opin Plant Biol 12:527–538

Werner T, Motyka V, Strnad M, Schmülling T (2001) Regulation of plant growth by cytokinin. Proc Natl Acad Sci U S A 98:10487–10492

Young H, Letham DS (1969) 6-(Substituted amino)purines: synthesis by a new method and cytokinin activity. Phytochemistry 8:1199–1203

Zatloukal M, Gemrotová M, Doležal K, Havlíček L, Spíchal L, Strnad M (2008) Novel potent inhibitors of *A. thaliana* of cytokinin oxidase/dehydrogenase. Bioorg Med Chem 16:9268–9275

Topolins and Related Compounds: Uses in Agriculture 22

Radoslav Koprna, Marek Zatloukal, and Karel Doležal

Contents

Abstract

Cytokinins are among the phytohormones essential for a number of developmental processes, a key one being cell division. In relation to crop plant productivity, these processes include tillering, grain setting, filling, nutrient uptake, pathogen resistance, and leaf senescence. There is a close positive correlation between cytokinin levels and cell division phase, but data from cytokinin application to field cereals are inconsistent due to the limited phases of cell division, among others. Further, despite extensive testing, cytokinin use as a commercial plant growth regulator falls short of prediction mainly due to the complexity of field trial effects. This chapter gives an overview of the results of pot and field experiments after exogenous application of a new aromatic cytokinin, 2-chloro-6-(3-methoxybenzylamino)purine (2ClMe*m*T), which we refer to as a regreening factor due to its marked antisenescent activity. Testing was done on winter wheat and spring barley. The effects on yield-producing properties was determined and

R. Koprna (✉) · M. Zatloukal · K. Doležal
Department of Chemical Biology, Faculty of Science, Palacky University, Olomouc, Czech Republic
e-mail: radoslav.koprna@upol.cz

N. Ahmad, M. Strnad (eds.), *Meta-topolin: A Growth Regulator for Plant Biotechnology and Agriculture*, https://doi.org/10.1007/978-981-15-9046-7_22

compared with crop spraying at different phenological stages. The data from several years were collected and statistically evaluated.

Keywords

6-Benzylaminopurine · Topolins · Cytokinin · Seed development · Crop yield · Agricultural applications

22.1 Introduction

Natural cytokinins can be chemically characterized as N^6-substituted adenines that affect several aspects of plant growth and development, which include cell division, nutrient uptake, leaf senescence, sink/source relationships, apical dominance, shoot initiation, differentiation and growth, phyllotaxis, and vascular, gametophytic, and embryonic development, as well as response to biotic and abiotic stress factors (Kieber and Schaller 2014). They are usually divided into isoprenoid cytokinins, represented by *cis*- and *trans*-zeatin (*c*Z, *t*Z), N^6-isopentenyladenine (iP), and dihydrozeatin (DHZ), the most abundant types; second, naturally occurring adenines with N^6-aromatic substituents, for example, 6-benzylaminopurine (BAP), kinetin (K), and topolins, less abundant and highly active; and synthetic diphenylureas, represented by thidiazuron (TDZ) and CPPU (4-chlorfenuron), which are the third category. Natural cytokinins occur as nucleobases, nucleosides, and nucleotides, and their homeostasis is maintained through biosynthesis and metabolism, conjugation, and degradation of the bioactive cytokinins (Werner and Schmülling 2009).

22.2 Cytokinins and Seed Development

Cytokinins, together with auxins and other phytohormones, control the growth and developmental processes that are critical to crop plant productivity. These growth phases include tillering, creation, and maintenance of the photosynthetic apparatus, deposition and filling of grains, and acquisition of basic resources, including carbon skeleton, mineral nutrients, and their effective assimilation and allocation. A close positive correlation between cell division phases and cytokinin levels has been demonstrated in developing fruits (e.g., Letham and Williams 1969; Bohner and Bangerth 1988; Lewis et al. 1996) and seeds, since the beginning of cytokinin research. Cereals (maize, wheat, rice, and barley) have also been shown to have sharp, transiently elevated endogenous levels of cytokinins immediately after anthesis (Morris et al. 1993). For example, in wheat, there is a sharp increase in endogenous cytokinin levels during the endosperm growth in the developing grains (Bennett et al. 1973), a phase crucial for determining the final sink size (Jameson et al. 1982). Furthermore, the complex regulation of endogenous cytokinins in leaves, spikes, and developing grains in wheat occurs with differential input of various members of the cytokinin gene family (Song et al. 2012). Similar findings have been reported in maize (Dietrich et al. 1995; Brugière et al. 2008) and rice

(Yang et al. 2003). In addition, the application of exogenous cytokinins prevented flower abortion/pod set and increased yields in lupin and soybean (e.g., Carlson et al. 1987; Atkins and Pigeaire 1993; Nagel et al. 2001; Nonokawa et al. 2012). On the other hand, the application of exogenous cytokinins to legumes has been less successful in the field (Dyer et al. 1987; Nagel et al. 2001), and therefore it is necessary to repeat these experiments and applications (Atkins et al. 2011; Nonokawa et al. 2012), but the agricultural use of cytokinins is likely to be unsuccessful due to high costs and impracticality. Yang et al. (2003) showed increased cell division and grain weight in field rice after application of kinetin for 5 days, starting on the second day after anthesis, on plants in variable stages of tiller development. Interestingly, in several cases, the inverse correlation between seed size and their number and increased cytokinin levels has been demonstrated (Sadras 2007; Paul-Victor and Turnbull 2009; Van Daele et al. 2012). In cereal field trials, there are highly inconsistent responses due to a narrow window of opportunities associated with the limited phase of cell division and activation of cytokinin degradation, their biosynthesis, etc. Taken together, the results suggest that cytokinins limit yield in both monocots and dicots, and their field applications are unlikely to be commercially viable (Jameson and Song 2016).

Historically, the increase in cereal yields was due to an increased harvest index (HI) rather than an increase in total biomass. However, the HI of many cereals is considered to reach a plateau, and therefore further increase in yield may most likely require an increase in plant biomass production through an increase in net photosynthesis production (Spano et al. 2003). This could be achieved by extending the operating time of the photosynthetic apparatus, delaying the senescence of the source leaves, extending the carbon uptake time, and delaying nitrogen remobilization. These are properties of typical plant development that are influenced by cytokinins (Thomas and Ougham 2014), and therefore investment in this area has considerable potential.

22.3 Aromatic Cytokinins in Agriculture and Biotechnology

Aromatic cytokinins are most commonly used in agriculture. One successful aromatic cytokinin (ARCK) is 6-benzylaminopurine (BAP). BAP affects plant growth and development processes, flower setting, and fruit quality by stimulating cell division (Gupta et al. 2012; Harrison and Kaufman 1980). The use of individual ARCKs increased grain size and their formation and subsequently improved the final yield of cereals. However, their positive effects were usually conditioned by other factors, such as the selected species, the method of cytokinin application, the plant developmental phase, and/or the environmental conditions of growth. In most studies, ARCKs were applied by irrigation or spraying, and treatments were performed in a wide range of developmental phases from germination to anthesis (Criado et al. 2009; Woodward and Marshall 1988; Koprna et al. 2016). Furthermore, in wheat, injections of BAP into the base of mother shoots during anthesis improved seed weight and yield when plants were grown under normal conditions;

however, this effect was genotype dependent during late sowing (Gupta et al. 2003). Seed priming with BAP or K stimulated grain production and yield in wheat and barley at high salinity (Afzal et al. 2013; Sarwat and El-Sherif 2007). In rice and maize, seed priming improved with K application, which also increased seed germination capacity (Wang et al. 1996). The effect of CK application on another yield-forming trait, i.e., tillering, was also studied in cereals. Langer et al. (1973) found that K promoted bud elongation in wheat. Recent studies on this crop also indicate BAP is an inductor of a more effective period of tillering (Alizadeh et al. 2010). Similar results have also been shown for other cereals, such as barley and oat, using various ARCKs (Harrison and Kaufman 1980; Suge and Iwamura 1993). Seed priming or spraying by cytokinins in the early developmental phases has also shown a significant stimulatory effect, especially under various abiotic stresses (Wang et al. 1996; Khan and Srivastava 1998). However, no effect on growth and tillering was observed in rice after BAP application, but an increase in yield was observed under salinity conditions (Liu et al. 2001). On the other hand, Zahir et al. (2001) reported a positive effect of K applied at several rice phenological phases, from early development to reproductive stages, on homogenous branch development and grain yield. Patel and Mohapatra (1992) also reported a positive effect of CK treatment on development of homogenous branch and grain yield in rice. Similar exogenous applications of ARCKs have also been tested on C4 cereals, such as corn and sorghum. Different CK treatment regimes applied in different phenological phases have been shown to improve germination, biomass production, and grain yield (Amin et al. 2007; Kaya et al. 2010; Pan et al. 1999). Furthermore, BAP-treated peach seedlings in general had larger tops, and therefore the leaf content would be expected to exhibit a reduced nitrogen percentage based on dry weight. BAP used on the leaves to some extent overcame the reduced growth of restricted plants, but its main effect was to affect the distribution of dry matter. BAP applications, in general, reduced mineral nutrient levels in leaves (Richards and Rowe 1977). The impact of the application of the cytokinin BAP in combination with other methods on the retention of green discoloration of broccoli heads and asparagus spears has shown positive results to maintain quality. This BAP application to the carpel delayed also chlorophyll degradation in the sepals, stimulated carpel growth, and, when mature heads of broccoli were used, stimulated emergence of petals (Clarke et al. 1994). A detailed analysis of the various ARCK applications can also be found in our review (Koprna et al. 2016).

22.4 Topolins: Their Biological Activity and In Vivo Plant Application

A new era began with the discovery topolin compounds and their relatives termed methoxytopolins (Strnad 1997; Tarkowská et al. 2003). The first biological testing of hydroxybenzyladenines (later, topolins) was carried out by the Japanese researchers who at first synthesized a wide range of kinetin analogues which included, among others, the *ortho*-, *meta*-, and *para*-substituted 6-(hydroxybenzylamino)purines

(Kuraishi 1959; Okumura et al. 1959). The series also contained *meta*-hydroxy-BAP, which was later termed *meta*-topolin (*m*T, Strnad et al. 1997). The compounds were only tested in the radish leaf discs expansion assay (Kuraishi and Okumura 1956). The effect of different BAP substituents on the promotion of growth of tobacco callus was analyzed by Iwamura et al. (1980). Keim et al. (1981) also showed that BAP, *o*T, and *m*T bind to CBF 1 protein with similarly high affinity. CBF 1 is a soluble protein isolated from wheat germ and binds preferentially aromatic cytokinins with relatively high affinity and specificity (Fox and Erion 1975). Further, Kaminek et al. (1987) reported significantly increased activity of *meta*-hydroxybenzyladenosine (later renamed *meta*-topolin riboside, *m*TR) over the standard BAP riboside (BAPR) in the tobacco callus and wheat leaf senescence assay. In the same year, Kaminek and Vaněk (1987) described the use of *m*T and *m*TR in agriculture and horticulture as shoot- and branch-stimulatory plant growth regulators. *Meta*-topolin riboside (*m*TR), tested under the name [(mOH)[9*R*]BAP or [(mOH)bzl^{6}Ado), was assessed by Czech scientists as a crop yield increasing substance on barley (Hradecká and Petr 1992) and wheat (Trčková et al. 1992; Hradecká and Petr 1992; Borkovec and Prochazka 1995; Trčková and Kamínek 2000). The application of cytokinins BAPR and *m*TR at 10^{-6} M concentration to 22 cultivars of 5 cereal species after flowering increased the mean values of relative plant growth rate, 1000-grain weight, number of grains per plant, and other yield-forming parameters (Hradecká and Petr 1992). The application of *m*TR to plants grown in a nutrient solution containing a low concentration of nitrate had almost no effect on the nitrogen content in the grains and caused a slight reduction in the rest of the plants. However, it increased significantly the total dry weight of both the grains and the whole plant. This resulted in increase in the total level of nutrients taken up by the cytokinin-treated plants. A similar trend was found in the case of phosphorus and, to a lesser extent, potassium and calcium (Trčková et al. 1992).

6-(2-Methoxybenzylamino)purine, (*ortho*-methoxytopolin, Me*o*T), and 6-(3-methoxybenzylamino)purine (*meta*-methoxytopolin, Me*m*T) and their 9-ribosides (Me*o*TR and Me*m*TR, respectively) were identified in *Populus* × *canadensis* cv *Robusta* leaves and in *Arabidopsis thaliana* plants (Tarkowská et al. 2003). These authors also evaluated the cytokinin activity of the four new phytohormones in three classical cytokinin bioassays. The bioassays are based on the stimulation of tobacco callus growth (tobacco callus assay), retention of chlorophyll in excised wheat leaves (wheat leaf senescence assay, WLSA), and dark induction of betacyanin synthesis in *Amaranthus* cotyledons (*Amaranthus* bioassay). It was very surprising that Me*m*T and Me*m*TR were highly active in the WLSA, and even *ortho*-analogues were highly active in this assay. All derivatives exhibited plus/minus double activity relative to BAP. In the other two bioassays, the phytohormones had equal or a slightly lower activity than BAP (Tarkowská et al. 2003). A systematic chemical approach made it possible to identify key substituents for achieving the high cytokinin and particularly high anti-senescence activity of BAP and BAPR derivatives. Among substituted BAPs, new aromatic cytokinin (ARCK) derivative 2-chloro-6-(3-methoxybenzyl)aminopurine (2ClMe*m*T) showing strong biological activities was discovered (Dolezal et al. 2006). This compound is highly active in classical CK bioassays (Koprna et al. 2016). The initial testing of its biological activities in field conditions was realized with winter wheat and spring barley as model cereals. The field plot trials were based on one single foliar application of 2ClMe*m*T (abbreviated as 2-Cl-3-MeOBAP). This was tested by spraying the plants in three phenological phases (tillering, stem elongation, and anthesis). The treatment had no effect in grain yield on wheat, while in barley increasing the grain yield was found following treatment applied at the stage of tillering and stem elongation by 7% and 4%, respectively (Koprna et al. 2016). Interestingly, the compound treatment showed positive effect on number of productive tillers. These result are in agreement with Suge and Iwamura (1993) who found tillering was the most effective period for CK application in barley. Also, other authors found possitive effect of CK application at grain yield via increased number od tillers in cereals (Sharif et al. 1980; Williams et al. 1980; Zahir et al. 2001).

22.5 New Aromatic Cytokinin Derivative 2-Chloro-6-(3-methoxybenzyl)aminopurine (2ClMe*m*T) in Field Trials

2ClMe*m*T was much more broadly tested in foliar treatments on cereals (spring barley and winter wheat) in the years 2012–2014. Application took place in autumn (wheat) in the 4–6 leaf phase (BBCH 14–16) and in spring (wheat and barley) in the tillering phase (BBCH 20–25), in stem elongation phase (BBCH 30–33), in flag leaf phase (BBCH 38–45), and in full heading phase (BBCH 55). The substance was tested under conditions of field trial plots; the sowing rate was 3.5 millions of germinated seeds per hectare. The experiments were performed in five repetitions (Olomouc locality) and four repetitions (Kroměříž locality), according to the methodology of cereal utility value testing and GEP (Good Experimental Practice). The foliar application was a dose of 300 liters per hectare at concentrations of 5 and 25 μM 2ClMe*m*T. Evaluation of strong (productive), medium, and weak (unproductive) offshoots was carried out at the end of the posting phase (BBCH 37–39), when the offshoot types were clearly visible. Evaluation of stem length and number of ears per sq. meter was performed at the phase end of heading (BBCH 59). The number of plants collected for this observation was 30 from each variant. Reading of ears per sq. was carried out for each repeat of the monitored variants from an area of 0.5 m^2. In the mixed sample from each variant, grain yield, the number of strong (productive), medium and weak (nonproductive) tillers, number of spikes per sq. meter, plant height, number of grains in spike, and thousand grain weight were measured. All tested variants of treatments were compared to non-treated controls. The results were statistically processed in the program PAST v. 2.17c. eStatistical significance was set at $p \leq 0.05$ and $p \leq 0.01$.

Application of 2ClMe*m*T on spring barley (Fig. 22.1) confirmed an increase in grain yield up to 8.18% over the control. After early application in the tillering stage, the number of productive tillers was moderately increased, while the number of medium tillers was considerably reduced, although this was greatly influenced by

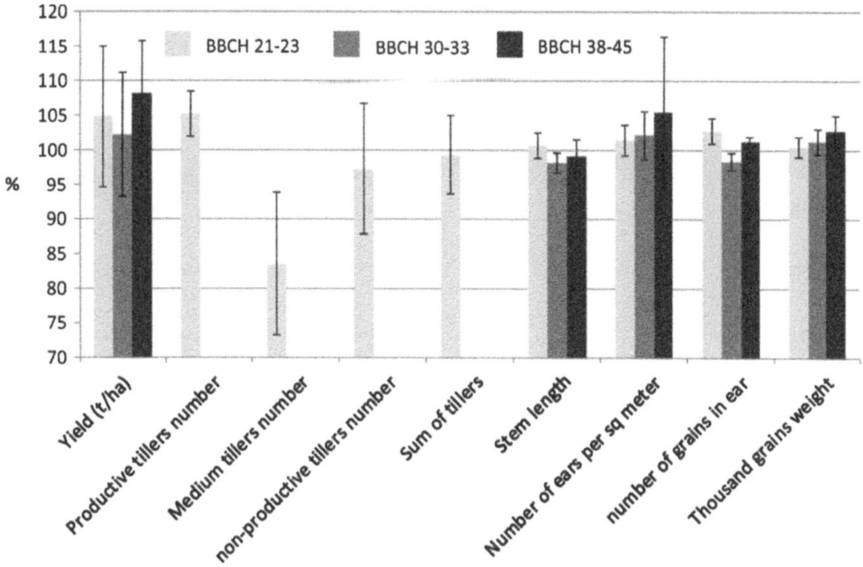

Fig. 22.1 Data from field experiments with 2ClMe*m*T in spring barley. Effect of the application on yield, number of tillers, stem length, weight of a thousand grains, number of grains in ear and number of ears per square meter

year of planting. In sum, the total number of tillers was unchanged. Only the number of each type of tiller was modified. Other plant parameters were almost unaffected, including stem length, number of ears per sq. meter, number of grains per ear, and thousand grain weight.

Application of 2ClMe*m*T on winter wheat (Fig. 22.2) at the stage of early tillering did not affect grain yield, but increased significantly the number of productive and medium tillers. Treatment at stage BBCH 55 (before flowering) increased the thousand grain weight. The main effect of 2ClMe*m*T was on productive tillers at the expense of medium tiller development. Sum of all tillers types was increased up to 10% over the control. There was only a marginal effect on winter wheat yield; no visible differences were observed when the compound was applied in different phenological phases.

It can be concluded from these results that in three consecutive field trials (2012–2014), the application of ARCK had a positive effect on yield through increased number of productive ears, but the effect was species dependent (only for spring barley), phenological phase dependent (at tillering and stem elongation stage, but not anthesis), and also year dependent. Previous studies also showed that field CK applications can increase seed weight, grain number, and yield in wheat, but the application was carried out at later phenological stages such as later booting or after ear emergence (Alizadeh et al. 2010; Wang et al. 2001). In barley, treatments of whole plants or ears at anthesis have been reported to increase grain weight and biological yield (Hosseini et al. 2008), whereas the increase in the production at

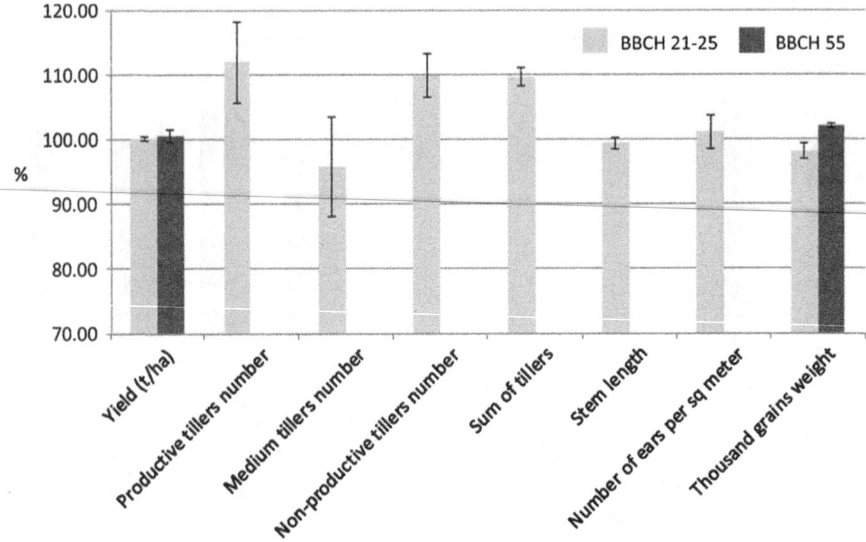

Fig. 22.2 Data from field experiments with 2ClMe*m*T in winter wheat. Effect of the application on yield, number of tillers, stem length, weight of a thousand grains, number of grains in ear, and number of ears per square meter

earlier phenological phases was related to stress conditions (Sarwat and El-Sherif 2007).

Cytokinin derivative 2ClMe*m*TR together with a fertilizer (N = 6%, K_2O = 5%, P_2O_5 = 10%, Cu = 0.1%, Mn = 0.5%) was applied by foliar treatment on winter wheat during the years 2016–2018 (Fig. 22.3). The substance was applied in terms BBCH 12–20 (autumn application) and BBCH 21–23 (spring application at the beginning of tillering). Grain yield exceeded the control variant by 5.08%; the number of productive tillers also increased (116.73% compared to control), while the number of other tillers was reduced (93.70% and 97.68%, respectively). The autumn application supported the development of productive tillers (106.75% of control), whereas medium, non-productive, as well as total number of tillers were decreased (91.18%, 79.21%, and 92.13%, respectively). The higher value of a thousand grain weight was observed in both tested variants (103.54 and 103.82%). Both applications also statistically significantly increased the grain yield, the treatment during tillering being more effective (grain yield increased by an average of 5.08%). Therefore, the most positive effect after application of cytokinin 2ClMe*m*T with fertilizer was observed at the beginning of tillering in the phenological stage BBCH 20–23.

According to the data, the effect of the cytokinin derivate in combination with foliar fertilizer had a more marked effect on increase in productive tillers at the expense of promoting the growth of other tillers (medium and non-productive). These changes in tiller structure led to an increase in grain yield and thousand grain weight, respectively.

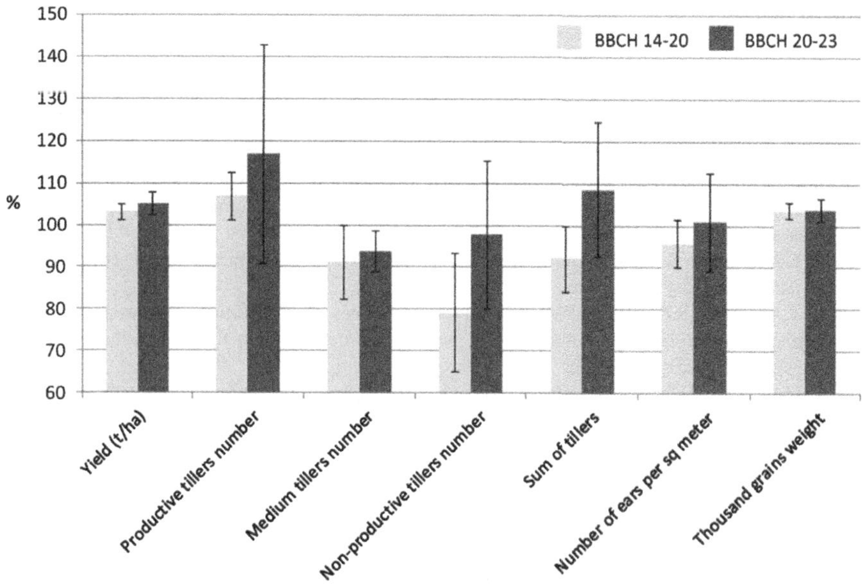

Fig. 22.3 Data from field experiments with 2ClMe*m*T application with nutrients in winter wheat. Effect of the application on yield, number of tillers, weight of thousand grains, and number of ears per square meter

22.6 Conclusion

Despite the large number of reports describing positive effects of exogenous treatments in various crops, the use of cytokinins (CKs) in agriculture appears to be confounded by a number of variables such as CK applied, crop, application time, and growth conditions. The data also show that aromatic CKs are preferred owing to their availability, lower price, greater stability in plants, resistance to degradation by cytokinin oxidase/dehydrogenase, and biological activity. More research on CK mode of action and the development of new modulators of key proteins controlling CK biosynthesis, degradation, and signaling would provide a more solid basis for the design and preparation of new targeted growth regulators with potential application in agriculture such as the novel aromatic cytokinin with regreening effects, described in this chapter.

Acknowledgments The authors of this review are grateful to Alexander Oulton, PhD., for a critical reading of the chapter. Supported by grant no. 18-07563S from the Czech Science Foundation and ERDF project "Plants as a tool for sustainable global development" (No. CZ.02.1.01/0.0/0.0/16_019/0000827).

References

Afzal I, Basra SMA, Cheema MA, Farooq M, Jafa MZ, Shahid M, Yasmeen A (2013) Seed priming: a shotgun approach for alleviation of salt stress in wheat. Int J Agric Biol 15:1199–1203

Alizadeh O, Haghighi BJ, Ordookhani K (2010) The effects of exogenous cytokinin application on sink size in bread wheat (*Triticum aestivum*). Afr J Agric Res 5:2893–2898

Amin AA, Rashad ESM, Hassanein MS, Zaki M (2007) Response of some white maize hybrids to foliar spray with benzyl adenine. Res J Agric Biol Sci 3:648–656

Atkins CA, Pigeaire A (1993) Application of cytokinins to flowers to increase pod set in *Lupinus angustifolius*. Aust J Agric Res 44:1799–1819

Atkins CA, Emery RJN, Smith PMC (2011) Consequences of transforming narrow leafed lupin (*Lupinus angustifolius* [L.]) with an *ipt* gene under control of a flower-specific promoter. Transgenic Res 20:1321–1332

Bennett MD, Rao MK, Smith JB, Bayliss MW (1973) Cell development in the anther, the ovule and the young seed of *Triticum aestivum* L. var. Chinese Spring. Philos Trans of Royal Soc B: Biol Sci 266:39–81

Bohner J, Bangerth F (1988) Cell number, cell size and hormone levels in semi-isogenic mutants of *Lycopersicon pimpinellifolium* differing in fruit size. Physiol Plant 72:316–320

Borkovec V, Prochazka S (1995) Proceedings of the Plant Growth Regulator Society of America 22nd, pp 360–364

Brugière N, Humbert S, Rizzo N, Bohn J, Habben JE (2008) A member of the maize isopentenyl transferase gene family, *Zea mays* isopentenyl transferase 2 (ZmIPT2), encodes a cytokinin biosynthetic enzyme expressed during kernel development. Plant Mol Biol 67:215–229

Carlson DR, Dyer DJ, Cotterman CD, Durley RC (1987) The physiological basis for cytokinin induced increases in pod set in IX93-100 soybeans. Plant Physiol 84:233–239

Clarke FC, Jameson PE, Christopher D (1994) The influence of 6-benzylaminopurine on post-harvest senescence of floral tissues of broccoli (*Brassica oleracea* var Italica). Plant Growth Regul 14:21–27

Criado MV, Caputo C, Roberts IN, Castro MA, Barneix AJ (2009) Cytokinin-induced changes of nitrogen remobilization and chloroplast ultrastructure in wheat (*Triticum aestivum*). J Plant Physiol 166:1775–1785

Dietrich JT, Kaminek M, Blevins DG, Reinbott TM, Morris RO (1995) Chnages in cytokinins and cytokinin oxidase activity in developing maize kernels and the effects of endogenous cytokinins on kernel development. Plant Physiol Biochem 33:327–336

Dolezal K, Popa I, Krystof V, Spíchal L, Fojtíková M, Holub J, Lenobel R, Schmülling T, Strnad M (2006) Preparation and biological activity of 6-benzylaminopurine derivatives in plants and human cancer cells. Bioorg Med Chem 14:875–884

Dyer DJ, Carlson DR, Cotterman JA, Sikorski JA, Ditson SL (1987) Soybean pod enhancement with synthetic cytokinin analogs. Plant Physiol 84:240–243

Fox JE, Erion JL (1975) A cytokinin binding protein from higher plant ribosomes. Biochem Biophys Res Commun 64:695–700

Gupta NK, Gupta S, Shukla DS, Deshmukh PS (2003) Differential responses of BA injection on yield and specific grain growth in contrasting genotypes of wheat (Triticum aestivum L.). Plant Growth Regul 40:201

Gupta S, Agarwal VP, Gupta NK (2012) Efficacy of putrescine and benzyladenine on photosynthesis and productivity in relation to drought tolerance in wheat (*Triticum aestivum* L.). Physiol Mol Biol Plants 18:331–336

Harrison MA, Kaufman PB (1980) Hormonal regulation of lateral bud (tiller) release in oats (*Avena sativa* L.). Plant Physiol 66:1123–1127

Hosseini MS, Poustini K, Ahmadi A (2008) Effects of foliar application of BAP on source and sink strength in four six-rowed barley (*Hordeum vulgare* L.) cultivars. Plant Growth Regul 54:231–239

Hradecká D, Petr J (1992) Spring barley kernel weight after treatment with a cytokinin. In: Sbornik Vysoke skoly zemedelske v Praze, fakulta agronomicka, Rada A: Rostlinná výroba, vol 54, pp 153–62
Hradecká D, Petr J (1992) The effect of cytokinins on the yield of some cereal. In: Kaminek M, Mok WS, Zažímalová E (eds) Physiology and biochemistry of cytokinins in plants. SPB Academic, The Hague, pp 245–247
Iwamura H, Fujita T, Koyama S et al (1980) Quantitative structure -activity relationship of cytokinin-active adenine and urea derivatives. Phytochemistry 19:1309–1319
Jameson PE, Song J (2016) Cytokinin: a key driver of seed yield. J Exp Bot 67:593–606
Jameson PE, McWha JA, Wright GJ (1982) Cytokinins and changes in their activity during the development of grains of wheat (*Triticum aestivum* L.). Z Pflanzenphysiol 106:27–36
Kaminek M, Vaněk T (1987) Agricultural and horticultural plant growth regulator compositions containing N^6-(m-hydroxybenzyl)adenosine and -adenine. From Czech. CS 238548. B1 19851113
Kaminek M, Vaněk T, Motyka V (1987) Cytokinin activities of N^6-benzyladenosine derivatives hydroxylated on the side-chain phenyl ring. J Plant Growth Regul 6:113–120
Kaya C, Tuna AL, Okant AM (2010) Effect of foliar applied kinetin and indole acetic acid on maize plants grown under saline conditions. Turk J Agric 34:529–538
Keim P, Erion J, Fox JE (1981) The current status of cytokinin-binding moieties. In: Guern J, Péaud-Lenoël C (eds) Metabolism and molecular activities of cytokinins. Proceedings in life sciences. Springer, Berlin
Khan MG, Srivastava HS (1998) Changes in growth and nitrogen assimilation in maize plants induced by NaCl and growth regulators. Biol Plant 41:93–99
Kieber JJ, Schaller GE (2014) Arabidopsis book. Am Soc Plant Biol 12:e0168. https://doi.org/10.1199/tab.0168. PMC3894907
Koprna R, De Diego N, Dundálková L, Spíchal L (2016) Use of cytokinins as agrochemicals. Bioorgan Med Chem 24:484–492
Kuraishi S (1959) Effect of kinetin analogs on leaf growth. Sci Pap Coll Gen Educ Univ Tokyo 9:67–104
Kuraishi S, Okumura FS (1956) The effect of kinetin on leaf growth. Bot Mag Tokyo 69:300–306
Langer RHM, Prasad PC, Laude HM (1973) Effects of kinetin on tiller bud elongation in wheat (*Triticum aestivum* L.). Ann Bot 37:565–571
Letham DS, Williams MW (1969) Regulators of cell division in plant tissues. VIII. The cytokinins of the apple fruit. Physiol Plant 22:925–936
Lewis DH, Burge GK, Hopping ME, Jameson PE (1996) Cytokinins and fruit development in the kiwifruit (*Actinidia deliciosa*). II. Effects of reduced pollination and CPPU application. Physiol Plant 98:87–195
Liu Z, Goto Y, Nishiyama I, Kokubun M (2001) Effects of foliar and root-applied benzylaminopurine on tillering of rice plants grown in hydroponics. Plant Prod Sci 4:220–226
Morris RD, Blevins DG, Dietrich JT et al (1993) Cytokinins in plant pathogenic bacteria and developing cereal grains. Aust J Plant Physiol 20:621–637
Nagel L, Brewster R, Riedell WE, Reese RN (2001) Cytokinin regulation of flower and pod set in soybeans (*Glycine max* (l.) Merr.). Ann Bot 88:27–31
Nonokawa K, Nakajima T, Nakamura T, Kokubun M (2012) Effect of synthetic cytokinin application on pod setting of individual florets within raceme in soybean. Plant Prod Sci 15:79–81
Okumura FS, Kotani Y, Ariga T, Kuraishi S (1959) Syntheses of kinetin-analogues. II. Bull Chem Soc Jpn 32:883–886
Pan B, Bai YM, Leibovitch S, Smith DL (1999) Plant growth promoting rhizobacteria and kinetin as ways to promote corn growth and yield in short season areas. Eur J Agron 11:179–186
Patel R, Mohapatra PK (1992) Regulation of spikelet development in rice by hormones. J Exp Bot 43:257–262
Paul-Victor C, Turnbull LA (2009) The effect of growth conditions on seed size/number trade-off. PLoS One 4:e6917

Richards D, Rowe RN (1977) Effects of root restriction, root pruning and 6-benzylaminopurine on the growth of peach seedlings. Ann Bot 41:729–740

Sadras VO (2007) Evolutionary aspects of the trade-off between seed size and number in crops. Field Crops Res 100:125–138

Sarwat MI, El-Sherif MH (2007) Increasing salt tolerance in some Barley genotypes (*Hordeum vulgare*) by using kinetin and benzyladenin. World J Agric Sci 3:617–629

Sharif R, Dale JE (1980) Growth-regulating substances and the growth of tiller buds in barley; effects of IAA and GA3. J Exp Bot 31:1191–1197

Song J, Jiang L, Jameson PE (2012) Co-ordinate regulation of cytokinin gene family members during flag leaf and reproductive development in wheat. BMC Plant Biol 12:78

Spano G, Di Fonzo N, Perrotta C, Platani C, Lawlor DW, Napier JA, Shewry PR (2003) Physiological characterization of 'stay green' mutants in durum wheat. J Exp Bot 54:1415–1420

Strnad M, Hanuš J, Vaněk T et al (1997) Meta-topolin, a highly active aromatic cytokinin from poplar leaves (*Populus x canadensis* Moencho cv. *Robusta*). Phytochemistry 42:213–218

Strnad M (1997) The aromatic cytokinins. Physiol Plant 101:674–688

Suge H, Iwamura H (1993) Effect of cytokinin and anticytokinin on the tillering of barley. Jpn J Crop Sci 62:595–600

Tarkowská D, Doležal K, Tarkowski P et al (2003) Identification of new aromatic cytokinins in Arabidopsis thaliana and Populus X canadensis leaves by LC-ESI-MS and capillary liquid chromatography/frit–fast atom bombardment mass spectrometry. Physiol Plant 117:579–590

Thomas H, Ougham H (2014) The stay-green trait. J Exp Bot 65:3889–3900

Trčková M, Kamínek M (2000) Nitrate uptake and nitrogen allocation in wheat is affected by exogenous cytokinins. In: Nitrogen in a sustainable ecosystem: from the cell to the plant. Backhuys Publishers, Leiden, pp 261–268

Trčková M, Kamínek M, Zmrhal Z (1992) Grain formation and distribution of nutrients in wheat plants after application of synthetic cytokinin N^6-(*meta*-hydroxybenzyl) adenosine. In: Kaminek M, Mok WS, Zažímalová E (eds) Physiology and biochemistry of cytokinins in plants. Kluwer Academic Publishing, The Hague, pp 241–244

Van Daele IN, Gonzalez I, Vercauteren L, de Smet D, Inzé D, Roldán-Ruiz I, Vuylsteke M (2012) A comparative study of seed yield parameters in Arabidopsis thaliana mutants and transgenics. Plant Biotechnol J 10:488–500

Wang Q, Zhang F, Smith DL (1996) Application of GA3 and kinetin to improve corn and soybean seedling emergence at low temperature. Environ Exp Bot 36:377–383

Wang Z, Cao W, Dai T, Zhou Q (2001) Effects of exogenous hormones on floret development and grain setin wheat. Plant Growth Regul 35:225–231

Werner T, Schmülling T (2009) Cytokinin action in plant development. Curr Opin Plant Biol 12:527–538

Williams RJ, Cartwright PM (1980) The effect of applications of a synthetic cytokinin on shoot dominance and grain yield in spring barley. Ann Bot 46:445–452

Woodward EJ, Marshall C (1988) Effects of plant growth regulators and nutrient supply on tiller bud outgrowth in barley (*Hordeum distichum* L.). Ann Bot 61:347–354

Yang J, Zhang J, Wang Z, Zhu Q (2003) Hormones in the grains in relation to sink strength and postanthesis development of spikelets in rice. Plant Growth Regul 41:185–195

Zahir AZ, Asghar HN, Arshad M (2001) Cytokinin and its precursors for improving growth and yield of rice. Soil Biol Biochem 33:405–408

The Pharmacological Activity of Topolins and Their Ribosides

23

Jiří Voller

Contents

Abstract

Cytokinins have diverse pharmacological activities including neuroprotective, immunomodulatory, and anticancer. They have been tested in patients as agents for oncology, neurology, and dermatology. The research has focused on isoprenoid cytokinins and on kinetin and N^6-benzyladenosine as representatives of aromatic cytokinins. The activities of topolins and their ribosides remain underexplored. Here we summarize the available data in the hope of encouraging interest in the therapeutic potential of these compounds.

Keywords

Aromatic cytokinins · Topolins · Cytokinin · Pharmacological activity · Cancer · Neurodegeneration

J. Voller (✉)
Laboratory of Growth Regulators, Institute of Experimental Botany of the Czech Academy of Sciences & Palacký University, Olomouc, Czech Republic
e-mail: jiri.voller@upol.cz

N. Ahmad, M. Strnad (eds.), *Meta-topolin: A Growth Regulator for Plant Biotechnology and Agriculture*, https://doi.org/10.1007/978-981-15-9046-7_23

Abbreviations

2OH3MeOBAR	N^6-(2-hydroxy-3-methoxybenzyl)adenosine
A_1R	A_1 adenosine receptor
$A_{2A}R$	A_{2A} adenosine receptor
A_{3R}	A_{3R} adenosine receptor
ADK	adenosine kinase
BA	N^6-benzyladenine
BAR	N^6-benzyladenosine
DNMT1	DNA-methyl transferase 1
DNPH1	2′-deoxynucleoside 5′-phosphate *N*-hydrolase 1
ENT1	Equilibrative nucleoside transporter 1
iP	N^6-isopentenyladenine
iPR	N^6-isopentenyladenosine
K	Kinetin
KR	Kinetin riboside
KRTP	Kinetin riboside-5′-triphosphate
*m*T	*meta*-Topolin
*m*TR	*meta*-Topolin riboside
NPC	Niemann-Pick disease type C
*o*T	*ortho*-Topolin
*o*TR	*ortho*-Topolin riboside
*o*TR5′MP	*ortho*-Topolin riboside-5′-monophosphate
*p*T	*para*-Topolin
*p*T	*para*-Topolin riboside
SCA3	Spinocerebellar ataxia type 3

The important role of cytokinins in plant cell growth and differentiation soon attracted the attention of researchers interested in those processes in human cells. Indeed, cytokinins and then their derivatives too were demonstrated to either promote or inhibit the division of mammalian cells depending on the cytokinin and experimental system used. Some also induce differentiation of various cell lines, including keratinocytes and certain leukemias. Many other pharmacological activities were reported, including neuroprotective, immunomodulatory, and anti-angiogenic (reviewed in Voller et al. 2017b, 2019; Kadlecová et al. 2019). Attention was focused mainly on N^6-isopentenyladenosine (iPR) and N^6-benzyladenosine (BAR) as candidate anticancer drugs and kinetin (K) for its ability to slow down the aging of human cells. The activities of topolins and their ribosides have been much less studied, however. Even those studies comparing the activities of several cytokinins typically use N^6-benzyladenine (BA) and/or K as sole representative(s) of aromatic cytokinins. Here we review the pharmacological activities of topolins and their ribosides in order to foster future research into their therapeutic potential.

23.1 Topolin Free Bases

Whereas certain cytokinin ribosides (CKRs) have been studied as candidate anticancer drugs, cytokinin bases have no or limited toxicity against human cell lines (Ishii et al. 2002; Mlejnek and Dolezel 2005; Voller et al. 2010). Long-term cultivation of human fibroblast with 80 μM K or *trans*-zeatin (*t*Z) even prevents detrimental changes typical for cells of high passage. Cytokinin bases also extend the lifespan of invertebrates. This was first demonstrated in diptera *Zapronius paravittiger* (Sharma et al. 1995) for K and later in nematode *Caenorhabditis elegans* for K, *meta*-topolin (*m*T), and *para*-topolin (*p*T) (Kadlecová et al. 2018). K, *trans*-zeatin, and Pyratine (kinetin 9-tetrahydropyran-2-yl) are active ingredients of clinically tested skin-improving products (McCullough et al. 2008; Ortiz et al. 2009; Tremaine et al. 2010; Wu et al. 2007). A recent 12-week clinical study on 39 subjects (Garcia et al. 2018) demonstrated the beneficial effects of topical *p*T (designated 4HBAP in this study). At a concentration of 0.1%, *p*T improved the appearance of fine and coarse wrinkles, reduced skin roughness, and hyperpigmentation. It also increased skin hydration and had a positive effect on both facial erythema and non-inflammatory acne lesions.

The mechanism of the (cyto)protective action of cytokinins is not fully understood. Research has traditionally focused on a direct or indirect (induction of defense mechanisms) antioxidant activity of K. It has also been proposed that K acts as a hormetin (Rattan et al. 2009) or as a hormetin precursor, prohormetin (Voller et al. 2017b). Molecular studies demonstrated that K after conversion into its riboside-5′-triphosphate acts as an ATP analog in PINK1 and CK2 kinases. The increased phosphorylation of their targets can have therapeutic effects in familial Parkinson's and Huntington's disease (Hertz et al. 2013; Bowie et al. 2018). Other cytokinins were seldom studied (Rattan and Sodagam 2005; Choi et al. 2009; Ji et al. 2010). The only study in a mammalian experimental system that included a topolin is Brizzolari et al. (2016). The authors demonstrated that K, iP, and *p*T, but not BA, were significantly active in the oxygen radical absorbance capacity (ORAC) assay at 2.5 and 5 μM concentrations. In the Trolox equivalence antioxidant capacity (TEAC) assay, only *p*T (0.5–5 μM) showed activity, probably due to phenolic hydroxyl moiety in the molecule. All four cytokinins quenched hydroxyl radicals generated in the 2-deoxyribose degradation assay. Evaluation of a larger cytokinin set, including other topolins, would be interesting. As antioxidant activity is not limited to hydroxylated derivatives, methoxylated derivatives of BA could also be included in such a study.

A recent exciting study reports that topolins are produced by *Mycobacterium tuberculosis*. They are formed by the activity of Rv1205, a homolog of the plant enzyme LONELY GUY, from respective riboside 5′-monophosphates. Cytokinin oxidase activity present in bacterial cells converts them to aldehydes that sensitize the bacteria to nitric oxide (Samanovic et al. 2018). Targeted perturbation of this pathway by, for example, cytokinin analogs, may present a novel therapeutic approach to tuberculosis treatment. As LONELY GUY homologs are present in the genomes of other bacterial pathogens, including *Staphylococcus aureus*, such an approach could have a significant impact on the treatment of bacterial infections.

23.2 Topolin Ribosides

Research into the pharmacological activity of topolin ribosides has focused mainly on the anticancer activity of *ortho*-topolin riboside (*o*TR) and the neuroprotective and neurological effects of *para*-topolin riboside (*p*TR). The only study that includes *meta*-topolin riboside (*m*TR) is our profiling of the anticancer activity of natural cytokinins where *m*TR showed no significant activity (Voller et al. 2010).

23.3 *ortho*-Topolin Riboside

iPR and BAR have been studied for their anticancer activities since the late 1960s (Grace et al. 1967; Fleysher et al. 1968); they even entered clinical trials in the next decade. Limited activity in a small and rather diverse study population together with stability issues (Mittelman et al. 1975) lead to a drop in interest in those compounds.

More than 40 years later, we carried out a systematic evaluation of the anticancer activity of almost all natural cytokinins known to date using a panel of diverse malignancies. We confirmed that cytotoxic activity is restricted to certain cytokinin ribosides (iPR, BAPR, KR) and their ribotides, whereas cytokinin bases show no or limited toxicity up to 100 μM. Moreover, we reported an especially strong cytotoxic activity of *ortho*-topolin riboside (*o*TR) with NCI60 GI50 values lower than those for other CKRs (Voller et al. 2010). *o*TR activity profile shares features typical for those of the "classical" cytokinin ribosides (Mlejnek and Dolezel 2005; Béres et al. 2010). First, it induces a pronounced ATP depletion. Second, vigorous conversion to 5′-monophosphates by adenosine kinase (ADK) is required for this effect; higher phosphates are present only in traces (Voller et al. 2010, 2017a). Mechanistic studies showed that *o*TR inhibited activation of the extracellular signal-regulated kinase-1/2 and the Akt pathway and induced apoptosis by the intrinsic pathway (Wang et al. 2012) in hepatocellular carcinoma SMMC-7721. In acute myeloid leukemia U937 cells, *o*TR induced a marker of endoplasmic reticulum stress and inhibited DNA-methyl transferase activity. The follow-up experiments, as well as a docking study, suggested that inhibition of DNA-methyl transferase 1 (DNMT1) may be responsible (Wang et al. 2018). It would be interesting to determine if DNMT1 or other methyltransferases also interact with *ortho*-topolin riboside-5′-monophosphate (*o*TR5′MP), the dominant species present in the cells. Direct interaction of *o*TR5′MP with putative oncogene DNPH1 (2′-deoxynucleoside 5′-phosphate *N*-hydrolase 1) was demonstrated by Amiable et al. (2013). The DNPH1 expression is regulated by ETV and Myc oncogenes (Lewis et al. 1997; Shin et al. 2008), and its rat homolog has a tumorigenic potential (Lewis et al. 2000). Its overexpression was associated with incidence of several malignancies (Rhodes et al. 2002; Shin et al. 2008; Uhlen et al. 2010) and with tumor grade in breast cancers (Shin et al. 2008). Moreover, the ability of its homolog to promote tumorogenesis was demonstrated in rats. *o*TR also induces differentiation of acute myeloid leukemia cell lines U937 and HL-60 (Wang et al. 2019). The only animal study published to date is a hollow-fiber assay with 12 NCI60 cancer cell lines. *o*TR administered by intraperitoneal injection (100 or 150 mg/kg/day for 4 days) induced at least 50% reduction of tumor mass in 16 out of 24 intraperitoneal implants. However, no tumor reduction over 50% was observed if *o*TR was given subcutaneously (Voller et al. 2010). This difference in activity suggests problems with bioavailability. No pharmacokinetic data have yet been published.

N^6-(2-Hydroxy-3-methoxybenzyl)adenosine (2OH3MeOBAR) is another natural CKR derived from *o*TR, whose cytotoxic activity was reported by our group (Doležal et al. 2007). This compound can efficiently induce apoptosis of some cell lines without disturbing intracellular ATP concentration levels. Accumulation of its ribotides in cytoplasm is limited in comparison with other cytotoxic cytokinin ribosides, and higher phosphates that could possibly incorporate into nucleic acids or interfere with nucleotide metabolism are formed (Voller et al. 2017a). Another class of metabolites distinctive for 2OH3MeOBAR are glucosides, derived from both the riboside and ribotides. Similar to *o*TR5′MP, 2OH3MeOBAR5′MP inhibits DNPH1. In the absence of ATP depletion, a relevant contribution of this mechanism to the cytotoxic activity seems more plausible.

In our opinion, a comparison of mechanistic effects of 2OH3MeOBAR with those of *o*TR and "classical" CKRs could help to delineate direct effects of CKRs and the secondary effects of energy perturbation. For example, it would be interesting to see if *ortho*-hydroxylated BARs had any impact on protein farnesylation. Bifulco's group discovered this effect for iPR (designated as i6A in their studies) (Laezza et al. 2006), and over the years, they explored its therapeutic potential in various cancers (Laezza et al. 2009, 2010; Ciaglia et al. 2016, 2018; Ranieri et al. 2018). They also demonstrated the anti-angiogenic (Pisanti et al. 2014) and immunomodulatory effects of iPR (Ciaglia et al. 2018). Similar studies for other cytokinin ribosides would be of great interest. Particularly interesting would be discovery of such activity for non-toxic CKRs like *m*TR, *p*TR, and zeatin ribosides, as some application of protein farnesylation inhibition may lie outside cancer therapy. iPR treatment, for example, normalizes the nuclear envelope of Hutchinson-Gilford progeria cells (Bifulco et al. 2013).

23.4 *para*-Topolin Riboside

para-Topolin riboside (*p*TR) has been studied mainly in connection with neuroprotection. It was identified as one of the cytoprotective substances in plant *Gastrodia elata* that is used in Chinese traditional medicine for the therapy of dizziness, headaches, and epilepsy (Huang et al. 2007). It is designated T1–11 or NHBA in the follow-up studies.

Huang et al. (2011) demonstrated that *p*TR binds human adenosine receptors $A_{2A}R$ (Ki = 2.62 μM) and activates rodent homologs in rat cells and mice. Although it also binds human A_3R (Ki = 0.1 μM), it may not activate it efficiently. It also interacts with nucleotide equilibrative transporter ENT1, as demonstrated in guinea pig protein (Ki = 0.54 μM). It inhibited adenosine uptake in a rat P12 cell line with maximum effect attained in low tens of micromoles. Docking studies further support the binding of *p*TR to human $A_{2A}R$ and ENT1. Inhibition of ENT1 can increase the adenosine levels in the extracellular space and thus contribute to the regulation of adenosinergic signaling in neural synapses. Indeed, *p*TR treatment increased the striatal adenosine level in mice (Huang et al. 2011). Both $A_{2A}R$ activation and ENT1 inhibition are being explored as treatment strategies for neurodegenerative conditions including Huntington's disease (Guitart et al. 2017; Kao et al. 2017). In Huntington's disease, expansion of the polyglutamine sequence makes huntingtin protein prone to aggregation and accumulation, which results in neuronal death. $A_{2A}R$ stimulation may counteract both aggregate accumulation and suppression of the proteasome activity by mutant huntigtin (Chiang et al. 2009). Indeed, *p*TR (0.05 mg/mL in drinking water ad libitum) showed positive effects on several symptoms of Huntington's disease in a transgenic mouse model R6/2. It prevented the rotarod performance deterioration and even extended the lifespan of the test animals. It may also affect brain atrophy; the difference was not statistically significant, however. At the molecular level, the treatment led to weaker accumulation of huntingtin, improved activity of proteasome, and increased levels of mRNA for brain-derived neurotrophic factor (BDNF) in the brains of the experimental animals.

*p*TR also showed a promising activity in a mouse model of spinocerebellar ataxia type 3 (SCA3) (Chou et al. 2015). SCA3 is another neurodegenerative condition caused by an expansion of the polyglutamine tract in the affected protein. *p*TR dosing (5–50 mg/kg/day in drinking water) started at the presymptomatic stage. At 50 mg/kg/day, *p*TR prevented neuronal death in pontine nuclei and attenuated cerebellar transcriptional downregulation. The authors suggest that *p*TR treatment improved the removal of ataxin 3 protein with the expanded polyglutamine sequence as it increased the chymotrypsin-like activity of the proteasome in the neuronal tissue. Doses of 5, 15, and 50 mg/kg/day alleviated motor symptoms. The addition of $A_{2A}R$ antagonist SCH58261 (5 mg/kg/day intraperitoneally) markedly interfered with the beneficial effect confirming the role of $A_{2A}R$.

Niemann-Pick disease type C (NPC) is a fatal neurovisceral lipid storage disorder usually caused by a loss-of-function mutation in NPC1 (intracellular cholesterol transporter 1). A proof-of-concept study demonstrated that stimulation of $A_{2A}R$ normalized the pathological phenotype of fibroblasts derived from an NPC patient, including cholesterol accumulation (Visentin et al. 2013). In a follow-up study (Ferrante et al. 2018) using a Balb/c $Npc1^{nih}$ mouse model of NPC, T1–11 (50 mg/kg/day in drinking water) attenuated cognitive impairment, protected Purkinje cells, and reduced accumulation of sphingomyelin in the liver. It also significantly extended the lifespan of the experimental animals. Importantly, the beneficial effects of CGS21680, a standard high affinity $A_{2A}R$ agonist with purine scaffold, were only limited. The authors hypothesize that that low penetration across the blood-brain barrier may be responsible. The ability of *p*TR to interact not only with $A_{2A}R$ but also with ENT1 may be an alternative explanation.

Another example demonstrating the potential therapeutic utility of dual $A_{2A}R$ and ENT1 inhibition is the observation that CGS21680 reduces ethanol-drinking behaviors in ENT1 knock-out, but not wild-type mice (Nam et al. 2013). Interestingly, *p*TR (0.1 mg/kg intraperitoneally) reduced ethanol-seeking behavior of wild-type C57BL/6J mice (Hong et al. 2019). The treatment had no significant effect on locomotor activity of the animals in the open field; however, a dose of 0.3 mg/kg decreased the activity significantly.

*p*TR may also have sedative and hypnotic activity. In mice strain ICR *p*TR (0.2 mg/kg intraperitoneally) reduced spontaneous locomotor activity and potentiated the effect of both a sub-hypnotic and hypnotic dose of pentobarbital. The sleep-promoting activity of *p*TR was confirmed by EEG analysis. It was mediated by the interaction with $A_{2A}R$ and A_1R receptors as the administration of inhibitors of either reduced the effect. The authors also demonstrated that *p*TR binds both A_1 and A_2AR mouse receptors in vitro. *p*TR also increased c-Fos expression in GABAergic neurons of the ventrolateral preoptic area of the hypothalamus sleep center. Importantly, *p*TR did not disturb sleep architecture as the administration of 1, and 5 mg/kg intraperitoneally decreased the time of wakefulness and increased the duration of non-REM sleep without significantly affecting REM sleep.

Overall, these data show that *p*TR has promising activity relevant to the treatment of some neurological and behavioral disorders. Dual affinity for $A_{2A}R$ and ENT1 seems to be critical for at least some of these effects, as an even more potent $A_{2A}R$ receptor agonist does not show comparable efficacy. Highly encouraging is also the fact that oral administration leads to effective concentrations beyond the blood-brain barrier as indicated by the reported CNS activities. Several important issues should be addressed, however.

First, the relevance of interspecies differences in target affinity and selectivity needs to be considered. Although the therapeutic activities of *p*TR in mouse models are mediated by $A_{2A}R$ activation, the role of other adenosine receptors cannot be excluded. Although *p*TR does not bind human A_1R, it binds the mouse homolog, and this interaction may contribute to its hypnotic activity. We also know that *p*TR binds human A_3R strongly, but the data for the mouse homolog are not available. We also lack experimental proof that *p*TR inhibits human ENT1.

Second, no data on the pharmacokinetics of *p*TR have been reported to date with an exception of a study of pTR metabolites in rat plasma and urine (Lei et al. 2011). What concentrations in plasma and target tissues are reached after intraperitoneal and oral administration? What is the plasmatic half-life of *p*TR? Are there any tissues accumulating *p*TR? We can expect that if such a hydrophilic compound ($X\log P = 0.8$, according to Pubchem) crosses the gastrointestinal wall and blood-brain barrier efficiently, a transporter is involved. At present, it cannot be excluded that *p*TR enters the cells through ENT1 if its inhibitory effects on adenosine transport are a result of competition. The transporter make-up of particular tissues may significantly influence *p*TR tissue and organ distribution.

Third, as *p*TR has hypnotic and sedative properties in mice, is it possible to reach therapeutic levels for other indications without affecting attention with a reasonable safety margin? This is a real concern, as, for example, reduction of ethanol-seeking behavior in mice was reported for a concentration of one-third of the dose that induced a negative effect on locomotor activity. Moreover, a double concentration has apparent sedative effects in another mouse strain. On the other hand, the differences in *p*TR receptor affinity for human and mouse adenosine receptors, including A_1R that is important for vigility, may prevent this possible side effect in humans.

Fourth, could an interaction of *p*TR with some intracellular targets contribute to the observed beneficial effects? Here may lie another possible explanation for the lack of effect of $A_{2A}R$ agonist CGS21680, especially if the potential intracellular target interacts with the phosphorylated form of the cytokinin. Although CGS21680 is also an adenosine derivative, it is an amide of ribofuranoic acid and thus lacks the 5′-hydroxy group necessary for phosphorylation.

Finally, *p*T is an antioxidant—could antioxidant activity contribute to the neuroprotective activity of *p*TR? On the other hand, if interaction with some intracellular targets contributes to *p*TR activity, can *p*T engage them as well?

23.5 Conclusions

Although the majority of researchers interested in drug discovery have focused on other cytokinins, the last 10 years brought many exciting observations of pharmacological activity of topolins and their ribosides. 2-Hydroxylated derivatives of N^6-benzyladenosine have promising anticancer activity. *o*TR shares hallmarks of other anticancer cytokinin ribosides including induction of severe ATP depletion, yet it is more active against NCI60 cancer cell line panel. It's even more active derivative with extra 3-methoxy group on the benzyl ring represents a compound with a novel and unique anticancer activity. *p*TR, a $A_{2A}R$ agonist, demonstrated promising activity in multiple models of neurodegeneration. Notably, it was superior to a more potent $A_{2A}R$ agonist, possibly because it also inhibits the ENT1 transporter. Such dual activity is exceptional among many reported adenosine receptor agonists. *p*T showed a skin-improving effect in a clinical trial. Although *p*T was tested as a cosmeceutical, the positive effect on acne and redness may justify further development for dermatology indications. Other topolins may also be suitable for cytoprotection and anti-aging studies—not only *p*T but also *m*T, extends the lifespan of *C. elegans*. We believe that topolins and their ribosides have diverse as yet unknown activities relevant to aging and disease treatment and recommend that researchers studying the pharmacological activities of cytokinins include them for comparison.

References

Amiable C, Pochet S, Padilla A et al (2013) N^6-substituted AMPs inhibit mammalian deoxynucleotide N-hydrolase DNPH1. PLoS One 8:e80755. https://doi.org/10.1371/journal.pone.0080755

Béres T, Zatloukal M, Voller J et al (2010) Tandem mass spectrometry identification and LC-MS quantification of intact cytokinin nucleotides in K-562 human leukemia cells. Anal Bioanal Chem 398:2071–2080. https://doi.org/10.1007/s00216-010-4126-5

Bifulco M, D'Alessandro A, Paladino S et al (2013) N6-isopentenyladenosine improves nuclear shape in fibroblasts from humans with progeroid syndromes by inhibiting the farnesylation of prelamin A. FEBS J 280:6223–6232. https://doi.org/10.1111/febs.12544

Bowie LE, Maiuri T, Alpaugh M et al (2018) N6-furfuryladenine is protective in Huntington's disease models by signaling huntingtin phosphorylation. Proc Natl Acad Sci 115:E7081–E7090. https://doi.org/10.1073/pnas.1801772115

Brizzolari A, Marinello C, Carini M et al (2016) Evaluation of the antioxidant activity and capacity of some natural N^6-substituted adenine derivatives (cytokinins) by fluorimetric and spectrophotometric assays. J Chromatogr B Anal Technol Biomed Life Sci 1019:164–168. https://doi.org/10.1016/j.jchromb.2015.12.047

Chiang MC, Chen HM, Lai HL et al (2009) The A2A adenosine receptor rescues the urea cycle deficiency of Huntington's disease by enhancing the activity of the ubiquitin-proteasome system. Hum Mol Genet 18:2929–2942. https://doi.org/10.1093/hmg/ddp230

Choi SJ, Jeong C-H, Choi S-G et al (2009) Zeatin prevents amyloid beta-induced neurotoxicity and scopolamine-induced cognitive deficits. J Med Food 12:271–277. https://doi.org/10.1089/jmf.2007.0678

Chou AH, Chen YL, Chiu CC et al (2015) T1-11 and JMF1907 ameliorate polyglutamine-expanded ataxin-3-induced neurodegeneration, transcriptional dysregulation and ataxic symptom in the SCA3 transgenic mouse. Neuropharmacology 99:308–317. https://doi.org/10.1016/j.neuropharm.2015.08.009

Ciaglia E, Abate M, Laezza C et al (2016) Antiglioma effects of N6-isopentenyladenosine, an endogenous isoprenoid end product, through the downregulation of epidermal growth factor receptor. Int J Cancer 140(4):959–972. https://doi.org/10.1002/ijc.30505

Ciaglia E, Laezza C, Abate M et al (2018) Recognition by natural killer cells of N6-isopentenyladenosine-treated human glioma cell lines. Int J Cancer 142:176–190. https://doi.org/10.1002/ijc.31036

Doležal K, Popa I, Hauserová E et al (2007) Preparation, biological activity and endogenous occurrence of N6-benzyladenosines. Bioorg Med Chem 15:3737–3747. https://doi.org/10.1016/j.bmc.2007.03.038

Ferrante A, Pezzola A, Matteucci A et al (2018) The adenosine A2A receptor agonist T1–11 ameliorates neurovisceral symptoms and extends the lifespan of a mouse model of Niemann-Pick type C disease. Neurobiol Dis 110:1–11. https://doi.org/10.1016/j.nbd.2017.10.013

Fleysher MH, Hakala MT, Bloch A, Hall RH (1968) Synthesis and biological activity of some N6-alkyladenosines. J Med Chem 11:717–720. https://doi.org/10.1021/jm00310a018

Garcia RL, McCullough JL, Reece B, Massino FJ (2018) A novel topical cytokinin improves coarse wrinkles and skin roughness without skin irritation. J Drugs Dermatol 17:772–779

Grace JT, Hakala MT, Hall RM, Blakeslee J (1967) N^6-Substituted adenine derivatives as growth inhibitors of human leukemic myeloblasts and S-180 cells. Proc Am Assoc Cancer Res 8:23–27

Guitart X, Chern Y, Ferré S (2017) Targeting the equilibrative nucleoside transporter ENT1 in Huntington disease. Oncotarget 8:12550–12551

Hertz NT, Berthet A, Sos ML et al (2013) A neo-substrate that amplifies catalytic activity of Parkinson's-disease-related kinase PINK1. Cell 154:737–747. https://doi.org/10.1016/j.cell.2013.07.030

Hong SI, Peyton L, Chern Y, Choi DS (2019) Novel adenosine analog, N6-(4-hydroxybenzyl)-adenosine, dampens alcohol drinking and seeking behaviors. J Pharmacol Exp Ther 371:260–267. https://doi.org/10.1124/jpet.119.261529

Huang N-K, Chern Y, Fang J-M et al (2007) Neuroprotective principles from Gastrodiaelata. J Nat Prod 70:571–574. https://doi.org/10.1021/np0605182

Huang N-K, Lin J-H, Lin J-T et al (2011) A new drug design targeting the adenosinergic system for Huntington's disease. PLoS One 6:e20934. https://doi.org/10.1371/journal.pone.0020934

Ishii Y, Hori Y, Sakai S, Honma Y (2002) Control of differentiation and apoptosis of human myeloid leukemia cells by cytokinins and cytokinin nucleosides, plant redifferentiation-inducing hormones. Cell Growth Differ 13:19–26

Ji C, Yang Y, Yang B et al (2010) Trans-Zeatin attenuates ultraviolet induced down-regulation of aquaporin-3 in cultured human skin keratinocytes. Int J Mol Med 26:257–263. https://doi.org/10.3892/ijmm-00000460

Kadlecová A, Jirsa T, Novák O et al (2018) Natural plant hormones cytokinins increase stress resistance and longevity of *Caenorhabditis elegans*. Biogerontology 19:109–120. https://doi.org/10.1007/s10522-017-9742-4

Kadlecová A, Maková B, Artal-Sanz M et al (2019) The plant hormone kinetin in disease therapy and healthy aging. Ageing Res Rev 55:100958

Kao YH, Lin MS, Chen CM et al (2017) Targeting ENT1 and adenosine tone for the treatment of Huntington's disease. Hum Mol Genet 26:467–478. https://doi.org/10.1093/hmg/ddw402

Laezza C, Notarnicola M, Caruso MG et al (2006) N^6-isopentenyladenosine arrests tumor cell proliferation by inhibiting farnesyl diphosphate synthase and protein prenylation. FASEB J 20:412–418. https://doi.org/10.1096/fj.05-4044lsf

Laezza C, Caruso MG, Gentile T et al (2009) N6-isopentenyladenosine inhibits cell proliferation and induces apoptosis in a human colon cancer cell line DLD1. Int J Cancer 124:1322–1329. https://doi.org/10.1002/ijc.24056

Laezza C, Malfitano AM, Di Matola T et al (2010) Involvement of Akt/NF-κB pathway in N6-isopentenyladenosine-induced apoptosis in human breast cancer cells. Mol Carcinog 49:892–901. https://doi.org/10.1002/mc.20666

Lei Y, Wang L, Cheng M, Xiao H (2011) Identification of major metabolites in rat urine and plasma of N6-(4-hydroxybenzyl) adenine riboside by LC/MS/MS. Biomed Chromatogr 25:344–352. https://doi.org/10.1002/bmc.1452

Lewis BC, Shim H, Li Q et al (1997) Identification of putative c-Myc-responsive genes: characterization of rcl, a novel growth-related gene. Mol Cell Biol 17:4967–4978

Lewis BC, Prescott JE, Campbell SE et al (2000) Tumor induction by the c-Myc target genes rcl and lactate dehydrogenase A. Cancer Res 60:6178–6183

McCullough JL, Garcia RL, Reece B (2008) A clinical study of topical Pyratine 6 for improving the appearance of photodamaged skin. J Drugs Dermatol 7(2):131–135. PMID: 18335649

Mittelman A, Evans JT, Chheda GB (1975) Cytokinins as chemotherapeutic agents. Ann N Y Acad Sci 255:225–234

Mlejnek P, Dolezel P (2005) Apoptosis induced by N^6-substituted derivatives of adenosine is related to intracellular accumulation of corresponding mononucleotides in HL-60 cells. Toxicol In Vitro 19:985–990. https://doi.org/10.1016/j.tiv.2005.06.023

Nam HW, Hinton DJ, Kang NY et al (2013) Adenosine transporter ENT1 regulates the acquisition of goal-directed behavior and ethanol drinking through A2A receptor in the dorsomedial striatum. J Neurosci 33:4329–4338. https://doi.org/10.1523/JNEUROSCI.3094-12.2013

Pisanti S, Picardi P, Ciaglia E et al (2014) Antiangiogenic effects of N6-isopentenyladenosine, an endogenous isoprenoid end product, mediated by AMPK activation. FASEB J 28:1132–1144. https://doi.org/10.1096/fj.13-238238

Ortiz A, Elkeeb L, Truitt A, Hindiyeh R, Aquino L, Tran M, Weinstein GJ (2009) Topical PRK 124 (0.125%) lotion for improving the signs and symptoms of rosacea. Drugs Dermatol 8(5):459–462

Ranieri R, Ciaglia E, Amodio G et al (2018) N^6-isopentenyladenosine dual targeting of AMPK and Rab7 prenylation inhibits melanoma growth through the impairment of autophagic flux. Cell Death Differ 25:353–367. https://doi.org/10.1038/cdd.2017.165

Rattan SIS, Sodagam L (2005) Gerontomodulatory and youth-preserving effects of zeatin on human skin fibroblasts undergoing aging in vitro. Rejuvenation Res 8:46–57. https://doi.org/10.1089/rej.2005.8.46

Rattan SIS, Fernandes RA, Demirovic D et al (2009) Heat stress and hormetin-induced hormesis in human cells: effects on aging, wound healing, angiogenesis, and differentiation. Dose-Response 7:90–103. https://doi.org/10.2203/dose-response.08-014.Rattan

Rhodes DR, Barrette TR, Rubin MA et al (2002) Meta-analysis of microarrays: interstudy validation of gene expression profiles reveals pathway dysregulation in prostate cancer. Cancer Res 62:4427–4433. https://doi.org/10.1007/978-1-4020-9654-9

Samanovic MI, Hsu HC, Jones MB et al (2018) Cytokinin signaling in Mycobacterium tuberculosis. MBio 9(3):e00989–e00918. https://doi.org/10.1128/mBio.00989-18

Sharma SP, Kaur P, Rattan SIS (1995) Plant growth hormone kinetin delays aging, prolongs the lifespan, and slows down development of the fruitfly Zaprionus paravittiger. Biochem Biophys Res Commun 216:1067–1071. https://doi.org/10.1006/bbrc.1995.2729

Shin S, Bosc DG, Ingle JN et al (2008) Rcl is a novel ETV1/ER81 target gene upregulated in breast tumors. J Cell Biochem 105:866–874. https://doi.org/10.1002/jcb.21884

Tremaine AM, Ortiz A, Elkeeb L, Tran M, Weinstein G (2010) Long-term efficacy and safety of topical PRK 124 (0.125%) lotion (Pyratine-XR) in the treatment of mildto-moderate rosacea. J Drugs Dermatol 9(6):647–650. PMID: 20645526

Uhlen M, Oksvold P, Fagerberg L et al (2010) Towards a knowledge-based human protein atlas. Nat Biotechnol 28:1248–1250. https://doi.org/10.1038/nbt1210-1248

Visentin S, De Nuccio C, Bernardo A et al (2013) The stimulation of adenosine A2A receptors ameliorates the pathological phenotype of fibroblasts from Niemann-Pick type C patients. J Neurosci 33:15388–15393. https://doi.org/10.1523/JNEUROSCI.0558-13.2013

Voller J, Zatloukal M, Lenobel R et al (2010) Anticancer activity of natural cytokinins: a structure-activity relationship study. Phytochemistry 71:1350–1359. https://doi.org/10.1016/j.phytochem.2010.04.018

Voller J, Béres T, Zatloukal M et al (2017a) The natural cytokinin 2OH3MeOBAR induces cell death by a mechanism that is different from that of the "classical" cytokinin ribosides. Phytochemistry 136:156–164. https://doi.org/10.1016/j.phytochem.2017.01.004

Voller J, Maková B, Kadlecová A et al (2017b) Plant hormone cytokinins for modulating human aging and age-related diseases. In: Rattan S, Sharma R (eds) Hormones in ageing and longevity. Springer International Publishing, Basel, pp 311–335

Voller J, Béres T, Zatloukal M et al (2019) Anti-cancer activities of cytokinin ribosides. Phytochem Rev 18:1101–1113

Wang L, Sun C, Wang Z-H, Guo G-Q (2012) Mechanism of apoptotosis induced by ortho-topolin riboside in human hepatoma cell line SMMC-7721. Food Chem Toxicol 50:1962–1968. https://doi.org/10.1016/j.fct.2012.03.053

Wang L, Zhao YH, Cheng J et al (2018) Ortho-topolin riboside induced cell apoptosis through ERS pathway and inhibited DNMT1 activity in acute myeloid leukemia cells. Biomed Res 29:2061–2067. https://doi.org/10.4066/biomedicalresearch.29-17-3499

Wang L, Cheng J, Lin F et al (2019) Ortho-topolin riboside induced differentiation through inhibition of STAT3 signaling in acute myeloid leukemia HL-60 cells. Turkish J Hematol 36:162–168. https://doi.org/10.4274/tjh.galenos.2019.2019.0020

Wu JJ, Weinstein GD, Kricorian GJ, Kormeili T, McCullough JL (2007) Topical kinetin 0.1% lotion for improving the signs and symptoms of rosacea. Clin Exp Dermatol 32:693–695. https://doi.org/10.1111/j.1365-2230.2007.02513.x

The manufacturer's authorised representative in the EU is Springer Nature Customer Service Centre GmbH, Europaplatz 3, 69115 Heidelberg, Germany. If you have any concerns regarding our products, please contact ProductSafety@springernature.com

Printed and bound by CPI Group (UK) Ltd, Croydon, CR0 4YY
30/09/2026
02225330-0011